# Quantitative PCR Protocols

# METHODS IN MOLECULAR MEDICINE™

## *John M. Walker,* SERIES EDITOR

METHODS IN MOLECULAR MEDICINE™

# Quantitative PCR Protocols

Edited by

## Bernd Kochanowski
## and Udo Reischl

*University of Regensburg, Germany*

Humana Press ✳ Totowa, New Jersey

# Preface

Since the polymerase chain reaction (PCR) was first developed in 1985, an enormous number of research reports have documented the versatility of this brilliant technique for in vitro amplification of nucleic acids. Although PCR has had a profound impact in many areas of research, contrary to expectation its routine application to the quantitation of nucleic acids has proven problematic in several aspects. The shortcomings are principally caused by the exponential nature of PCR, whereby small variations in amplification efficiency may dramatically affect the yield of amplification product. Even minimal temperature deviations that occur between adjacent wells of a thermocycler or day-to-day variations in the efficiency of nucleic acid preparation can lead to significant differences in the extent of amplification between otherwise identical samples.

However, knowing more about the intrinsic limitations of PCR is the first step towards surmounting the shortcomings associated with this promising methodology. With the introduction of appropriate standards of known amount, which are co-amplified with the sample using the same primers, it is increasingly feasible to address biological or diagnostic questions that are difficult or impossible to answer using any other experimental approach.

The techniques and experimental strategies described in *Quantitative PCR Protocols* are representative of those most generally applicable to routine work at present. Apart from a brief description of the principles of quantitative PCR, the book describes both established and novel strategies, each of which has been applied successfully to such problems as the analysis of eukaryotic gene expression, the quantitation of viral loads in clinical specimens, reporter gene expression, and quantitative oncogene analysis.

Particular emphasis is placed on the underlying principles of the design of competitive or noncompetitive standards, as well as the optimization of the amplification process; these are crucial in any successful quantitative application. Basic problems with the interpretation of the results are addressed as well. Some duplication of important topics has been introduced purposely to offer the reader several approaches to the same problem. It is hoped that this collection of detailed protocols, providing comprehensive and up-to-date information, will be especially useful to researchers and to students needing to become familiar with the principles of quantitative PCR, and guiding them to set up test systems tailored to their specific practical needs. Since approaches to the amplification of nucleic acids in a quantitative manner and to the technology involved in product detection are subject to continual improvement, *Quantitative PCR Protocols* does not attempt the impossible task of treating every variety of experimental approach in the field. Rather, it depicts a kind of cross-section of realistic possibilities for the user's conception of still more refined assays.

Quantitative PCR—myth or reality? At the present moment the truth lies somewhere in-between and, since the usefulness of quantitation mainly depends on the particular application, only the future will show which assays will prove most useful in individual diagnostic situations.

We are especially indebted to Prof. Hans Wolf and Prof. Wolfgang Jilg for giving us the opportunity to gain substantial experience in the field. Without their confidence and continuous support many things would not have been possible. We also thank Prof. John Walker for his encouragement and Humana Press for their excellent assistance during the assembly of this volume. Finally, we are grateful to all of the contributing authors for their constantly high level of motivation and enthusiasm and, last but not least, for providing such good manuscripts.

*Bernd Kochanowski*
*Udo Reischl*

# Contents

# Contributors

AFTAB A. ANSARI • *Department of Pathology, Emory University School of Medicine, Atlanta, GA*

G. ALAN BEARD • *Elcatech, Winston-Salem, NC*

JACQUES BONNET • *Laboratoire d'Immunologie Moleculaire, Université de Bordeaux II, France*

ELIZABETH BONNEY • *Inserm Unit, Necker Hospital, Paris, France*

DAVID B. CORRY • *Department of Medicine, University of California, San Francisco, CA*

GEORGE J. DOELLGAST • *Department of Biochemistry, Bowman Gray School of Medicine, Winston-Salem, NC*

MICHAEL J. FASCO • *Wadsworth Center, New York State Department of Health, Albany, NY*

FRANCK GRISCELLI • *Inserm Unit, Necker Hospital, Paris, France*

MEINHARD HAHN • *Institüt für Biochemie, Justus-Liebig-Universität Giessen, Germany*

SABINE HERBLOT • *Laboratoire d'Immunologie Moleculaire, Université de Bordeaux II, France*

MARK HOLODNIY • *AIDS Research Center, Veterans Affairs Health Care System, Palo Alto, CA*

NATHAN IYER • *Department of Microbiology and Immunology, Bowman Gray School of Medicine, Winston-Salem, NC*

WOLFGANG JILG • *Institute for Medical Microbiology and Hygiene, University of Regensburg, Germany*

ERIC DE KANT • *Department of Anatomy and Embryology, LUMC, Leiden, The Netherlands*

MARLYSE C. KNUCHEL • *Centers for Disease Control and Prevention, Atlanta, GA*

BERND KOCHANOWSKI • *Institute for Medical Microbiology and Hygiene, University of Regensburg, Germany*

THOMAS KÖHLER • *Institute of Chemical Chemistry, Leibzig, Germany*

LOUIS S. KUCERA • *Department of Microbiology and Immunology, Bowman Gray School of Medicine, Winston-Salem, NC*

JERZY KULSKI • *Centre for Molecular Immunology and Instrumentation, University of Western Australia, Subiaco, Australia*

OLIVIER LANTZ • *Inserm Unit, Necker Hospital, Paris, France*

RICHARD M. LOCKSLEY • *Department of Medicine, University of California, San Francisco, CA*

FRANÇOIS MALLET • *Unité Mixte de Recherche, CNRS-BioMérieux, Ecole Normale Supérieure de Lyon, France*

ALFRED PINGOUD • *Institüt für Biochemie, Justus-Liebig-Universität Giessen, Germany*

ADRIAN PUNTSCHART • *Department of Anatomy, University of Bern, Switzerland*

LUC RAEYMAEKERS • *Laboratorium voor Fysiologie, KULeuven Campus Gasthuisberg, Leuven, Belgium*

UDO REISCHL • *Institute for Medical Microbiology and Hygiene, University of Regensburg, Germany*

STEPHEN H. RICHARDSON • *Department of Microbiology and Immunology, Bowman Gray School of Medicine, Winston-Salem, NC*

BENOÎT ROUSSEAU • *Laboratoire d'Immunologie Moleculaire, Université de Bordeaux II, France*

MARGARET SHEEHAN • *Elcatech, Winston-Salem, NC*

PAUL D. SIEBERT • *Clontech Laboratories, Palo Alto, CA*

ANU SUOMALAINEN • *Department of Human Molecular Genetics, National Public Health Institute, Helsinki, Finland*

ANN-CHRISTINE SYVÄNEN • *Department of Human Molecular Genetics, National Public Health Institute, Helsinki, Finland*

YASSINE TAOUFIK • *Inserm Unit, Necker Hospital, Paris, France*

MICHAEL VOGT • *Department of Anatomy, University of Bern, Switzerland*

CHANG-NING WANG • *Biotronics Corporation, Lowell, MA*

# I

## Reviews

# 1

# Quantitative PCR

*A Survey of the Present Technology*

**Udo Reischl and Bernd Kochanowski**

## 1. Introduction

The polymerase chain reaction (PCR) is a powerful tool for the amplification of trace amounts of nucleic acids, and has rapidly become an essential analytical tool for virtually all aspects of biological research in experimental biology and medicine. Because the application of this technique provides unprecedented sensitivity, it has facilitated the development of a variety of nucleic acid-based systems for diagnostic purposes, such as the detection of viral *(1)* or bacterial pathogens *(2)*, as well as genetic disorders *(3)*, cancer *(4)*, and forensic analysis *(5)*. These recently developed systems open up the possibility of performing reliable diagnosis even before any symptoms of the disease appear, thus considerably improving the chances of success with treatment. For many routine applications, particularly in the diagnosis of viral infections, the required answer is the presence or the absence of a given sequence in a given sample. Therefore, PCR is in able for the early diagnosis of HCV infection *(6)*, HSV encephalitis *(7)*, or HIV infection of babies of HIV-positive mothers *(8)*. On the other hand, since even minute amounts of DNA are detected, the medical interpretation of positive results for widespread infectious agents like CMV *(9)* or HHV6 *(10)* turned out to be rather difficult.

Nevertheless, with the continuous development of PCR technology, there is now a growing need, especially in areas, such as therapeutic monitoring *(11–13)*, quality control, disease diagnosis *(14)*, and regulation of gene expression *(15)*, for the quantitation of PCR products, and thereby deducing the number of template molecules present in a sample prior to amplification.

From: *Methods in Molecular Medicine, Vol 26: Quantitative PCR Protocols*
Edited by: B. Kochanowski and U. Reischl © Humana Press Inc., Totowa, NJ

In contrast to a simple positive/negative determination, inherent features of the amplification process may constrain the use of PCR in cases where an accurate quantitation of the input nucleic acids is required. Although the theoretical relationship between the amount of starting template nucleic acid and the amount of PCR product can be demonstrated under ideal conditions, this does not always apply for most typical biological or clinical specimens. Dealing with PCR-based quantification of nucleic acids, one has always to keep in mind that any parameter that is capable of interfering with the exponential nature of the in vitro amplification process might ruin the in sic quantitative ability of the entire procedure. Even very small differences in the kinetic and efficiency of individual amplification steps will have a large effect on the amount of product accumulated after a limited number of cycles.

Inherent factors that will lead to tube-to-tube or sample-to-sample variability are, for example, thermocycler-dependent temperature deviations, the presence of individual DNA polymerase inhibitors in clinical samples, pipeting variations, or the abundance of the target sequence in the specimen of interest *(16,17)*. Various approaches have been developed in the last few years to circumvent these problems, but the extremely desirable goal of truly quantitative PCR has still proven elusive.

Here we would like to present an overview on the current methodology and to address the advantages as well as the limitations of individual protocols. Since the number of applications is increasing with the volumes of relevant journals, this article should provide a knowledge base for investigators to become familiar with quantitative PCR-based assays and even guide them in setting up their own assay systems. For ease of presentation, a brief summary of statistical aspects of the amplification reaction will be given, followed by a more detailed overview of detection strategies and procedures, and an appraisal of their value in the quantitation of PCR products.

## 2. Strategies to Obtain a Quantitative Course of Amplification: How to Make an Exponential Reaction Calculable

### 2.1. Theoretical Framework of PCR

It is well known that the PCR educt is amplified during the PCR procedure in an exponential manner. (Note: throughout the text, we will use the term "PCR educt" for the target of interest prior to amplification, whereas the term "PCR product" refers to the corresponding amplification products.) A mathematical description for the product accumulation within each cycle is:

$$Y_n = Y_{n-1} \cdot (1 + E_v) \text{ with } 0 \leq E_v \leq 1. \tag{1}$$

$E_v$ represents the efficiency of the amplification, $Y_n$ the number of molecules of the PCR product after cycle $n$, and $Y_{n-1}$ the number of molecules of the PCR product after cycle $n-1$. To calculate the number of molecules of the PCR product after a given number of cycles from the starting amount of PCR educt, this recursive equation has to be solved. Since $E_v$ stays constant for a limited number of cycles during the exponential phase of the amplification reaction, this is only possible within this particular period. Therefore, the accumulation of the PCR product can be approximately described by **Eq. 2**:

$$Y = X \cdot (1 + E_e)^n \tag{2}$$

$Y$ represents the number of molecules of the PCR product, $X$ the PCR educt molecules, $n$ the number of cycles, and $E_e$ the efficiency with a value between 0 and 1. **Equation 2** is valid only for a restricted number of cycles, usually up to 20 or 30. Then the amplification process slows down to constant amplification rates, and finally it reaches a plateau where the target is not amplified any more. For **Eq. 1** this would result in a steady decline of $E_v$, until the value reaches 0. The over all efficiency ($E$) of the amplification process is dependent on the primer/target hybridization, the relative amount of the reactants, especially the DNA polymerase/target quotient, and it may vary with the position of the sample in the thermocycler or the presence of coisolated DNA polymerase inhibitors in different clinical samples. The number of cycles for which **Eq. 2** holds true is partly determined by the amount of PCR educt. Target strand reannealing and enzyme saturation events are leading to a decline of $E_v$ *(16,17)*.

As described later, is it easy to quantitate the PCR product, but because of varying efficiencies ($E_e$) and varying numbers of cycles ($n$) for which **Eq. 2** is valid, the result does not necessarily represent the amount of PCR educt. As already mentioned, inherent tube-to-tube and sample-to-sample variations are potential causes. At least three procedures of a PCR setup are described in the following paragraphs that have been devised to rule out those variabilities. The measures that have to be carried out are dependent on the  desired precision. In general, it is much easier to determine relative changes than to quantitate absolute numbers of the PCR educt. For measuring RNA copy numbers, the varying efficiencies of the reverse transcription process have to be normalized, and for low copy numbers of the PCR educt, stochastical problems have to be taken into account *(18)*.

## 2.2. PCR-Based Quantification with External Standards

A serial dilution of a known amount of standard, often a plasmid, can be amplified in parallel with the samples of interest. Provided that a linear PCR product/PCR educt relation for the standard dilution series is observed, the relative amount of PCR educt for samples in the same PCR run can be deduced. A typical example is shown in **Fig. 1**. Using replicates, this method may provide fairly accurate results and even rule out tube-to-tube variations, but it is

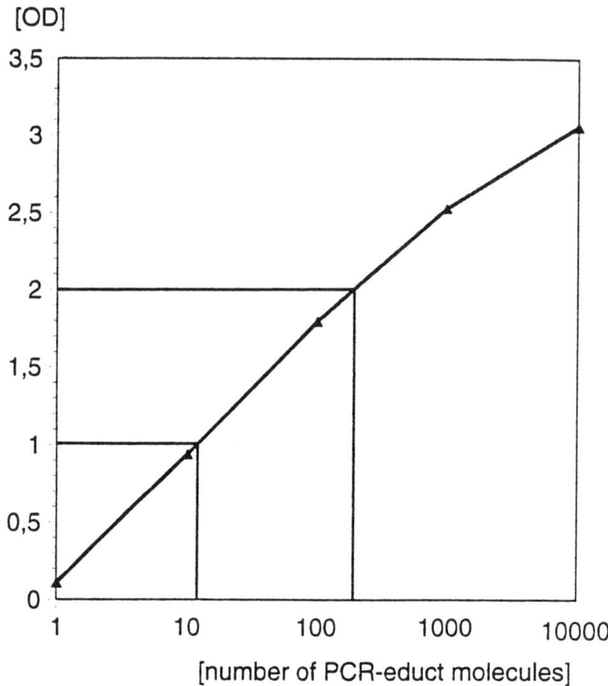



Fig. 1. ELOSA-based PCR quantification of HBV amplification products according to the external standard procedure. As a reference, a standard plasmid dilution series was subjected to PCR amplification. The biotin-labeled PCR product was hybridizised with a digoxigenin-labeled probe, bound to streptavidin-coated microtiter plates and subsequently quantitated using <DIG>:HRP conjugate and 2.2'-azino-di {2-ethyl-benzthiazolin-sulfonat] *(6)*. An examplary curve is shown—with the variation that the ELOSA-derived value for 1 molecule of PCR educt is not positive in every experiment (for statistical reasons). It is shown that two samples with OD values of 1.0 and 2.0 would correspond to 15 and 200 mol of PCR educt/vol, respectively.

not capable to rule out sample-to-sample variations. A potential and always lurking drawback to this simple procedure is the sensitivity of the PCR for small variations in the setup. Because of resulting differences in the efficiency, they may devastate precision and reproducibility. Therefore, if a quantification with external standard is established, precision (replicates in the same PCR run) and reproducibility (replicates in separate PCR runs) has to be analyzed to understand the limitations within a given application.

Keeping **Eq. 2** in mind, it is clear that quantification with this procedure must be done in the exponential phase, which is also dependent on the relative

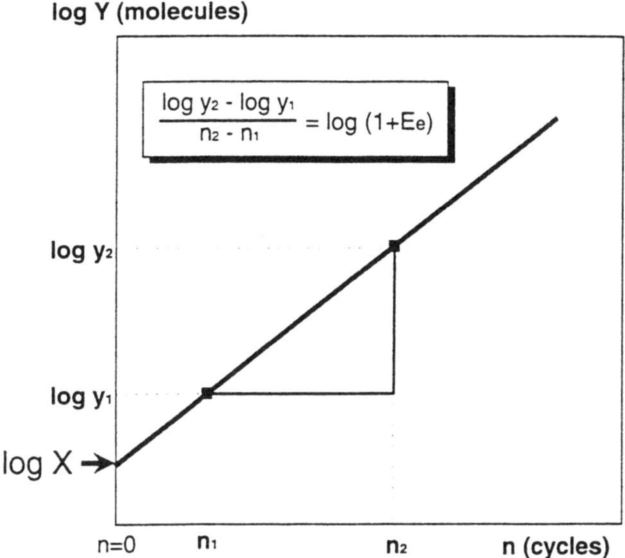



Fig. 2. Determination of the number of molecules of the PCR educt (*X*) from the amount of PCR product after cycle number *n*1 and *n*2 (*Y*1 and *Y*2, respectively (30). *X* can be calculated according to **Eq. 3**.

amount of the PCR educt. Rigorous analyses have to be performed to demonstrate that with increasing number of cycles, the results do not change.

A more sophisticated application for PCR quantification is the determination of the amount of PCR product molecules with increasing number of cycles. After the transformation of **Eq. 2** to:

$$\log(Y) = \log(X) + \log(1 + E_e) \cdot n \tag{3}$$

a linear relationship between the PCR product $\log(Y)$ and $n$ can be drawn, provided $E_e$ remains constant. Then the PCR educt $\log(X)$ can be tentatively determined as the *y*-intercept, which can be extrapolated from the slope $\log(1 + E_e)$ as shown in **Fig. 2**. In this case, no external standards are needed, although well-defined positive controls seem essential. A possible problem with this procedure is the fact that within the first few cycles of the PCR, the efficiency (*E*) is much lower than between cycles 10 and 30 *(18)*. In spite of this theoretical problem, it seems nevertheless possible to gain realistic results *(19)*.

This procedure has the advantage that different amplification efficiencies ($E_e$) of the samples will be detected, if the absolute number of PCR product molecules can be determined. In our hands, quantification with external standards proved to be sufficient to gain primarily quantitative results of DNA targets isolated from acellular clinical samples. The isolated DNA is then subjected to competitive PCR, where less competitors are necessary (*see* **Subheading 2.4.**).

Because of higher sensitivity, PCR-based quantification with an external standard has been recently used in connection with nested PCR, but since the major problem of nested PCR is connection, there is a greatest risk if no internal control is used. If one of the recently developed highly sensitive detection methods *(see below)* is applied for the detection of the first-round products, nested PCR can be avoided at most of the common applications.

A variation of this procedure is the limited dilution analysis of the PCR educt. The PCR analysis is performed with a dilution series of the educt *(10,20–22)*. The least positive sample is thought to contain the same amount of PCR educt as the last positive sample of a dilution series of a known standard. This procedure also has been used in conjunction with nested PCR.

Limited dilution analysis has the disadvantage that efficiencies of different PCR runs may vary, so that the reproducibility could be low. Another problem that emerges is the Gaussian distribution of a low number of PCR educts within a sample. Therefore, each dilution has to be analyzed repeatedly for a correct identification of the least positive sample.

### 2.3. Quantification with Noncompetitive Internal Standards

Depending on the extraction procedure applied, nucleic acids isolated from cellular material usually contain a lot of nontarget DNA or RNA. The presence of cellular nucleic acids facilitates the coamplification of one of these cellular targets with the target of interest within the same PCR tube (multiplex PCR). This second cellular target shares neither the primer binding sites nor the region in between with the target of interest. For DNA-PCR, almost any gene would do. Typical targets, for example, are pyruvate dehydrogenase *(23)*, proenkephalin *(24)*, or $\beta$-actin *(25)*. For RNA-PCR, the task turns out to be more difficult. Here a cellular mRNA has to be selected that has an even level of transcription and is in dent of different degrees of cellular activation. A lot of mRNAs have been evaluated for this purpose. First attempts had been performed with mRNA for HLA, $\beta$-actin, DHFR, or GPDH *(26–29)*. More recently, the mRNA of histone H3.3 or the 14S rRNA has been used as a cellular target *(30,31)*.

To our knowledge, no comparison of the different internal standards has been published so far, and it is still unknown if all of them fulfill the criteria of an even and undisturbed transcription. Since this is the crucial point of the entire procedure, more attention should be paid to it. Since, for example, HLA-antigens, and thus the corresponding mRNA, are downregulated by Epstein-Barr virus (EBV) *(32)*, they should not be used as internal standards for the quantification of EBV mRNA. It is also known that β-actin mRNA levels are increasing with the malignant transformation of cells *(33)*.

The main advantage of this procedure is its simplicity and the fact that no profound molecular biology is needed. Replicates rule out tube-to-tube and to some extent sample-to-sample variations, although individual inhibitors of the polymerase may be missed. On the other hand, this method bears some pitfalls that should be kept in mind. The efficiency of the reverse transcription for the internal standard and the target of interest may vary, and more disturbing, it may even vary dramatically for the same target *(34)*. Therefore, it seems to be very cumbersome to use this procedure for RNA-PCR. Quantitation during the exponential phase of the amplification process makes it possible to determine relative changes of the primary target, but if it is not checked that both targets are showing the same amplification efficiency $(E)$ within a given number of cycles, absolute quantification is not possible.

Quantification with a noncompetitive internal standard has been reviewed in detail by Ferre *(34)*. He demonstrated, as reasoned above, that the procedure is useful for monitoring relative changes of nucleic acid targets. He stated, nevertheless, that several replicates have to be applied and that, owing to a given precision, at least a twofold change of the PCR educt is required to detect a relative change. Therefore, each new setup of the assay requires a complete reevaluation of the parameters discussed above.

## 2.4. Competitive PCR

For competitive PCR, an internal standard has to be constructed that competes with the primary target for enzyme, nucleotides, and primer molecules. The competitor bears the same primer binding region, but the sequence in between is modified in such a way that amplification products derived from the competitor and the target of interest can be differentiated, for example, by gel-electrophoreses, enzyme-linked oligonucleotide sorbent assay (ELOSA), or HPLC. As long as the number of molecules of both PCR educts are equal, it is theoretically possible to use a competitor within a nested PCR assay *(35)*. *In praxi*, for each application, it has to be demonstrated that it really works in conjunction with nested PCR. We observed, for example, that a reduction of the cycle number within the second PCR did increase the capability power of the nested PCR procedure for quantification purposes.

For initial attempts, competitors were used that differ from the wild-type target only by a point mutation. In most cases, these point mutations are introduced in such a way that an additional restriction enzyme recognition site is created within the competitor nucleic acid *(36,37)*. Following restriction enzyme cleavage, the resulting products of competitor and primary target can be easily separated by electrophoresis on an agarose gel and quantitated by hybridization with a labeled probe or with the help of a labeled PCR primer. Although these competitors are showing a very high degree of similarity to the wild-type product, this procedure is no longer regarded as a quantitative one. This is owing to the fact that the amplification products have to be diluted and that a second enzymatic step is necessary. In particular, if the amplification products of the competitor are not cut completely by the restriction enzyme, a false quantification results.

More recently, deletions of a part of the wild-type sequence or insertions of foreign sequences are used for the *de novo* construction of competitors, which are analyzed by gel electrophoresis *(38)*. Reviewing the literature, it seems obvious that there are no general rules or strategies for the construction of these modifications *(39–43)*. Often a critical analysis of precision and reproducibility is found, but a more detailed evaluation of the amplification efficiencies ($E_e$) of the wild-type target and the competitor has, to our knowledge, in most cases not been performed. Usually it is demonstrated that these applications allow a relative quantification, and it is assumed that an absolute quantification can also be performed. Computer simulations confirmed recently that different amplification efficiencies ($E_e$) of the wild-type target and the competitor may allow a very precise relative quantification, although an absolute quantification is out of reach *(44)*. For absolute quantification, it is therefore most important to demonstrate that $E_e$ of the wild-type target and the competitor are equal. It may be also very helpful to evaluate the competitor on samples with a known amount of wild-type target molecules.

Competitors for microtiter plate-based assays do not need to have a different length, since they are differentiated from wild-type amplification products by sequence. Therefore, specific sequences may be deleted or inserted, and both targets can be detected separately by hybridization procedures. Again, the amplification efficiencies of both target and competitor have to be equal to allow absolute quantification; otherwise, only relative quantification is possible.

For quantitating single clinical samples, one has to perform several competitive PCR assays with a constant amount of the target of interest and varying amounts of competitor. That is owing to the fact that only equimolar amounts of competitor and the target of interest result in a reliable quantification. It is likely that the number of competitive PCR assays needed is reduced by the application of ELOSA-based assays (B. K., unpublished results and *41*).

In general, since competitive PCR is capable of ruling out tube-to-tube and sample-to-sample variations, it seems to be the method of choice for accurate PCR quantification. If the criteria mentioned above are taken into account, we consider this procedure appropriate for absolute quantification and for quantification of low copy targets.

## 3. Detection and Quantitative Measurement of PCR Products

### 3.1. Labeling of PCR Products

By itself, the amplification of a target nucleic acid is not an analytical procedure. To detect the presence and specifity of amplified DNA and, if necessary, to quantitate the amount of specific PCR products present in the reaction mixture, the amplification system has to be linked to an appropriate detection system. For this purpose, the amplification products have to be equipped with any kind of label that can be detected subsequently either in a direct or indirect way. For many years, the most commonly used methods for the detection of PCR-amplified DNA were based on radioactive labels. Because of the difficulties encountered in the handling of such radioactive isotopes, a variety of highly sensitive nonradioactive indicator systems have been developed. Suitable nonradioactive labels include helix-intercalating dyes, like ethidium bromide or bis-benzimide *(45)*, covalent bound dyes (e.g., fluorescein) or enzymes (e.g., horseradish peroxidase [HRP]) *(46)*, and alkaline phosphatase *(47)* as well as distinct reporter molecules, such as digoxigenin or biotin. For detailed reviews on the variety of direct and indirect nonradioactive bioanalytical indicator systems, *see* **refs.** *48* and *49*.

Since the PCR is based on the oligonucleotide-primed *de novo* synthesis of template-complementary DNA by the enzymatic action of a DNA polymerase, nonradioactive reporter molecules can be easily incorporated into the amplification products either in the presence of labeled deoxyribonucleotide (dNTP) an logs and/or labeled primer oligonucleotides present in the amplification mixture *(50,51)*. Labeled deoxyribonucleotides are commercially available in the form of digoxigenin- or biotin-dUTP (e.g., Boehringer Mannheim GmbH, Mannheim, Germany). Primer oligonucleotides can be precisely labeled at their 5'-end during their chemical synthesis using digoxigenin-, biotin- or fluorescein-phosphoramidite components. Labeling with photodigoxigenin, a photoreactive compound that binds covalent to amino groups upon UV irradiation *(52)*, results in a statistical distribution of digoxigenin molecules along the oligonucleotide.

Bifunctional conjugates, like antidigoxigenin antibody fragments (<DIG>) or streptavidin (SA), covalently linked to the customary enzymes HRP or alkaline phosphatase (AP) were commonly used for the detection of labeled PCR products in an ELISA-type reaction. The high stability of these enzymes, their wide application in diagnostic assays, and the development of appropriate detection systems are factors that have contributed to their suitability as reporter enzymes. Once a digoxigenin-labeled amplification product is fixed on a solid phase, incubation with <DIG>:AP conjugate, for example, resulted in a tight attachment of the antibody portion to the digoxigenin residues, and the enzyme portion of the bifunctional conjugate is capable of catalyzing subsequent color reactions that yield optical, luminescing *(53)* or fluorescing signals *(54)*, depending on the substrate used. Since the resulting signal can be precisely quantified by appropriate instrumentation, this strategy has recently be come well established in the field of quantitating PCR products. The use of enzymes for signal generation can also be considered an amplification method, since many product molecules are produced per enzyme molecule.

Detection strategies for amplification products can generally be divided in two parts. On the one hand, there are assay systems that are capable of detecting the presence or the absence of amplification products, and on the other hand, there are assay systems that are specific for amplification products with a given sequence. Although the border between these assay formats is vague, for ease of presentation, we decided to divide this chapter into nonsequence-specific and sequence-specific detection systems, and to outline the individual principles with the help of selected examples.

### 3.2. Nonsequence-Specific Detection Systems

A lot of PCR applications are already optimized with regard to the buffer $MgCl_2$ condition, temperature profile, and so forth, and are leading to well-defined amplification products without the formation of any byproducts that are different in size. Under suitable conditions, the relative amount of amplification products in these cases is strictly dependent on the amount of starting material present in the amplification mixture. Therefore, quantification of the PCR products by physical or enzymatic means is almost sufficient for a rough determination of the amount of the PCR educt (*see* **Subheading 2.1.**).

### 3.2.1. Gel Systems

Applicable formats include well-established laboratory techniques, like agarose or polyacrylamide gel electrophoresis, and subsequent quantitative detection of ethidium bromide-stained amplification products using gel scanners or suitable computer-assisted video equipment. A quantitative detection of radioactive labeled amplification products can be accomplished either by autoradiography or by Cherenkov counting of excised gel pieces. A recent development in the field is the application of automated DNA sequencers for the quantification of fluorescence-labeled nucleic acids (e.g., Applied Biosystems 373A DNA sequencer in combination with the GeneScan software [Applied Biosystems, a division of Perkin-Elmer, Foster City, CA]). With the help of these instruments, the gel-associated lack of sequence specificity can be nearly overcome by an accurate size determination in the basepair range and ultimate detection sensitivities in the femtomole range of individual dye-labeled amplification products. Since these instruments can differentiate up to four distinct fluorescence dyes, internal or external standards can be applied and analyzed in parallel within the same gel lane as the amplification products, thus reducing the possibility of lane-to-lane artifacts. Since automated DNA sequencers and fluorescence-labeled primers are still expensive, at present, this promising technique is main restricted to research applications.

### 3.2.2. HPLC

Direct HPLC of PCR products using, for example, a 2.5-μm nonporous polymer-based an ion-exchange column, a 12- to 25-min gradient cycle time and UV absorbance detection have been shown to meet the analytical criteria for practical PCR product quantitation *(55)*. PCR samples can be injected onto the column directly after amplification without further purification, and the sensitivity is adequate to provide the detection of unlabeled amplification products in the femtogram range (this corresponds to around $10^3$ molecules of PCR educt). The detection limit of labeled amplification products may be lower and will depend on the availability of suitable detector systems (e.g., the use of fluorescence-labeled primers in conjunction with a fluorescence detection device *[56]*) The size-differentiating features of HPLC even allow the use of internal standards different in size to align variations in amplification efficiency more precisely. If the amplification parameters are well adjusted, the linear form of the graph of PCR product output vs log (template input) leads to a calibration curve that comes up to four decades of target concentration into one decade of HPLC-quantitated PCR product concentration.

### 3.2.3. Solid-Phase Assays

In general, the attachment of amplification products to a solid phase is advantageous to carrying out several measures in parallel and under comparable conditions. The most widely used and convenient solid-phase plastic support medium for this kind of bioanalytical assay is the 96-well microtiter plate. These plates lend them selves to some degree of automation, such as the use of plate washers and, for colorimetric enzyme assays, the use of multichannel spectrophotometric plate readers. Since many proteins adsorb passively to polystyrene by hydrophobic interaction, it is possible to coat microtiter plates with molecules like streptavidin. This results in a solid-phase medium that is capable of the specific capture of biotin or, in practice, biotinylated molecules. Streptavidin-precoated plates are already available from different manufacturers and are well suited for setting up quantitative assays for biotinylated PCR products. Double-labeling of PCR products with biotin and reporter molecules like digoxigenin can be employed for a subsequent quantification in microtiter plate-based assay formats. The simultaneous incorporation of biotin and digoxigenin into the amplification products can either be achieved in the presence of both digoxigenin- and biotin-deoxyribonucleotide analogs (DIG-/bio-dUTP) in the amplification mixture or, in a more specific manner, in the presence of a biotinylated primer 1 and DIG-dUTP or a biotinylated primer 1 and a digoxigeninylated primer 2. Since the absolute concentration of labeled deoxyribonucleotide analogs in the reaction mixture has a significant influence on the *Taq* DNA polymerase activity (reduced elongation rate), the optimal concentration of DIG-/bio-dUTP has to be determined individually (*see* **Fig. 3**).

In a typical assay format, double-labeled amplification products were ethanol-precipitated (to remove unincorporated label) and subsequently incubated in streptavidin-coated microtiter plates for at least 2 h at room temperature with occasional shaking. Following several wash steps, incubating with <DIG>:AP conjugate and a substrate solution results in the generation of a quantitative color or fluorescence signal, depending on the substrate used. The solid-phase capture of labeled amplification products mediated by the streptavidin–biotin interaction allows for the accomplishment of the indicator reaction in  solution; this is essential for quantitatively determining the concentration of target molecules. Similar to standard ELISA, the colorimetric detection of the PCR products makes this quantification procedure suitable for screening a large number of samples. Taken into account that this convenient assay for mat is not sequence-specific and the signal measured is assembled from all amplification products, whether specific or not, present in the amplification mixture, its use is limited to PCR reactions that lead to a well-defined product. On the other hand, these intrinsic limitations can be easily overcome by the sequence-specific hybridization of biotin-labeled probes to digoxigenin-labeled amplification  products and a subsequent detection in form of double-labeled hybrids according to the ELOSA principle (*see* **Subheading 3.3.2.**).

In general, streptavidin-mediated solid-phase capture of biotin-labeled target molecules in solution turned out to be an effective, versatile, and easy to handle assay format and will certainly evolve into a key technology in the field of quantitative PCR.

### 3.2.4. SPA Assay

The scintillation proximity assay (SPA) is based on a similar concept. This assay relies on the use of fluomicrospheres as the solid phase, coated with acceptor molecules that are capable of binding labeled ligands in solution *(57)*. In a typical application of this technique, one of the PCR primers is labeled with biotin, and tritiated nucleotides are incorporated during the amplification reaction. Once the amplification procedure is complete, streptavidin-coated SPA beads (Amersham International, UK) are added to capture the biotinylated

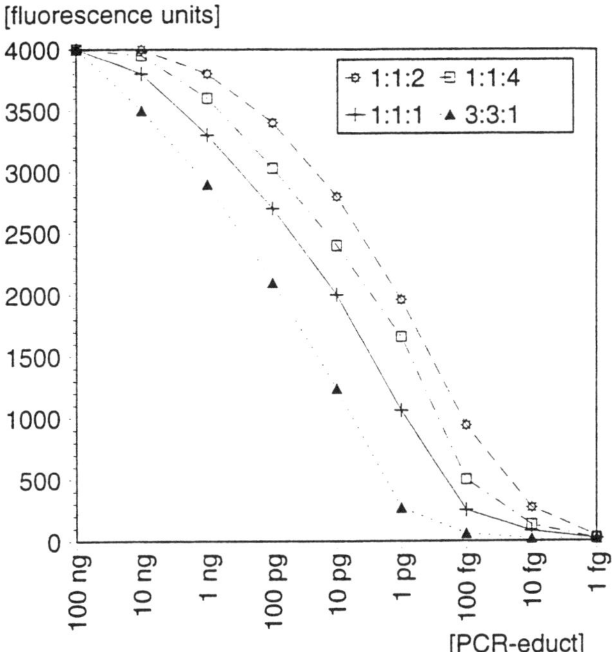



Fig. 3. Quantification of biotin/digoxigenin double-labeled HBV amplification products (543 bp) in streptavidin-coated microtiter plates using <DIG>:AP conjugate and 4-methyl-umbilliferyl-phosphate as flourescing substrate (plot is based on average values). The molar ratio of DIG-dUTP:dTTP in the reaction mixture is indicated.

PCR products. This capture event brings the tritium close enough to the microsphere so that the fluor incorporated within it is excited to emit a pulse of light that is measurable in a conventional scintillation counter. On the other hand, the majority of unincorporated tritium molecules are too far away from the SPA beads to enable the transfer of energy. Compared to color-developing assays, the SPA format has a broader linear detection range. Using unlabeled primers in combination with a postamplification hybridization with biotinylated probes complementary to an internal sequence of the amplicon, this quantitative assay format can also be configured to be sequence-specific. For example, this system has been successfully applied to the quantification of cytomegalovirus DNA in blood specimens and was capable of detecting changes in the level of viral DNA within a three-log dynamic range and a detection limit of $4 \times 10^4$ molecules of PCR educt *(22)*.

### 3.2.5. Transcription-Mediated Detection

Quantitative measurement of specific mRNA species can be achieved by a combination of RT-PCR and a subsequent in vitro transcription reaction. In the course of this straightforward method, a T7 RNA polymerase promoter sequence is incorporated during the PCR reaction at the 5'-end of the amplification products, and following the amplification reaction, an in vitro transcription reaction is carried out in the presence of labeled ribonucleotides. The linear transcription reaction greatly increases the amount of amplified product and, there by, gains an additional dimension of sensitivity for the detection of low-abundance mRNA. Using the expression of an endogenous gene as a denominator for normalization of the quantitative data, an internal control is provided for the amount of intact RNA successfully isolated and converted to cDNA. This method is aimed at measuring the relative rather than the  absolute levels of gene expression by determining a ratio between PCR products of the desired target gene and an endogenous internal standard gene in separate reactions, and then comparing it with the same ratio in another sample. Using  serial dilutions of the cDNA samples, a less than twofold difference in gene expression can be discriminated even if the absolute amount of input mRNA or cDNA is not known *(58)*.

## 3.3. Sequence-Specific Detection Systems

Although the theory of PCR is straightforward, the primers are frequently annealing to nontarget sequences, especially in complex template mixtures, and this so-called mispriming significantly lowers the purity of the amplified target portion in the final product. Therefore, probe-based methods remain a key feature of current detection  systems primarily because of the additional information and sequence specifity they provide. Probes have been adjusted to nonisotopic colorimetric systems by labeling them with reporter molecules, such as digoxigenin, biotin, or distinct enzymes, or with dye molecules capable of emitting light (chemiluminescence). In the field of quantitative PCR, probes were mainly bound to the wells of microtiter plates since this format has certain advantages for reproducible results and automation.

### 3.3.1. Dot-Blot Procedures

Classically, hybridization assays are carried out by dot-blot or Southern blot procedures, in which the amplified target is denatured, immobilized on a nitrocellulose or a nylon membrane surface, and then hybridized with an appropriate labeled DNA probe. Even if a laboratory is not equipped with an ELISA plate reader, sequence-specific detection and quantification of amplification products can be carried out within a simple dot-blot format. After spotting and immobilization of the PCR products on a nylon membrane, the dot-blot meth-

odology utilizes the sequence-specific hybridization of labeled oligonucleotides to indicate the presence or absence of specific amplified sequences. The
reverse dot-blot procedure is based on sequence-specific oligonucleotide
probes immobilized on a nylon membrane via link age of poly-T tails and subsequent hybridization with denatured labeled amplification products that are in
solution. Since the target is not directly bound to the membrane surface, the
reaction kinetics in this assay essentially approach a liquid phase, which allows
a rapid hybridization reaction. If biotinylated probes or biotinylated amplification products (in the case of the reverse dot-blot) are used, the nonradioactive
detection is usually carried out with streptavidin–AP conjugates producing a
colored dot. Under ideal conditions and in comparison with samples with
known concentration, the color intensity represents the relative amount of specific amplification products. In the case of reverse dot-blot procedures, a more
accurate quantification of amplification products can be achieved by providing
membrane strips with a series of dots containing a shading amount of probe.
Apart from quantitative applications, this format offers the practical advantage
of detecting multiple alleles within a given amplification product simultaneously (HLA-DQA genotyping *[59]*) or different pathogens in a single
hybridization reaction.

### 3.3.2. Solid-Phase Capture

The adaption of the solid-phase capture technique to microtiter plates or
paramagnetic beads results in the most convenient assay formats for the
sequence-specific detection and quantification of PCR products in routine practice. With respect to the basic principle, they were recently named ELOSA.
Although individual strategies have been developed, these assay formats share
a common principle: molecules that support the solid-phase capture and molecules that mediate the subsequent detection are located on different strands of
nucleic acids. In contrast to the double-labeling of PCR products mentioned
above, double-labeled molecules are formed within these assays exclusively
on the hybridization of labeled probes to labeled PCR products.

Providing the sequence-specific detection of distinct amplification products
in a complex mixture, this post-PCR hybridization event is also crucial for
most of the quantitative procedures. In principle, there are two different
hybridization-based concepts for the capture and subsequent detection of
amplification products on a solid phase.

### 3.3.2.1. Immobilized Capture Probe

Oligonucleotides representing a characteristic part of the amplified
sequence, so-called capture probes, are attached either covalently or via
biotin:streptavidin linkage to a solid phase, and labeled PCR products are

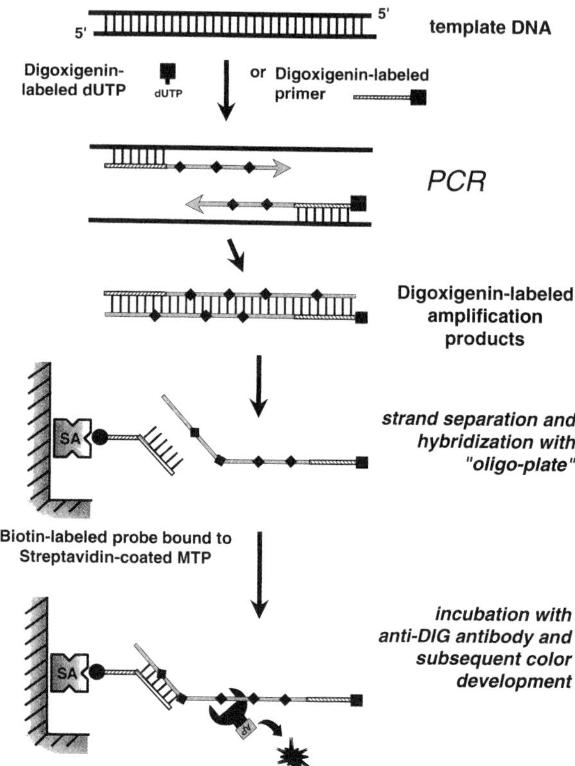



Fig. 4. Immobilized capture probe. Following strand separation, a sequence-specific detection of digoxigenin-labeled amplification products is carried out by hybridization to immobilized probes.

hybridized using stringent conditions. Following several wash steps, the amount of specific amplification products can be determined by a label-mediated detection reaction (*see* **Fig. 4**).

### 3.3.2.2. IMMOBILIZED AMPLIFICATION PRODUCT

Biotin-labeled PCR products are attached to a streptavidin-coated solid phase and subsequently hybridized with a labeled probe complementary to internal sequences of the specific amplification product (*see* **Fig. 5**). Another possibility is the covalent binding of aminated amplification products to carboxylated wells of microtiter plates *(60)*.

Although ingenious protocols have been developed (e.g., **ref.** *61*), for reliable results, it is advisable to denature the double-stranded PCR products via

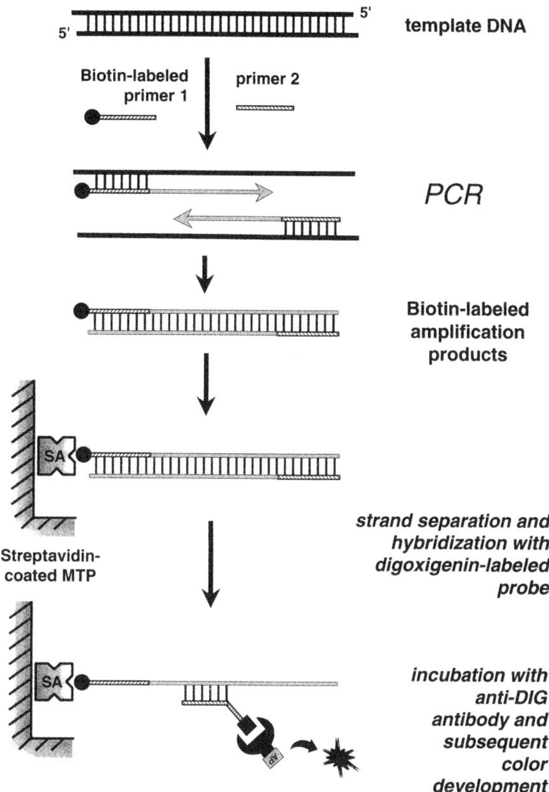


Fig. 5. Immobilized amplification product. 5'biotin-labeled amplification products are immobilized on a streptavidin-coated microtiter plate. Following strand separation, a sequence-specific detection is carried out with the help of digoxigenin-labeled probes.

heat or alkali treatment before hybridization with a specific probe. As an in-solution assay, there is a constant diffusion of target and probe that speeds up the reaction kinetics and allows for a rapid hybridization reaction. The sensitivity mainly depends on the label used for the subsequent detection of the hybrids. Within these kinds of experiments, the use of digoxigenin labels and <DIG>:AP or <DIG>:HRP conjugate is recommended in combination with substrates yielding an optical, luminescing, or fluorescing signal. The detection can be automated using ELISA readers, and usually sensitivities in the attomole range of PCR educt are obtained. Although this detection format does not offer the utmost sensitivity levels, in our hands, it proved to be sufficient for the majority of quantitative applications (*see* **Fig. 1**). Furthermore, this hybridization format opens up the simultaneous quantitation of amplification products and internal standards that are equal in length, but differ in distinct nucleotide sequences. For example, target molecules and internal standards are coamplified using a set of biotinylated primers, attached to a streptavidin-coated solid phase, and subsequently hybridized to specific probes bearing different labels. After separate quantitation of the amount of each label, the initial concentration of the target molecules can be determined precisely in comparison to the internal standard. Apart from reporter molecules, like digoxigenin, chemiluminescent probes or distinct antibodies can be used as well for the hybridization-mediated detection of specific amplification products.

### 3.3.3. Electrochemiluminescence

A recently developed assay, the QPCR System from Perkin Elmer Instruments (Foster City, CA), utilizes the analytical capabilities of an electrically initiated chemiluminescent reaction (electrochemiluminescence) to provide sensitive and reproducible DNA quantitation at the attomole level. Again, the convenient assay for mat of streptavidin:biotin-mediated solid-phase capture of the amplification products to magnetic beads is applied in combination with a sequence-specific oligonucleotide probe labeled with Tris (2,2'-bipyridine) ruthenium (II) chelate (TBR). In contrast to commonly applied acridinium esters *(62)*, the high stability of ruthenium bipyridyl labels allows their incorporation during oligonucleotide synthesis *(63)*. Following hybridization, the bead-bound sample is supplemented with a tripropylamine solution (TPA) and is delivered to the detection cell of the electrochemiluminescence device. As the increasing voltage of the electrode reaches a specific level, a simultaneous oxidation of both the TPA and TBR occurs. The oxidized TPA is converted to an unstable highly reducing intermediate that reacts with oxidized TBR converting it to the excited state form. The excited-state species relaxes back to the ground state with the emission of light at 620 nm. Since the intensity of the emitted light is directly proportional to the amount of TBP labels present in the detection cell, the initial amount of specific amplification products can be quantitatively determined by measuring and integrating the light intensity at 620 nm. This system provides linear responses over more than three orders of magnitude (which corresponds to a dynamic range of at least four logs of initial PCR-educt copy numbers), sensitivities down to 70 attomoles *(64)*, and can be easily automated. In comparison to ELOSA techniques, no error-prone enzymatic steps are involved in these electrochemiluminescence procedures. Nevertheless, the impact of this theoretical advantage in practice has to be determined.

### 3.3.4. DNA Immunoassay

The availability of a monoclonal antibody (MAb) recognizing selectively double-stranded DNA has permitted the development of a novel enzyme immunoassay capable of detecting specific hybridization events. This methodology was adapted to the "immobilized capture probe format" mentioned above and has been termed "DNA Enzyme Immuno Assay" (DEIA; Sorin Biomedica, S.p.A. Saluggia, Italia) *(65)*. When DNA:DNA hybrids are formed between the capture probe and specific amplification products, the monoclonal anti-dsDNA an body is added and, as in conventional diagnostic ELISA systems, the presence and amount of DNA–anti complexes are indicated subsequently by a colorimetric reaction developed with the help of a secondary enzyme-conjugated antibody (murine anti-IgG:POD). A comparable assay format is based on the hybridization of biotinylated PCR products with unlabeled RNA probes and a subsequent detection of the resulting hybrids with the help of an enzyme-labeled antibody specific for DNA:RNA hybrids. These immunoassays can be used for the detection of any type of amplified DNA and eliminate the need for labeling DNA or primers. The DEIA assay has already been successfully applied to detect the presence of the gene coding for HBV core antigen and HLA typing. The possibility of crossreactions and the cost of these MAb are limiting the as says potential large-scale application at present.

### 3.3.5. Primer Elongation Assay

The single nucleotide primer extension assay (SNuPE) represents one of the most practicable assay formats for the identification and quantification of point mutations (e.g., allelic variants in DNA or RNA) and the measurement of specific mRNA levels. This post-PCR assay consists of the enzymatic extension by one base of an oligonucleotide primer hybridized just 5' to the position of mismatch in the presence of only one labeled dNTP specific for either the wild-type or a variant sequence (*see* **Fig. 6**). Here a previous solid-phase capture of amplification products is not absolutely required, since the introduction of the label by the template-dependent elongation of a perfect matching primer is specific for a given sequence within the amplification products. Nevertheless, a selective ethanol precipitation and agarose gel purification of the PCR products should be carried out prior to the assay, since the complete removal of dNTPs present in the initial amplification mixture is an essential prerequisite to obtaining quantitative results. A major advantage of the method is its usefulness for quantitative measurement over a wide range. Furthermore, a given transcript can be detected in up to 1000-fold excess of RNA from other alleles, depending on which nucleotides differ. This method can be easily adapted for

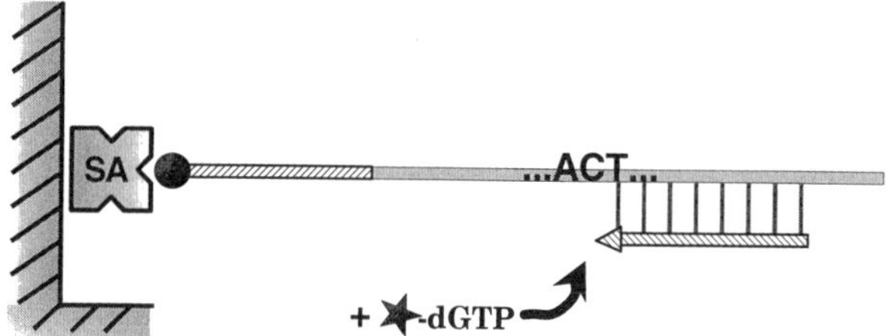



Fig. 6. Primer elongation assay. A distinct oligonucleotide primer is hybridized with its 3'-end immediately next to the base of interest within a denatured amplification product and subsequently elongated in the presence of corresponding labeled deoxyribonucleotide by the enzymatic action of a DNA polymerase.

quantitation of absolute amounts of a specific transcript by the addition of an internal standard *(66)*. Under optimal conditions, the background is below 1%, but varies significantly with the different kind of mismatches *(67)* (*see* Chapter 15).

## 4. Future Prospects

Techniques allowing for a precise quantification of minute amounts of nucleic acids derived from in vitro amplification techniques will undoubtedly have a substantial future impact on the practice of molecular biology and laboratory medicine. Especially in the field of medical diagnosis, techniques are desirable that are capable of providing the absolute amount of distinct nucleic acids rather than providing relative amounts. In the case of HIV infection, for example, absolute measurements of particular RNA levels will provide a means for following the progression of viral infection and monitoring the efficacy of therapeutic intervention *(11)*.

In the last few years, much effort has been spent in the development of detection systems with ultimate sensitivity. Since the overall performance of an analytical system is mainly dependent on its weakest part, some still unpredictable features of the real amplification procedure should be investigated in more detail. These investigations will provide further insight into the complex kinetics and may result in more robust amplification systems showing improved reliability. In contrast to the original purpose of PCR (the detection of as few target molecules as possible), for quantitative aspects, more stress should be placed on novel strategies that could improve the performance (e.g., linearization or enlargement of the exponential phase of the amplification procedure) rather than improving the overall sensitivity. For quantitative aspects, it is more important to differentiate between 100 and 500 molecules of PCR educt rather than to detect single PCR educt molecules. Emphasis should also be placed on the identification of suitable noncompetitive internal standards, which are not dependent on cell-cycle or cell-activation events. Another important aspect is the design of competitors, that are as similar as possible to the target of interest. This object can be achieved by the application of hybridization-based detection formats.

A promising application in the field of basic research and medical diagnostics is the quantification of distinct mRNA levels with the aim of elucidating gene regulation, virus replication, or immunological responses. Since this knowledge is an essential prerequisite for causative therapy and therapy monitoring, quantitative RT-PCR will evolve as a key technology in this field. The introduction of a thermostable DNA polymerase from Thermus thermophilus (rTth), which has both reverse transcription and DNA polymerase activities under certain reaction conditions, may eliminate the need for reopening the reaction tubes in the course of a RT-PCR and therefore reducing carryover contaminations.

Similar to techniques for the in vitro amplification of nucleic acids, the spread and acceptance of individual assays for the quantification of amplification products will ultimately be limited by cost, sensitivity, and specifity. For a list of actual applications, *see* **refs. *67–86***.

## 5. General Considerations

For standard PCR conditions, quantification should be carried out during the exponential phase of amplification. For this reason, it is important to optimize individual parameters of the entire amplification process carefully, so that the over all amplification can be controlled and the "plateau" phase avoided.

A quantitative PCR assay consists of three elements. Therefore, potential variations in the performance of the initial sample preparation should also be ruled out carefully, in addition to refining the amplification and detection procedures.

Standards used for the quantification of the sample should be chosen carefully to ensure reliable and accurate results. Here we recommend the use of recombinant plasmids, which can be easily created from individual amplification products with the help of the TA cloning kit (Invitrogen BV, NV Leek, The Netherlands).

For absolute quantification, the amplification efficiencies of the target of interest and the internal standard, whether competitive or noncompetitive, have to be determined. Internal standards should coamplify with the target of interest in equal efficiency.

To enhance the statistical validity of the data, it is recommended to carry out the assays several times.

Streptavidin-precoated microtiter plates from different manufacturers vary significantly in their ability to bind biotin-labeled amplification products, and there are no rules governing the choice of plate. Generally, a precise comparison should be carried out on sample plates using a well- defined dilution series of biotin-labeled amplification products.

## Acknowledgment

We gratefully acknowledge the support of Professor H. Wolf and Professor W. Jilg, giving us the opportunity to evaluate some of the latest quantitative procedures in our diagnostic laboratory.

## References


1. Clementi, M., Menzo, S., Bagnarelli, P., Manzin, A., Valenza, A., and Varaldo, P. E. (1993) Quantitative PCR and RT-PCR in virology. *PCR Methods Appl.* **2,** 191–196.
2. Tompkins, L. S. (1992) The use of molecular methods in infectious diseases. *N. Engl. J. Med.* **327,** 1290–1297.
3. Mazza, C., Mantero, G., and Primi, D. (1991) DNA enzyme immunoassay: A rapid and convenient colorimetric method for diagnosis of cystic fibrosis. *Mol. Cell. Probes* **5,** 459–466.
4. Zhang, X. Y. and Ehrlich, M. (1994) Detection and quantification of low numbers of chromosomes containing bcl-2 oncogene translocation using semi-nested PCR. *Biotechniques* **16,** 502–507.
5. Arnheim, N., White, T., and Rainey, W. E. (1990) Application of PCR: Organismal and population biology. *BioScience* **40,** 174–182.
6. Alter, M. J., Margolis, H. S., Krawczynski, K., et. al. (1992) Natural history of community-acquired hepatitis C in the United States. *N. Engl. J. Med.* **327,** 1899–1905.
7. Boerman, R. H., Peters, A. C. B., Arnoldus, E. P. J., Raap, A. K., van Loon, A. M., Bloem, B. R., and van der Ploeg, M. (1992) Polymerase chain reaction detection of herpes simplex virus in cerebrospinal fluid, in *Diagnosis of Human Viruses by Polymerase Chain Reaction Technology* (Becker, Y. and Darai, G., eds.), Springer-Verlag, Berlin, Germany, pp. 119–133.
8. Krivine, A., Yakudima, A., Le May, M., Pena-Cruz, V., Huang, A. S., and McIntosh, K. A. (1990) Comparative study of virus isolation, polymerase chain reaction, and antigen detection in children of mothers infected with human immunodeficiency virus. *J. Pediatr.* **16,** 372–376.
9. Weber, B., Nestler, U., Ernst, W., Rabenau, H., Braner, J., Birkenbach, A., Scheuermann, E.-H., Schoeppe, W., and Doerr, H. W. (1994) Low correlation of human cytomegalie DNA amplification by polymerase chain reaction with cytomegalovirus disease in organ transplant recipients. *J. Med. Virol.* **43,** 187–193.

10. Cone, R. W., Hackman, R. C., Huang, M.-L. W., Bowden, R. A., Meyer, J. D., Metcalf, M., Zeh, J., Ashley, R., and Corey, L. (1993) Human herpesvirus 6 in lung tissue from patients with pneumonitis after bone marrow transplantation. *N. Engl. J. Med.* **329,** 155–161.

11. Saksela, K., Stevens, C., Rubinstein, P., et al. (1994) Human immunodeficiency virus type 1 mRNA expression in peripheral blood cells predicts disease progression independently of the numbers of CD4+ lymphocytes. *Proc. Natl. Acad. Sci. USA* **91,** 1104–1108.

12. Futscher, B. W., Blake, L. L., Gerlach, J. H., Grogan, T. M., and Dalton, W. S. (1993) Quantitative polymerase chain reaction analysis of mdr1 mRNA in multiple myeloma cell lines and clinical specimens. *Anal. Biochem.* **213,** 414–421.

13. White, T. J., Madej, R., and Pershing, D. H. (1992) The polymerase chain reaction: Clinical applications. *Adv. Clin. Chem.* **29,** 161–196.

14. Bitsch, A., Kirchner, H., Dupke, R., and Bein, G. (1993) Cytomegalovirus transcripts in peripheral blood leukocytes of actively infected transplant patients detected by reverse transcription-polymerase chain reaction. *J. Inf. Dis.* **167,** 740–743.

15. O'Garra, A. and Vieira, P. (1992) Polymerase chain reaction for detection of cytokine gene expression. *Curr. Opinion Immunol.* **4,** 211–215.

16. Ruano, G., Brash, D. E., and Kidd, K. K. (1991) PCR: The first few cycles. *Amplifications* **7,** 1–4.

17. Sardelli, A. D. (1993) Plateau effect—understanding PCR limitations. Amplifications **9,** 1–5.

18. Wages, J. M., Jr. and Fowler, A. K. (1993) Amplification of low number sequences. *Amplifications* **11,** 1–3.

19. Wiesner, R. J. (1992) Direct quantification of picomolar concentrations of mRNAs by mathematical analysis of a reverse transcription/exponential polymerase chain reaction assay. *Nucleic Acids Res.* **20,** 5863–5864.

20. Drouet, E., Michelson, S., Denoyel, G., and Colimon, R. (1993) Polymerase chain reaction detection of human cytomegalovirus in over 2000 blood specimens correlated with virus isolation and related to urinary virus excretion. *J. Virol. Methods* **45,** 259–276.

21. Singer-Sam, J., Robinson, M. O., Bellve, A. R., Simon, M. I., and Riggs, A. D. (1990) Measurement by quantitative PCR of changes in HPRT, PGK-**1,** PGK-**2,** APRT, MTase, and Zfy gene transcripts during mouse spermatogenesis. *Nucleic Acids Res.* **18,** 1255–1259.

22. Rawal, B. K., Booth, J. C., Fernando, S., Butcher, P. D., and Powles, R. L. (1994) Quantification of cytomegalovirus DNA in blood specimens from bone marrow transplant recipients by the polymerase chain reaction. *J. Virol. Methods* **47,** 189–202.

23. Rolfs, A., Schuller, I., Finckh, U., and Weber-Rolfs, I. (1992) *PCR. Clinical Diagnostics and Research.* Springer-Verlag, Berlin, Germany, p. 11.

24. Wackym, P. A., Simpson, T. A., Gantz, B. J., and Smith, J. H. (1993) Polymerase chain reaction amplification of DNA from archival celloidin-embedded human temporal bone sections. *Laryngpscope* **103,** 583–588.

25. Hruza, C., Dobianer, K., Beck, A., et al. (1993) Her-2 and Int-2 amplification estimated by quantitative PCR in Paraffin-embedded ovarian cancer tissue samples. *Eur. J. Cancer* **29,** 1593–1597.
26. Aoki, S., Yarchoan, R., Thomas, R. V., Pluda, J. M., Marczyk, K, Broder, S., and Mitsuya, H. (1990) Quantitative analysis of HIV-1 proviral DNA in peripheral blood mononuclear cells from patients with AIDS and ARC: Decrease of proviral DNA con tent following treatment 2',3'-dideoxyinosine (ddI). *AIDS Res. Hum. Retrovir.* **6,** 1331–1339.
27. Yamamura, M., Uyemura, K., Deans, R. J., Weinberg, K., Rea, T. H., Bloom, B. R., and Modlin, R. L. (1991) Defining protective responses to pathogens: Cytokine profiles in leprosy lesions. *Science* **254,** 277–279.
28. Mohler, K. M. and Butler, L. D. (1990) Differential production of IL-2 and IL-4 mRNA in vivo after primary sensitization. *J. Immunol.* **145,** 1734–1739.
29. Alard, P., Latz, O., Sebagh, M., Calvo, C. F., Weill, D., Chavanel, G., Senik, A., and Charpentier, B. (1993) A versatile ELISA-PCR assay for mRNA quantification from a few cells. Biotechniques **15,** 730–737.
30. Siebert, P. D. (1993) Quantitative PCR. Methods and Applications **3,** Clontech Laboratories, Inc., Palo Alto, CA.
31. Leonard, M., Brice, M., Engel, J. D., and Papayannopoulou, T. (1993) Dynamics of GATA transcription factor expression during erythroid differentiation. *Blood* **82,** 1071–1079.
32. Jilg, W., Volz, R., Markert-Hahn, C., Mairhofer, H., and Wolf, H. (1991) Expression of class I hisocompatibility complex antigens in Epstein-Barr Virus-carrying lymphoblatoid cell lines and Burkitt lymphoma cells. *Cancer Res.* **51,** 27–32.
33. Ostrowski, L. E., Krieg, P., Finch, J., Cress, A. E., Nagle, R., and Bowden, G. T. (1989) *Carcinogenesis* **10,** 1439–1444.
34. Ferre, F. (1992) Quantitative or semi-quantitative PCR: Reality versus myth. *PCR Methods Appl.* **2,** 1–9.
35. Yun, Z., Lundeberg, J., Johansson, B., Hedrum, A., Weiland, O., Uhlen, M., and Sönnerborg, A. (1994) Coloremetric detection of competitive PCR products for quantification of hepatitis C viremia. *J. Virol. Methods* **47,** 1–14.
36. Becker-André, M. (1993) Absolute levels of mRNA by polymerase chain reaction-aided transcript titration assay. *Methods Enzymol.* **218,** 420–445.
37. Gilliland, G., Perrin, S., and Bunn, H. F. (1990) Analysis of cytokine mRNA and DNA: Detection and quantification by competitive polymerase chain reaction. *Proc. Natl. Acad. Sci. USA* **87,** 2725–2729.
38. Gilliland, G., Perrin, S., and Bunn, H. F. (1992) Competitive PCR for quantitation of mRNA, in *PCR Protocols: A Guide to Methods and Applications* (Innis, M. A., Gelfand D. H., Sninsky, J. J., and White, T. J, eds.), Academic, San Diego, CA, pp. 60–69.
39. Huang, S.-K., Essayan, D. M., Krishnaswamy, G., Yi, M., Kumai, M., Su, S.-N., Xiao, H.-Q., Lichtenstein, L. M., and Liu, M. C. (1994) Detection of allergen- and mitogen-induced human cytokine transcripts using a competitive polymerase chain reaction. *J. Immunol. Methods* **168,** 167–181.

40. Kumar, U., Thomas, H. C., and Monjardino, J. (1994) Serum HCV RNA levels in chronic HCV hepatitis measured by quantitative PCR as say; correlation with serum AST. *J. Virol. Methods* **47,** 95–102.

41. Lehtovaara, P., Uusi-Oukari, M., Buchert, P., Laaksonen, M., Bengtström, M., and Ranki, M. (1993) Quantitative PCR for hepatitis B virus with colorimetric detection. *PCR Methods Appl.* **3,** 169–175.

42. Scheuermann R. H. and Bauer S. R. (1993) Polymerase chain reaction-based mRNA quantification using an internal standard: Analysis of oncogen expression. *Methods Enzymol.* **218,** 446–473.

43. Zipeto, D., Baldanti, F., Zella, D., Furione, M., Cavicchini, A., Milanesi, G., and Gerna, G. (1993) Quantification of human cytomegalovirus DNA in peripheral blood polymorphonuclear leukocytes of immunocompromised patients by polymerase chain reaction. *J. Virol. Methods* **44,** 45–56.

44. Raeymaekers, L. (1993) Quantitative PCR: Theoretical considerations with practical implications. *Anal. Biochem.* **214,** 582–585.

45. Haff, L. A. and Mezei, L. M. (1989) Measurement of PCR amplification by fluorescence. *Amplifications* **1,** 8–10.

46. Pollard-Knight, D., Read, C. A., Downes, M. J., Howard, L. A., Leadbetter, M. R., Pheby, S. A., McNaughton, E., Syms, A., and Brady, M. A. (1990) Non-radioactive nucleic acid detection by enhanced chemiluminescence using probes directly labelled with horseradish peroxidase. *Anal. Biochem.* **85,** 84–89.

47. Murakami, A., Tada, J., Yamagata, K., and Takano, J. (1989) Highly sensitive detection of DNA using enzyme linked DNA probe. Colorimetric and fluorimetric detection. *Nucleic Acids Res.* **17,** 5587–5595.

48. Kessler, C. (1991) The digoxigenin:anti-digoxigenin (DIG) technology—a survey on the concept and reaction of a novel bioanalytical indicator system. *Mol. Cell. Probes* **5,** 161–205.

49. Pollard-Knight, D. (1990) Current methods in nonradioactive nucleic acid labeling and detection. *Technique* **2,** 113–132.

50. Rüger, R., Höltke, H.-J., Reischl, U., Sagner, G., and Kessler, C. (1990) Use of the polymerase chain reaction for non-radioactive labeling specific DNA sequences with Digoxigenin. *J. Clin. Chem. Clin. Biochem.* **28,** 566–568.

51. Reischl, U., Rüger, R., and Kessler, C. (1994) Nonradioactive labeling and high-sensitive detection of PCR products. *Mol. Biotechnol.* **1,** 229–240.

52. Mühlegger, K., Huber, E., von der Eltz, H., Rüger, R., and Kessler, C. (1990) Nonradioactive labeling and detection of nucleic acids: IV. Synthesis and properties of the nucleotide compounds of the digoxigenin system and of photodigoxigenin. *Biol. Chem. Hoppe Seyler* **371,** 953–965.

53. Höltke, H.-J., Sagner, G., Kessler, C., and Schmitz, G. (1992) Sensitive chemiluminescent detection of digoxigenin-labeled nucleic acids: A fast and simple protocol and its applications. *Biotechniques* **12,** 104–113.

54. Bush, C. E., Di Michele, L. J., Peterson, W. R., Sherman, D. G., and Godsey, J. H. (1992) Solid-phase time-resolved fluorescence detection of human immunodeficiency virus polymerase chain reaction amplification products. *Anal. Biochem.* **202,** 146–151.

55. Katz, E. D., Bloch, W., and Wages, J. (1992) HPLC quantitation and identification of DNA amplified by the polymerase chain reaction. *Amplifications* **8**, 10–13.

56. Oefner, P. J., Huber, C. G., Puchhammer-Stöckl, E., Umlauft, F., Grünewald, K., Bonn, G. K., Kunz, C. (1994) High-performance liquid chromatography for routine analysis of hepatitis C virus cDNA/PCR product. *Biotechniques* **16**, 898–908.

57. Bosworth, N. and Towers, P. (1989) Scintillation proximity assay. *Nature* **341**, 167,168.

58. Horikoshi, T., Danenberg, K., Volkenandt, M., Stadlbauer, T., and Danenberg, P. V. (1993) Quantitative measurement nuclear oncogenes, such ant of relative gene expression in human tumors, in *Methods in Molar Biology*, vol. 15: *PCR Protocols: Current Methods and Applications* (White, B. A., ed.), Humana, Totowa, NJ, pp. 177–188.

59. Saiki, R. K., Walsh, D. S., and Erlich, H. A. (1989) Generic analysis of amplified DNA with immobilized sequence-specific oligonucleotide probes. *Proc. Natl. Acad. Sci. USA* **86**, 6230–6234.

60. Kohsaka, H., Taniguchi, Atsuo, Richman, D. D., and Carson, D. A. (1993) Microtiter format gene quantification by covalent capture of competitive PCR products: Application to HIV-1 detection. *Nucleic Acids Res.* **21**, 3469–3472.

61. Khudyakov, Y. E., Gaur, L., Singh, J., Patel, P., and Fields, H. A. (1994) Primer specific solid-phase detection of PCR products. *Nucleic Acids Res.* **22**, 1320–1321.

62. Arnold, L. J., Hammond, P. W., Wiese, W. A., and Nelson, N. C. (1989) Assay formats involving acridinium ester-labelled DNA probes. *Clin. Chem.* **35**, 1588–1594.

63. Kenten, J. H., Casadei, J., Link, J., Lupold, S., Willey, J., Powell, M., Rees, A., and Massey, R. (1991) Rapid electrochemiluminescence as says of polymerase chain reaction products. *Clin. Chem.* **37**, 1626–1632.

64. Wages, J. M., Dolenga, L., and Fowler, A. F. (1993) Electrochemiluminescent detection and quantitation of PCR-amplified DNA. *Amplifications* **10**, 1–6.

65. Mantero, G., Zonaro, A., Albertini, A., Bertolo, P., and Primi, D. (1991) DNA enzyme immunoassay: General method for detecting products of polymerase chain reaction. *Clin. Chem.* **37**, 422–429.

66. Singer-Sam., J. and Riggs, A. D. (1993) Quantitative analysis of messenger RNA levels: Reverse transcription-polymerase chain reaction single nucleotide primer extension assay. *Methods Enzymol.* **225**, 344–351.

67. Singer-Sam, J, (1994) Quantitation of specific transcripts by RT-PCR SNuPE assay. *PCR Methods Application* **3**, S48–S50.

68. Schnell, S. and Mendoza, C. (1997) Theoretical description of the polymerase chain reaction. *J. Theor. Biol.* **188**, 313–318.

69. Guitierrez, R., Garcia, T., Gonzales, I., Sanz, B., Hernandez, P. E., and Martin. R. (1997) A quantitative PCR-ELISA for the rapid enumeration of bacteria in refrigerated raw milk. *J. Appl. Microbiol.* **83**, 518–523.

70. Luo, W., Aosia, F., Ueda, M., Yamashita, K., Shimizu, K., Sekiya, S., and Yano, A. (1997) Kinetics in parasite abundance in susceptible and resistant mice infected with an avirulent strain of *Toxoplasma gondii* by using quantitative competitive PCR. *J. Parasitol.* **83,** 1070–1074.

71. Zhang, J. and Byrne, C. D. (1997) A novel highly reproducible quantitative competitive RT PCR system. *J. Mol. Biol.* **274,** 338–352.

72. Deane, M., Gor, D., Macmahon, M. E., Emery, V., Griffith, P. D., Cummins, M., and Prentice, H. G. (1997) Quantification of CMV viraemia in a case of transfusion-related graft-versus-host disease associated with purine analogue treatment. *Br. J. Haematol.* **99,** 162–164.

73. Caballero, O. L., Menezes, C. L., Costa, M. C., Fernandes, S. C., Anacleto, T. M., de Oliveira, R. R., Viotti, E. A. Tavora, E. R., Vilaca, S. S., Sabbaga, E., de Paula, F. J., Tavora, P. F., Villa, L. L. and Simpson, A. J. (1997) Highly sensitive single-step PCR protocol for diagnosis and monitoring of human cytomegalovirus infection in renal transplant recipients. *J. Clin. Microbiol.* **35,** 3192–3197.

74. Revillion, F., Hornez, L., and Peyrat, J. P. (1997) Quantification of c-erbB-2 gene expression in breast cancer by competitive RT-PCR. *Clin. Chem.* **43,** 2114–2120.

75. Afgani, B., Lieberman, J. M., Duke, M. B., and Stutman, H. R. (1997) Comparison of quantitative polymerase chain reaction, acid fast bacilli smear, and culture results in patients receiving therapy for pulmonary tuberculosis. *Diagn. Microbiol. Infect. Dis.* **29,** 73–79.

76. Brun-Vezinet, F., Boucher, C. Loveday, C. Descamps, D., Fauveau, V., Izopet, J., Jeffries, D., Kay, S., Krzyanowski, C., Nunn, A., Shuurman, R., Seigneurin, J. M. Tamalet, C., Tedder, R., Weber, J., and Weverling, G. J. (1997) HIV-1 viral load, phenotype, and resistance in a subset of drug-naive participants from the Delta trial. The National Virology Groups. Delta Virology Working Group and Coordinating Committee. *Lancet* **350,** 983–990.

77. Bia, X., Hosler, G., Rogers, B. B., Dawson, D. B., and Scheuermann, R. H. (1997) Quantitative polymerase chain reaction for human herpes virus diagnosis and measurement of Epstein-Barr virus burden in posttransplant lymphoproliferative disorder. *Clin. Chem.* **43,** 1843–1849.

78. Yeh, C. T., Shyu, W. C., Sheen, I. S., Chu, C. M., and Liaw, Y. F. (1997) Quantitative assessment of hepatitis C virus RNA by polymerase chain reaction and a digoxigenin detection system: comparison with branched DNA assay. *J. Virol. Methods* **65,** 219–226.

79. Righetti, P. G. and Gelfi, C. (1997) Capillary electrophoresis of DNA for molecular diagnostics. *Electrophoresis* **18,** 1709–1714.

80. Harting, I. and Wiesner, R. J. (1997) Quantification of transcript-to-template ratios as a measure of gene expression using RT-PCR. *Biotechniques* **23,** 450–455.

81. Hullin, R., Asmus, F., and Steinbeck, kG. (1997) Competitive RT-PCR for studying gene expression in micro biopsies. *Mol. Cell. Biochem.* **172,** 89–95.

82. Paffard, S. M., Miles, R. J., Clark, C. R., and Price, R. G. (1997) Amplified enzyme-linked-immunofilter assays enable detection of $50–10^5$ bacterial cells within 1 hour. *Anal. Biochem.* **248,** 265–268.

83. Rowe, D. T., Qu, L., Reyes, J., Jabbour, N., Yunis, E., Putnam, P., Todo, S., and Green, M. (1997) Use of quantitative competitive PCR to measure Epstein-Barr virus genome load in the peripheral blood of pediatric transplant patients with lymphoproliferative disorders. *J. Clin. Microbiol.* **35,** 1612–1615.

84. Imbert-Marcille, B. M., Cantarovich, D., Ferre-Aubineau, V., Richet, B., Soulillou, J. P., and Billaudel, S. (1997) Usefulness of DNA viral load quantification for cytomegalovirus disease monitoring in renal and pancreas/renal transplant recipients. *Transplantation* **63,** 1476–1481.

85. Favre, N., Bordmann, G., and Rudin, W. (1997) Comparison of cytokine measurements using ELISA, ELISPOT and semi-quantitative RT-PCR. *J. Immunol. Methods 204,* 57–66.

86. Payan, C., Veal, N., Crescenzo-Chaigne, B., Belec, L., and Pillot, J. (1997) New quantitative assay of hepatitis B and C viruses by competitive PCR using alternative internal sequences. *J. Virol. Methods* **65,** 299–305.

# 2

# General Principles of Quantitative PCR

## Luc Raeymaekers

### 1. PCR Amplification

Polymerase chain reaction (PCR) is a repetitive amplification process by which in each step (designated $i$), the copy number of product already accumulated during the previous step ($Pi_{-1}$) is multiplied by a factor that depends on the efficiency $Ei$ of the DNA synthesis during that step. $Ei$ is a measure of the relative increment of the amount of product in one step, defined as:

$$Ei = (Pi - Pi_{-1})/Pi_{-1} \tag{1}$$

Because the copy number can at most double in one step, $Ei$ has a value between 0 and 1. The absolute value of the increase of the copy number in one cycle is:

$$Pi - Pi_{-1} = Pi - 1 \times Ei \tag{2}$$

The amount of product accumulated after n cycles is obtained by summation of the one-step increments:

$$P_{\mathrm{n}} = P_{\mathrm{o}} + \sum_{i=1}^{n} P_{i\text{-}1} \times E_i \tag{3}$$

in which $P_0$ denotes the starting amount of template.
A mathematically equivalent expression for **Eq. 3** is the iterated product:

$$P_{\mathrm{n}} = P_0 \times \prod_{i=1}^{n} (1 + E_i) \tag{4}$$

Because the values of $Ei$ are not known *a priori*, it follows from these equations that the absolute amount of $P_0$ cannot be determined from one single measurement of $P_{\mathrm{n}}$. However, the problem of quantitative PCR can, in principle, easily be solved in different ways. The procedures that have been applied

*From: Methods in Molecular Medicine, Vol 26, Quantitative PCR Protocols*
*Edited by B. Kochanowski and U. Reischl ©Humana Press Inc., Totowa, NJ*

can be divided into two main categories, which will be designated as kinetic methods and coamplification methods. Another useful distinction that can be made is between absolute quantification, i.e., the determination of $P_0$ in terms of number of molecules, and relative quantification, i.e., the measurement of the ratio of $P_0$ in various samples. Obviously, relative quantification requires less stringent controls than absolute quantification.

## 2. Methods of Quantification

### 2.1. Kinetic Methods

If the efficiency is constant in each cycle (its value denoted by $E$ without subscript), **Eq. 4** can be rewritten as:

$$P_n = P_0 \times (1 + E)^n \tag{5}$$

It is possible to determine the value of $E$ by taking samples at several consecutive or nonconsecutive cycles during the exponential phase of the PCR and by measuring the amount of product $Pi$ in each sample. The collection of more than two samples is necessary, and preferably as many as possible to ascertain that the efficiency remains constant; in other words, the PCR had not yet reached the stage at which the efficiency starts to decrease. If consecutive samples have been taken, the value of $E$ can be determined from **Eq. 1**. Otherwise, the following more general equation can be used, obtained by rearranging **Eq. 5** and replacing $n$ by the parameter $j$ (the number of cycles in the sampling interval), $Pn$ by $Pj$ (the amount of product sampled at the higher number of cycles), and replacing $P_0$ by $Pi$-$j$ (the amount of product sampled at the lower number of cycles):

$$E = -1 + (P_i /Pi\text{-}j)^{1/j} \tag{6}$$

Once the efficiency has been determined, $P_0$ is calculated from the measured amount of product and the cycle number according to **Eq. 7**, which is a rearrangement of **Eq. 5**:

$$P_0 = Pi/(1 + E)i \tag{7}$$

Alternatively, $P_0$ can be determined by plotting the logarithm of the measured values of $Pi$ as a function on $n$, according to the logarithmic form of **Eq. 5**:

$$\log Pi = \log P_0 + n \times \log (1 + E) \tag{8}$$

The value of $P_0$ can be read on the graph where $n$ equals zero, or $P_0$ can be calculated by performing a linear regression analysis of **Eq. 8** *(1)*. It should be noted that for each of these procedures, it is very important to obtain an accurate value of $E$. Because of the exponential nature of the PCR, small differences in the value of $E$ result in appreciable differences in the amount of product. For example, in two separate runs starting from the same copy number of template, but in one condition amplifying with an efficiency of one and in the other one of 0.8, the quantity of the resulting products will differ by a factor of 24 after 30 cycles and by a factor of 68 after 40 cycles. Another crucial consideration in this respect is that for kinetic PCR, the method used to quantify the PCR products should not only give a signal that is linear with the quantity of product, but also the signal should not be compressed, or, when it occurs, the degree of compression should be small and accurately known. Any compression of the signal that is not taken into account results in an artifactual underestimation of the efficiency. Therefore, the construction of a standard curve based on a dilution series of the template of interest should be recommended for all quantitative PCR applications, because it represents an additional control on the efficiency of amplification and on the range of concentrations that can reliably be quantified. The methodology used in the recently marketed ABI PRISM™ 7700 system (Perkin Elmer, Foster City, CA) can be considered as based on kinetic PCR in that the apparatus continuously measures the amount of product during the run (avoiding the complications of frequent opening and sampling of the PCR tubes). The software of this system does not extrapolate the amplification plot to the start of the PCR, but instead calculates the threshold cycle where the amplification plot crosses some signal threshold.

## 2.2. Coamplification Methods

### 2.2.1. Principle of the Method

These methods involve the coamplification of the sequence of interest, together with a second control sequence, which is either a known quantity of a related cDNA, or a constitutively expressed control gene that is used as a reference (for review *see 2–5*). The main advantages of this technique are that the results are not affected by tube to tube variations in amplification efficiency, and it is not necessary to restrict the PCR to the exponential phase. Reliable quantification is still possible if the PCR extends into the linear phase or even into the saturation phase, provided that it is ascertained that the amplification efficiency is the same for both templates throughout the PCR, including the final cycles. Quantitative coamplification PCR rests on the assumption that the product ratio of the target and standard sequences reliably reflects the ratio of their initial copy numbers. Therefore, it is a prerequisite for this method that the amplification efficiency $E$ is identical for both sequences. In describing coamplification PCR, we will make use of **Eq. 4** given above, replacing the symbol $P$ either by $T$ (the quantity of target sequence) or by $S$ (the quantity of standard sequence), and denoting the amplification efficiency of the target and standard sequences respectively by $E^T$ and $E^S$.

$$T_n = T_0 \times \prod_{i=1}^{n} (1 + E_i^T) \qquad (9)$$

$$S_n = S_0 \times \prod_{i=1}^{n} (1 + E_i^S) \tag{10}$$

Because it is a prerequisite of the method that the efficiency for target and standard should be the same in each cycle, even when the efficiencies are decreasing during the plateau phase, $E_i^T = E_i^S$ for all values of $i$. It follows that in these conditions when making the ratio of **Eqs. 9** and **10**, the iterated product terms cancel out, such that the following equation is valid:

$$(Tn/Sn) = (T_0/S_0) \tag{11}$$

**Eq. 11** allows to calculate $T_0$ from the known quantities $S_0$, $Sn$, and $Tn$:

$$T_0 = (Tn \times T_0)/Sn \tag{12}$$

If the absolute value of $S_0$ is not known, relative comparison of $T_0$ between different samples is still possible by adding the same quantity or defined dilutions of $S_0$ to each sample.

### 2.2.2. Coamplification of the Target and an Unrelated Sequence, Such as a Control Gene

Reliable quantification by this method requires stringent controls, because the target and the standard sequences usually are unrelated, both with respect to the primer binding sites and the intervening sequence. This situation increases the chance that both templates are amplified with different efficiencies, especially during the later linear phase of the PCR. An additional difficulty resides in the fact that there is often a vast difference in the initial copy number of both templates. Without precautions, the PCR may for one template run into saturation, whereas the other one is still being amplified (*see also* **Subheading 3.1.**). These difficulties are more easily solved by using a standard sequence that resembles the target sequence, as described in the next section.

### 2.2.3. Coamplification of the Target and a Closely Related Sequence: Competitive PCR

A minimum requirement for reliable competitive PCR is the identity of the primer-binding sites (however, some mismatches appear to be tolerated, *see* **6**). To ensure equal amplification efficiency of target and standard under all circumstances, a close resemblance of the intervening sequence (length, base composition—if possible—even sequence) is recommended as well. Although quantification can be done by running a single PCR tube and applying **Eq. 12**, it is recommended to add an amount of standard that does not differ too much from the amount of target. In practice, reliable quantification requires the analysis of several PCR tubes in parallel, each containing the same quantity $T_0$ of the target sequence to be quantified but differing in the initial amount of standard sequence added ($S_0$). The range of the dilution series of $S_0$ preferentially should encompass the quantity $T_0$ *(7,8)* (except when the copy number of $T_0$ is so small that statistical considerations become important: *see* **Subheading 3.3.**). The most convenient way to analyze the data is to construct a standard curve by plotting the logarithm of the product ratio of target and standard vs the logarithm of the quantity of standard sequence added to the tube (log $S_0$) *(8,9)*. From **Eq. 11** one derives:

$$\log (Tn/Sn) = \log T_0 - \log S_0 \tag{13}$$

It is clear from **Eq. 13** that such a standard curve should be a straight line with a slope of $-1$. At the point of equivalence of *Tn* and *Sn*, log $(Tn/Sn) = 0$ and log $T_0 = \log S_0$. At this point on the graph, the value of $T_0$ that is to be determined equals that of $S_0$ (**Fig. 1**).

## 3. Sources of Error in Quantitative PCR

### 3.1. The Impact of the Plateau Phase on Quantification

The exponential phase of the reaction extends over a limited number of cycles because of the accumulation of product. If several PCR tubes, each containing a different initial amount of template, are run in parallel, and if the amplification is extended beyond the exponential phase into the saturation phase, initial differences in the amount of template will be compressed, because tubes containing more starting material will reach the saturation phase sooner than tubes containing a smaller amount. This phenomenon results in a systematic bias against the more abundant PCR products. Therefore, relative quantifications between different samples without coamplification of a resembling standard sequence requires suitable controls on the purely exponential nature of the PCR in all the tubes to be compared. The same precautions apply to the method of coamplification of the sequence of interest with an unrelated, constitutively expressed sequence, such as actin. These "housekeeping" genes are often expressed at much higher levels than the target sequence. The product corresponding to such standard sequence may accumulate up to concentrations that inhibit the amplification, whereas the efficiency of amplification of the target sequence is little diminished. One of the advantages of competitive PCR, at least in theory, is its insensitivity to the effect of saturation of the PCR. However, an interference of saturation with the quantification cannot be fully excluded for some templates, as will be explained in the **Subheading 3.2.**

### 3.2. Standard Curves of Competitive PCR Having a Slope Different from −1

In the original description of the method of constructing a log–log standard curve to evaluate competitive PCR, the predicted property that the slope should

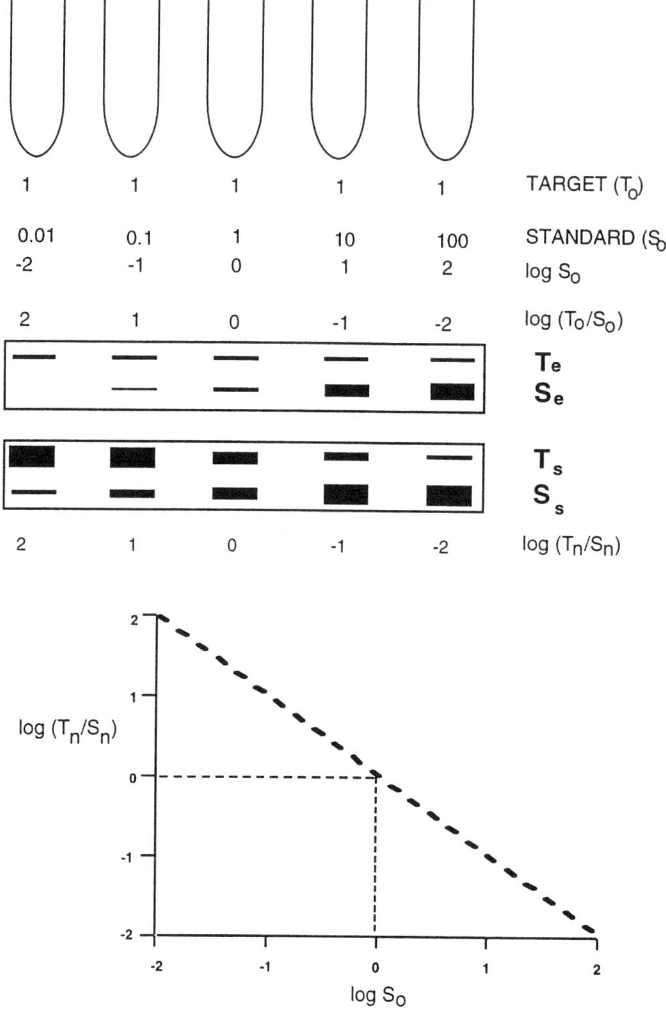



Fig. 1. Idealized overview of a competitive PCR experiment. A series of PCR tubes are spiked with the same but unknown copy number (in the Example 1 relative units) of target sequence ($T_0$) and with a dilution series (in the example from 0.01 to 100) of a known copy number of the standard sequence ($S_0$). Schematic gel patterns of the PCR products are shown obtained at the end of the exponential phase ($T_e$ and $S_e$) and after full saturation of the PCR ($T_s$ and $S_s$). The graph shows the standard curve constructed from the quantified gel bands. The value of $T_0$ corresponds to that of $S_0$ at the point of equivalence, i.e., where log ($T_n /S_n$) = 0.

equal −1 has not been mentioned *(8)*. As a consequence of this omission, this paper and many papers published afterwards show standard curves that do not conform to theory. Although this fact does not necessarily imply that the quantifications based on these curves are grossly wrong (*see* explanations below), it is obvious that such errors should be avoided in the future.

Because many nonorthodox standard curves have been published (an incomplete scan of the literature before October 1996 yielded more than 20 papers), it is important to consider possible causes. There are at least three types of explanations:

1. If the PCR is run into saturation, a systematic bias against the more abundant PCR products may occur if their sequences differ significantly *(10)*. The consequence of this phenomenon is that the ratio of the products ($T_n/S_n$) is smaller than the ratio of the initial copy number ($T_0/S_0$) when $T_n$ greater than $S_n$ (or larger when $T_n$ less than $S_n$). As a result, the slope of the standard curve will be smaller in absolute value than 1. There is no shift of the position of the point of equivalence, so that the quantification based on the position of this point is correct.

2. Systematic errors may arise in some methods of quantification of the PCR products. It has been observed that ethidium bromide-stained bands yield a tilted standard curve when analyzed on agarose gels but not on polyacrylamide gels *(11)*. Also in this situation, the point of equivalence remains at the same position. The effect is probably a result of the higher background staining in agarose gels.

3. A slope deviating from −1 may be caused by the unequal amplification efficiencies of target and standard. The shift of the slope is accompanied by a shift of the point of equivalence, resulting in erroneous quantification. It can been shown by computer simulation *(12)*, and it can also be seen intuitively from **Fig. 1** that a deviation of the slope only occurs if the difference between the amplification efficiencies of target and standard varies among the PCR tubes of the dilution series on which the standard curve is based. (A difference between $E^T$ and $E^S$ that is identical in all the tubes in all cycles results in a parallel shift of the graph, thus maintaining the slope = 1 property but resulting in a shift of the point of equivalence.) A possible cause that could be responsible for such a phenomenon is up until now speculative. However, it is reasonable to suppose that in some cases two similar sequences that amplify with the same efficiency during the exponential phase may start to amplify with different efficiencies during the later linear stages of the PCR. Small differences in the relevant properties of the templates may not show up in conditions when the DNA polymerase and all substrates are abundantly available, and the concentration of the reaction products is still below inhibiting levels. These differences may become important, however, if the binding of substrates or of the polymerase, the rate of template annealing, or the rate of strand dissociation become rate-limiting. Because the different samples constituting the standard curve contain different copy numbers of template, each tube will spend a different number of cycles in the nonexponential phase of the PCR and will be differentially affected by the difference between $E^T$ and $E^S$.

## 3.3. Stochastic Effects in the Quantification of Small Numbers of Molecules

It should be noted that the equations given in the first part of this chapter are valid only if the magnitude of the influence of statistical variations on the outcome of the PCR can be neglected. Statistical considerations become important when the number of template molecules is small and when the efficiency is significantly smaller than one because the amount of PCR product that is produced in one cycle depends on molecular fluctuations. For example, starting from a single copy of a DNA sequence that is amplified with an efficiency of 0.8, the probability that one copy remains after the first cycle is 20%. In theory, the final copy number after $n$ cycles may be any number between 1 and $1.8n$. As a consequence, the analytical equations given above do not apply when the initial copy number is low. A more rigorous description of the process of PCR in these conditions should be based on the theory of branching processes. Thorough mathematical descriptions of PCR reactions in these conditions have been published *(13,14)*. However, simulations of such PCR trajectories can be easily implemented on computer because the distribution of $Pn$ is the binomial distribution with parameters $Pn_{-1}$ and E *(14)**. The expected outcome of the PCR in different conditions and of the confidence intervals have been calculated *(14)*. As expected, the confidence interval of the estimation of the initial copy number of the target is larger for a lower initial copy number. The uncertainty also increases with decreasing amplification efficiency. For instance, when the initial copy number is 100, the relative uncertainty (ratio of uncertainty over true value) is 10% for $E = 0.9$, and 25% for $E = 0.5$ (when the number of cycles is >20). The relative uncertainty computed

---

*The following is an example of the simulation in the Mathcad program (version 6.0) of 500 PCR runs of 20 cycles starting from one copy of template that is amplified with an efficiency of 0.6. *P* denotes the copy number, *m* denotes the amplification factor (= 1 + p), and p is the probability of duplication, which for large copy numbers corresponds to the amplification efficiency. *rnd*(1) generates a random number between 0 and 1.

PROGRAM:     p:=0.6 n (number of cycles) :=20 trials:=500 $_j$:=1..trials $P_{0,j}$:=1

$i$:=1..$n$ u$_{i,j}$:= rnd(1) $P_{i,j}$:=$P_{i-1,j}$ + qbinom(u$_{i,j}$,$P_{i-1,j}$,p) $m_{n,j}$:=$P_{i,j}/P_{i-1,j}$

RESULTS (shown in the format required by the program):
mean[$(P^T)$ less than 20 greater than] = $1.222 \times 10^4$ (the mean number of copies at 20 cycles),
stdev[(P$^T$)<20>] = $5.984 \times 10^3$
mean[(m$^T$)<20>] = 1.6 stdev[(m$^T$)<20>] = 0.007

As described by Peccoud and Jacob *(14)*, the mean value of *m* is an estimation of the real amplification factor that converges to the real value as $i$ tends to infinity. For a limited number of cycles, an estimation of $P_0$ can be calculated for each run from the mean value of *mi* according to **Eq. 7**. This value of $P_0$ in this particular simulation of 500 runs was $1.01 \pm 0.497$ (mean ± stdev).

with one single initial copy is 255% for $E = 0.5$, whereas it is 99% for $E = 0.9$. Because the uncertainty increases with decreasing initial copy number, it follows that the accuracy of coamplification PCR of a very low copy number of target will be higher when using a larger copy number of standard than by using a copy number of standard that is similar to that of the target *(13)*. Stochastic effects also may be important when PCR is used in combination with the limiting dilution technique. This method requires that many samples contain one or a few templates *(15,16)*. It follows that the method is reliable only if the efficiency equals or is very close to 1.

## *3.4. Other Confusions about Quantitative PCR*

Besides the problem concerning the slope of the standard curve of competitive PCR, the reader should be warned about some other illegitimate—but nevertheless published—simplifications in order to avoid a chain-reaction multiplication of errors in the literature.

It has been stated that reliable quantification is possible with competitive PCR, even when the efficiencies of the target (*ET*) and the standard (*ES*) are different, assuming that the ratio *ET/ES* is a constant value *(17)*. This statement should be made more precise in that it applies only to relative quantification and not to absolute quantifications, as can be seen from **Eq. 14**, which itself is based on **Eqs. 8** and **13**:

$$\log (T_\mathrm{n}/S_n) = \log (T_0/S_0) + n \times \log ([1 + E^T]/[1 + E^S]) \qquad (14)$$

Absolute quantification is not possible if the term at the right is not equal to zero. The deviation of the quantification from the real value equals the value of the right term, which increases with the difference in amplification efficiency and with the number of cycles. In theory, relative quantification, i.e., comparing $T_0$ in different samples, is still possible as long as $n$ and $E^T/E^S$ remain constant. However, it seems that reliable relative quantification in these circumstances is possible only in theory, as it would require too many controls to be feasible in practice.

It has been stated that the ratio $T_0/S_0$ is proportional to the ratio of the slope of the line relating $T_n$ to the number of cycles $n$, divided by the slope of a similar graph for $S_n$, if both slopes are determined during the linear phase of the PCR, i.e., close to saturation *(18)*. Obviously, there is neither a theoretical nor a practical reason why this should the case. On the contrary, one would expect the inverse because a sample containing more starting material would run closer to saturation and consequently show a less steep increase of the amount of product as a function of $n$.

A simple PCR method for relative quantitation has been proposed as an alternative to other methods, such as competitive PCR *(19)*. The authors describe a method consisting in making a series of progressive dilutions by mixing the two samples to be compared in different ratios. According to the authors, the alignment of the quantities of amplified product in each tube along a line would demonstrate that the amplification efficiency in each tube was equal, allowing direct comparison between the two samples. It is clear that this method does not eliminate the trap of running the PCR close to or into saturation, thereby compressing the difference between the amount of product in the two samples. A linear regression line may be obtained, even in conditions of near-saturation, particularly if one allows for errors—even small errors—in the quantification of the products. Furthermore, this paper contains several mistakes in the calculations, and it also shows graphs relating the amount of PCR product to the number of cycles according to which the amplification factor would be much larger than two, which is theoretically impossible.

## Acknowledgement

I thank Dr. G. Droogmans of the Laboratory of Physiology for the help with the computer program.

## References


1. Wiesner, R. J. (1992) Direct quantification of picomolar concentrations of mRNAs by mathematical analysis of a reverse transcription/exponential polymerase chain reaction assay. *Nucleic Acids Res.* **20,** 5863–5864.
2. Ferre, F. (1992) Quantitative or semi-quantitative PCR: reality versus myth. *PCR Methods Appl.* **2,** 1–9.
3. Cross, N. C. (1995) Quantitative PCR techniques and applications. *Br. J. Haematol.* **89,** 693–697.
4. Reischl, U. and Kochanowski, B. (1995) Quantitative PCR. A survey of the present technology. *Mol. Biotechnol.* **3,** 55–71.
5. Raeymaekers, L. (1995) A commentary on the practical applications of competitive PCR. *Genome Res.* **5,** 91–94.
6. Kwok, S., Kellogg, D. E., McKinney, N., Spasic, D., Goda, L., Levenson, C., and Sninsky, J.J. (1990) Effects of primer-template mismatches on the polymerase chain reaction: human immunodeficiency virus type 1 model studies. *Nucleic Acids Res.* **18,** 999–1005.
7. Wang, A. M., Doyle, M. V., and Mark, D. F. (1989) Quantitation of mRNA by the polymerase chain reaction. *Proc. Natl. Acad. Sci. USA.* **86,** 9717–9721.
8. Gilliland, G., Perrin, S., Blanchard, K., and Bunn, H. F. (1990) Analysis of cytokine mRNA and DNA: detection and quantitation by competitive polymerase chain reaction. *Proc. Natl. Acad. Sci. USA* **87,** 2725–2729.
9. Siebert, P. D. and Larrick, J. W. (1992) Competitive PCR. *Nature.* **359,** 557,558.

10. Mathieu Daude, F., Welsh, J., Vogt, T., and McClelland, M. (1996) DNA rehybridization during PCR: the 'Cot effect' and its consequences. *Nucleic Acids Res.* **24,** 2080–2086.
11. Bouaboula, M., Legoux, P., Pessegue, B., Delpech, B., Dumont, X., Piechaczyk, M., Casellas, P., and Shire, D. (1992) Standardization of mRNA titration using a polymerase chain reaction method involving co-amplification with a multispecific internal control. *J. Biol. Chem.* **267,** 21,830–21,838.
12. Raeymaekers, L. (1993) Quantitative PCR: theoretical considerations with practical implications. *Anal. Biochem.* **214,** 582–585.
13. Nedelman, J., Heagerty, P., and Lawrence, C. (1992) Quantitative PCR with internal controls. *Comput. Appl. Biosci.* **8,** 65–70.
14. Peccoud, J. and Jacob, C. (1996) Theoretical uncertainty of measurements using quantitative polymerase chain reaction. *Biophys. J.* **71,** 101–108.
15. Villarreal, X. C., Grant, B. W., and Long, G. L. (1991) Demonstration of osteonectin mRNA in megakaryocytes: the use of the polymerase chain reaction. *Blood.* **78,** 1216–1222.
16. Sykes, P. J., Neoh, S. H., Brisco, M. J., Hughes, E., Condon, J., and Morley, A. A. (1992) Quantitation of targets for PCR by use of limiting dilution. *Biotechniques.* **13,** 444–449.
17. Zachar, V., Thomas, R. A., and Goustin, A. S. (1993) Absolute quantification of target DNA: a simple competitive PCR for efficient analysis of multiple samples. *Nucleic Acids Res.* **21,** 2017–2018.
18. Santagati, S., Bettini, E., Asdente, M., Muramatsu, M., and Maggi, A. (1993) Theoretical considerations for the application of competitive polymerase chain reaction to the quantitation of a low abundance mRNA: estrogen receptor. *Biochem. Pharmacol.* **46,** 1797–1803.
19. Nicoletti, A. and Sassy-Prigent, C. (1996) An alternative quantitative polymerase chain reaction method. *Anal. Biochem.* **236,** 229–241.

# 3

# Effects of Collection, Processing, and Storage on RNA Detection and Quantification

**Mark Holodniy**

## 1. Introduction

Historically, clinicians and researchers have relied upon the development of clinical endpoints or the use of surrogate markers in the evaluation of disease pathogenesis and in response to various therapeutic agents. In addition, microbiologic methods of detecting various pathogens have usually required the culture of an agent. The majority of bacterial pathogen culture methods have been standardized and identification has become relatively straightforward. Nonetheless, a wide variety of unculturable pathogens have been identified. Recent advances in molecular diagnostics have provided clinicians with the ability to measure directly infectious agents. Certain viral pathogens such as human immunodeficiency virus (HIV) are detectable by standard culture techniques whereas others such as hepatitis C virus (HCV) are not.

The inherent biologic and interlaboratory variability, time, expense, and sample-processing requirements of quantitative HIV culture techniques have made HIV culture outside of specific clinical trials or pathogenesis-based studies extremely problematic. Therefore, it is necessary to search for standard and reproducible molecular techniques to quantify viruses from various tissues and blood. Early in the HIV epidemic, it was discovered that infectious virus could be cultured in cell-free plasma. With the advent of molecular amplification techniques, it was determined that cell-free HIV RNA from viral particles could be detected and quantitated in plasma *(1,2)*. Over the next several years, many studies were published that demonstrated the quantity of cell-free HIV in plasma or serum. These studies also demonstrated the clinical utility of RNA quantitation in determining clinical progression of a patient, response to vari-

From: *Methods in Molecular Medicine, Vol 26: Quantitative PCR Protocols*
Edited by: B. Kochanowski and U. Reischl © Humana Press Inc., Totowa, NJ

ous antiretroviral treatment regimens, and predicted clinical response based on virologic responses *(3,4)*. Various biologic phenomena were also found to alter plasma HIV RNA levels. Immunizations with influenza and pneumococcal vaccines *(5,6)* and concomitant infections such as bacterial pneumonia or opportunistic infections *(7,8)* have been found to increase HIV RNA levels.

Over the past few years, a tremendous amount of work has been concerned with a second blood-borne pathogen, HCV. Various molecular techniques have identified the importance of plasma HCV RNA genotype and quantification as a marker for clinical progression, and antiviral response in HCV disease *(9–11)*. In order to validate whether viral RNA is useful for clinical practice and monitoring in patients with HIV or HCV disease, a thorough understanding of assay performance and sample processing and collection issues must be undertaken. This chapter reviews current issues that affect collection, processing, and storage of blood, blood plasma, serum, or tissue, and that can affect the ability to quantitate and/or detect viral RNA in these clinical samples. The majority of the literature that has been published regarding this topic involves HIV and HCV.

## 2. Methods of RNA Quantification

Commercial assays can now measure plasma RNA viral load by either target or signal-associated amplification of virion associated HIV or HCV RNA. Available target amplification assays of RNA include reverse transcription-polymerase chain reaction (RT-PCR; Amplicor HIV-I Monitor, Roche, Branchburg, NJ) and nucleic-acid signal-based amplification (NASBA; Organon Teknika, Durham, NC). A signal-associated amplification assay called branched DNA (bDNA, Chiron, Emeryville, CA) has also been developed. All of these assays measure the same target and biologic phenomena. However, specific assays have slightly different sensitivity thresholds and performance characteristics. The specific technology and procedures of these assays are reviewed elsewhere *(12–14)*. However, a brief discussion that defines the characteristics, reproducibility, and intra/interassay variability is necessary. In addition, it is important to identify the inherent patient-related biologic variability of these viral agents as measured by these assays. When interpreting data from individual samples, performance characteristics of these assays must be considered and understood.

Recent data suggests that in clinically stable HIV-infected patients, there is short-term biologic stability of HIV RNA in plasma. The studies indicate that there are approximately $0.1–0.3$ $\log_{10}$ copy/mL diurnal, daily, or monthly variations in virus levels *(15)*. The observed intraassay standard deviation (SD) for the three commercially available assays ranges from $0.12–0.33$ $\log_{10}$ copy/mL

*(16)*. The total variability taking into account both assay and inherent patient or biologic variability is no more than approx 0.5 $\log_{10}$ copy/mL. Therefore, changes of greater than 0.5 $\log_{10}$ U/mL are considered biologically significant. In addition, fluctuations may be greater at the sensitivity threshold of an assay. For example subjects with extremely low copy numbers are reported to have a variability as great as 1 $\log_{10}$ copy/mL (10-fold) when assessed by RT-PCR *(17)*.

Several recent publications describe performance related issues. In the study by Schuurman et al. *(18)*, three separate laboratories performed all three of the commercially available assays on the same set of plasma samples. In reconstruction panels, the bDNA assay gave the highest reproducibility with standard deviations for all dilutions ranging from 0.05–0.12 $\log_{10}$ copy/mL. When clinical samples were analyzed, the mean interlaboratory difference was 0.18 $\log_{10}$ copy/mL. However, this was not believed to be of any clinical significance. In another study by Yen-Lieberman et al. *(19)*, 65% of 41 laboratories with several different RNA quantitation methods employed a common set of standards and were able to achieve < 0.15 log10 copy/mL variability. Regression analyses indicated that differences among laboratories using the same kit were greater than differences among the population average regressions for the same commercial kits. Finally, Van Damme et al. *(20)* compared NASBA and RT-PCR on the same samples and found that the majority of samples differed by <0.5 $\log_{10}$ copy/mL. Taken together, these three comparative studies indicate that assay variability, secondary to the individual assay type or operator, affects RNA measurement to a small degree and needs to be factored in with the effects of collection and processing activities on RNA quantitation.

## 3. Effect of Blood Components and Anticoagulants

Components within blood or tissues that could inhibit enzymatic amplification of genetic material are also critical to our understanding of the use of molecular diagnostic techniques for viral quantitation. Various compounds, including heme at a concentration of 0.8 m$M$ or greater, can inhibit DNA polymerase, or, more specifically, *Taq* DNA polymerase used in amplification techniques. Panaccio et al. *(20a)* also found that as little as 1% v/v blood in a PCR reaction inhibited *Taq* polymerase. Other chemicals commonly used in molecular biology can also inhibit reactions, such as ethylenediaminetetraacetic (EDTA), sodium dodecyl sulfate (SDS), or guanidinium hydrochloride *(21,22)*.

In 1990 and 1991, it was noted by several groups that samples amplified successfully either from serum, EDTA, or citrate-containing blood were found to be inhibited when collected in the presence of heparin. Gustafson also found that blood samples collected in the presence of heparin were shown to yield decreased quantities of cellular DNA *(23)*. Buetler et al. *(24)* were the first to describe heparin interference with PCR. They found that the inhibitory effect of heparin on PCR could not be removed by boiling the samples. They did find in their cellular DNA experiments that incubating the DNA with heparinase II prior to PCR could reverse the inhibitory effect. Our group described the inhibitory effect of heparin on the ability to quantitate HIV RNA in the plasma of HIV-infected patients by RT-PCR. When compared to blood collected in EDTA or acid citrate dextrose (ACD), heparinized plasma produce markedly decreased signal in our RT-PCR enzyme-linked immunosorbent assay (ELISA) assay *(25)*. We also confirmed Buetler's observation that genomic DNA from peripheral blood mononuclear cells in both whole blood and after extraction also demonstrated an inhibitory effect when collected in the presence of heparin. We showed that as little as 0.05 U of heparin/reaction could suppress the presence of a visible band of PCR product as seen by gel electrophoresis and ethidium bromide staining. A partial reversal of the inhibitory effect could be achieved by incubation of the samples with 10-fold greater concentrations of *Taq* polymerase (25 U/reaction). By adding approx 7.5 U of heparinase to the reaction, a significant amount of the inhibitory effect of heparin could be reversed. Nakamura et al. *(25a)* used a two-step PCR procedure that increased the sensitivity of their assay so that as little as one molecule of HIV could be detected. Although their assay had good sensitivity in serum, they noted that their assay could not overcome the inhibitory effect of heparin on RT-PCR.

Further data by Imai et al. *(26)* demonstrated that heparin above the concentration of $1 \times 10^{-4}$ U/mL inhibited RT in a dose-dependent manner. Heparin did not inhibit RT at concentrations below $1 \times 10^{-2}$ U/mL. Imai also suggested that because calcium is required for heparinase activity and because it will inhibit the RT and subsequent PCRs, calcium must be removed prior to RT. Calcium was removed in their assay using glycoletherdiamine-tetraacetic acid (EGTA). EGTA concentrations above 5 m*M* were also found to inhibit PCR, but not reverse transcription. Thus, if this procedure is used to remove calcium interference, the EGTA concentration should be below 5 m*M*.

Izrael et al. *(27)* described the inhibitory effect of heparin, both on murine leukemia virus (MLV) reverse transcriptase and *Taq* DNA polymerase. They found that 0.1 U/g of RNA (0.005 U/L reaction) could completely inhibit RT and *Taq* DNA polymerase. Tsai et al. *(28)* described the use of heparinase treatment for RT-PCR assays of tissues containing high levels of heparin from mast cells. RNA was initially extracted from peritoneal cells containing approximately 4% mast cells. These results clearly showed that mRNA transcripts could not be detected in heparin-rich cellular samples unless a heparinase was added prior to the RT-PCR reaction.

Panaccio et al. *(29)* described a PCR protocol for amplification of DNA directly from whole blood, termed "formamide low temperature" (FoLT) PCR. They showed that EDTA and sodium heparin containing blood produced a greater signal than blood containing lithium or fluoride heparin. The process involved the use of formamide at 18% v/v. They hypothesized that the presence of formamide and lower temperature PCR reduced the amount of protein coagulation and thereby allowed more DNA template to be accessible for amplification. Burckhardt *(30)* demonstrated that up to 80% v/v of whole blood could be used in PCR reactions when *Taq* polymerase was required and was anticoagulant-dependent. Satsangi et al. *(31)* described different strategies to reverse the effects of heparin. Lymphocytes collected in preservative-free heparin were extracted using proteinase K digestion and phenol chloroform extraction. Across a broad range of DNA concentrations, including pretreatment with chelex and spermine, different DNA polymerases, buffer systems and various magnesium concentrations and/or albumin, little of the amplification was accomplished. They were, however, successful in producing PCR product after preincubation of their DNA sample with heparinase II (0.2 U/reaction for 45 min; reaction temperature not specified). Thus, despite the data provided by Poli et al. *(32)* that chelex could overcome the effects of heparin, Satsangi et al. could not reproduce these results with their specific PCR methodology. Interestingly, Di Cioccio et al. *(32a)* had described the inhibitory effects of heparin on cellular DNA polymerases years before.

Several studies have recently been published that compare the quantitation of both HIV and HCV RNA in serum vs plasma collected in the presence of different anticoagulants. Aoki-Sei described a comparison between RNA copy number obtained from heparinized plasma vs that from serum. Although heparinized plasma was used in their RT-PCR assay, they found a significantly higher (mean eightfold, range 1.2–15.3-fold) RNA level in heparinized plasma than in serum *(33)*. The ability to detect and quantitate viral particles in heparinized plasma was facilitated by the fact that a 1 h ultracentrifugation step preceded the extraction process. The ultracentrifugation step was apparently sufficient to remove heparin containing plasma and also leave a virus pellet that did not contain heparin as an inhibitor of RT-PCR.

The effect of anticoagulant and blood-collection tube type was further assessed by our group using bDNA technology *(34)*. We found that blood-collection tubes containing EDTA as an anticoagulant produced the highest HIV RNA copy number. Although HIV RNA could be detected in plasma containing either heparin, ACD, or EDTA as an anticoagulant, the RNA levels found in tubes containing heparin using bDNA were approximately 30% lower than those compared to EDTA. RNA levels in serum were significantly lower *(50% or greater)* than those in plasma containing any of the anticoagulants.

Cell preparation tubes (CPT) and plasma preparation tubes (PPT) containing different anticoagulants were also analyzed. These new type of blood-collection tubes contained a gel barrier and ficoll hypaque for cell separation in CPT tubes. Through density centrifugation, red blood cells (RBC) and neutrophils were removed from the plasma and trapped below the gel barrier. Mononuclear cells remained above the gel barrier in the plasma. PPT tubes removed all cells from plasma, including platelets. These tubes yielded HIV RNA levels that reflected the intrinsic anticoagulant, i.e., EDTA containing plasma was higher than citrate or heparin containing plasma from both CPT and PPT tubes. Part of this may be related to an anticoagulant dilution effect. Tubes containing EDTA were either plastic and spray-coated internally with anticoagulant or were glass and contained only 50 µL of liquid anticoagulant. Tubes with ACD contained approximately 1.5 mL of anticoagulant as a diluent. CPT tubes with citrate as an anticoagulant contained an additional 0.8 mL of ficoll hypaque for cell separation. Thus, a plasma sample and, hence, RNA copy number could be diluted by as much as 20% in these kinds of tubes when compared to standard EDTA blood-collection tubes.

This was confirmed by Todd et al. *(35)*, who also used bDNA. Plasma samples collected in ACD and heparin tubes yield approximately 12% and 38% lower HIV RNA levels, respectively, than those collected in EDTA. Serum again yielded RNA levels that were approximately threefold lower than those from plasma containing EDTA. Their study also showed that there was no significant effect of hemoglobin (1 mg/mL), increased lipemia, or bilirubinemia on HIV RNA levels. Panels of antiretroviral drugs added to plasma also had no effect on RNA levels tested with the bDNA assay. A recent study by Dickover et al. *(36)* using RT-PCR (Amplicor HIV-I monitor) found that RNA levels were decreased in ACD tubes when compared to EDTA. This probably reflects the large volume of anticoagulant and diluent in ACD tubes. Heparinized samples were also evaluated with heparinase pretreatment prior to RT-PCR. Utilizing this procedure, they were able to show that heparinized plasma contained approximately 30% less RNA when compared to EDTA. It is impossible to determine whether the RNA levels in plasma containing heparin were lower as a result of residual heparin interfering with the RT-PCR reaction and incomplete heparinase inactivation, or if other factors were involved which related to plasma containing heparin. In another study by Izopet et al. *(37)*, plasma samples collected in EDTA and sodium citrate tubes were assessed using the Roche RT-PCR assay. Results from their study also indicate that RNA measurements from tubes containing EDTA were significantly higher than those from samples taken in plasma containing sodium citrate. Again, as already shown by others, when compared to samples containing serum, HIV RNA concentrations were significantly higher in both plasma types when compared to serum.

In addition to analyzing the effect of anticoagulant on plasma RNA quantitation, several groups have assessed the impact that other constituents contained within plasma have on RNA quantification. Controversy exists as to the amount of virion associated RNA that is truly cell-free or is platelet-associated. Two case reports published in 1993 indicated discordant results in determining whether virion-associated RNA was in fact platelet-associated or not *(38,39)*. Zhu and colleagues *(40)* studied three patients in which platelet-rich and platelet-poor plasma were analyzed for HIV RNA levels. They concluded that approximately 0.5–3% of HIV RNA was platelet-associated, and that the number of platelets containing a single virus was on the order of $1/10^5$ platelets. We performed a more rigorous analysis on 10 subjects in which the amount of virus contained in platelet-rich vs platelet-depleted plasma was compared in both ACD and PPT tubes *(41)*. In both ACD and PPT tubes, the pelleted platelet debris fraction contained approximately 5% of the HIV RNA when compared to data acquired from plasma from both tubes processed under normal centrifugation parameters. Thus, the contribution to RNA signal imparted by platelet fraction appears to be minimal, although 5% of the signal needs to be accounted for when processing parameters are considered.

Some work has also been published regarding the effect of anticoagulants on quantitation of HCV. Wang et al. *(42)* qualitatively studied the signal for HCV obtained by RT-PCR in serum, and plasma containing sodium citrate or heparin. Furthermore, in this HCV assay, they were unable to obtain any PCR product from heparinized plasma as depicted by gel electrophoresis and ethidium-bromide staining. They also demonstrated qualitatively that both serum and citrated plasma could detect the HCV fragment. Willems et al. *(42a)* were able to detect HCV RNA in peripheral blood mononuclear cells (PBMCs) collected in heparinized blood and in an accompanying serum and EDTA samples. They were unsuccessful in detecting HCV RNA from plasma containing heparin. They also confirmed Izraeli's *(27)* observation on the inhibitory effect of heparin individually on both RT and PCR steps. A study by Manzin et al. *(43)* compared an unspecified plasma type (anticoagulant not stated) vs serum. A 2–9-fold reduction in HCV RNA was seen in serum when compared to the plasma samples. The loss of signal seen in serum did not appear to be predictable. Because the intersubject variation of this particular assay was determined to be 0.22 logs, the difference was considered to be biologically significant. In a study by Conrad et al. *(44)* they compared HCV RNA quantitation in serum and EDTA PPT tubes as measured by RT-PCR. In general, HCV copy number did not vary by more than 10% between the fresh-frozen serum (used as a control) and PPT tube plasma held out to 120 h after collection. Thus, it appeared that plasma collected in PPT tubes held for up to 5 d maintained stability of HCV RNA signal.

In summary, both HIV and HCV demonstrated marked differences in RNA levels from different plasma- or serum-containing samples when quantitated by RT-PCR or bDNA assays. However, the data published with respect to clinical use of the NASBA assay has demonstrated that there did not appear to be an anticoagulant or serum effect in terms of reducing RNA levels in different plasma or serum medium. Vandamme et al. *(45)* utilized the NASBA assay and found that signal was equivalent when EDTA and heparin plasmas were used from the same clinical sample. Two additional patients samples were studied that included serum. They again found the signals to be comparable. Although this data was reported in the text, no data was shown. In addition, they used an RT-PCR assay to compare their results to NASBA. By using the Boom extraction procedure *(46)* prior to RT-PCR, they were able to generate signals that were also equal in EDTA and heparin (data stated, but not shown). In another study published by VanGemmen et al. *(47)*, reconstruction experiments in which wild-type HIV RNA was added to citrate, heparin, and EDTA plasma or serum showed there was no significant difference in the quantitation within the four sample types. No clinical samples were collected in the various plasmas or evaluated in this study. However, other potential interfering substances such as lipids and hemoglobin were also investigated in reconstruction experiments and did not appear to affect quantitative RNA measurements.

Thus, it appears that the routine use of RT-PCR for infectious agents in plasma collected in the presence of heparin will be inhibited. If heparinase is added after extraction and prior to RT-PCR, then a significant increase in signal can be obtained. If a modification of RNA extraction is used in which RNA is extracted using guanidinium followed by capture and elution with silica, then inhibitory effects of heparin are potentially removed. Both the NASBA assay and a modified RT-PCR assay now take advantage of this procedure in order to utilize heparin containing samples (*see* **Subheading 5.**). Signal amplification using branch chain technology has the capability of producing signal in any of the medium discussed with some inhibitory effect seen in samples containing heparin relative to EDTA. Because bDNA does not require enzymatic amplification of the nucleic-acid target, there are no enzymes in the system that can be inhibited by heparin. Why heparin reduces the signal seen when compared to other types of plasma is unknown. Plasma containing EDTA appears to offer the best results for all assays. Thus, sensitivity considerations should be taken into account if different plasma media are used for blood collection.

## 4. Sample Storage and Processing

It is also important to consider whether RNA measurement is stable in samples collected and maintained, either on the bench top at room temperature or perhaps in the refrigerator, until plasma is separated and then frozen. In addition, the stability of RNA signal overtime in the freezer also needs to be taken into account when samples are pulled, thawed, and then analyzed for RNA quantitation. Our group first demonstrated that significant decay in HIV RNA signal occurred over time in samples. We compared citrated CPT and ACD tubes. When whole blood was held in a standard ACD tube at ambient temperature on the bench top for 24 h, over a 50% reduction in signal was seen. However, 95% of the signal was retained after 24 h of holding spun plasma in the CPT tube *(48)*. Further decay over 48 and 72 h in the CPT tubes was seen. After 72 h at room temperature, 60% of the 2-h value was present vs 40% of the ACD tube value held as whole blood. This work was performed using a RT-PCR assay. Similar results were obtained using bDNA assay and standard heparinized and CPT collection tubes. However, signals were somewhat lower and the retention of RNA signal after 24 h even in the CPT tubes was only 75% when compared to baseline. The signal in CPT tubes was significantly greater than the standard heparinized collection tube, which retained only 53% of the baseline signal at 24 h.

When using heparin as an anticoagulant, CPT tubes with clarified plasma still maintained 50% of their baseline signal after 72 h compared to only 30% of signal in whole blood held at room temperature in the heparinized standard blood-collection tubes. This data was further confirmed by Dickover et al. *(36)* using a slightly different experimental design. Blood was collected in tubes containing EDTA, ACD, and heparin. In the first 6 h HIV RNA levels decreased by 11, 20, and 32% respectively. From 6–48 h, RNA levels decreased in all anticoagulants, but at a slower rate of loss. After 48 hours using RT-PCR (Amplicor HIV monitor), heparin tubes demonstrated a 50% reduction in signal when compared to baseline. Somewhat different results were obtained by Vandamme et al. *(20)*, whereby using the NASBA assay and blood collected in the presence of EDTA, they showed relatively stable plasma RNA levels when whole blood was held over 48 h. Although the RNA load fluctuated over 0.6 logs, there was no correlation with the time of processing. Similar results were apparently obtained using an RT-PCR assay, but data was not shown. Thus, it appears that blood containing EDTA either in terms of separated plasma or in terms of whole blood appears to produce longer lasting and more stable RNA signal in terms of collection time.

We have also shown stability of HIV RNA signal in EDTA plasma held at 4°C for up to 5 d (Holodniy, personal observation). An early study by Coombs et al. *(49)* using an immunocapture RT-PCR assay and qualitative results (using isotopic hybridization and gel electrophoresis) indicated that plasma RNA signal was stable after multiple freeze-thaws and storage at ambient temperature up to 6 d. They also concluded that there was no apparent difference between PCR signal generated from heparinized or citrated plasma even after nine cycles of freeze-thawing. Thus, the immunocapture method could evaluate signal generated from heparinized plasma by capturing the viral particles and separating them from the plasma containing heparin. However, results of this study are not conclusive because of the qualitative nature of the data presented.

Todd et al. *(35)* also showed the effects of multiple freeze-thaw cycles on RNA quantitation using the first generation bDNA assay. Five patient plasma samples were subjected to 1–3 freeze-thaw cycles, and samples were stored frozen at –20 or –80°C, and then thawed to ambient temperature. Their data indicated that the difference in RNA quantitation between samples subjected to one and three freeze-thaw cycles at –80°C was not significantly different. However, samples that were subjected to three freeze-thaw cycles at –20°C did display a significant decline in RNA levels. This suggests that there is less stability of RNA when kept at –20°C vs –80°C.

Aoki-Sei also showed that in five patient samples, one freeze thaw did not significantly alter RNA signal *(33)*. Todd et al. *(35)* also evaluated the stability of samples for RNA signal when stored at –80°C or –20°C for over a 1-yr period. RNA levels were found to decline significantly after 12 wk when stored at –20°C in a nonfrost-free freezer. Identical samples held at –80°C did not appear to decay. Using an RT-PCR assay, Winters et al. *(50)* earlier found that samples held at –70°C had consistent RNA levels over 6 mo. They compared storage techniques by either storing plasma neat or in the presence of guanidinium isothiocynate. There did not appear to be a difference in the stability with the addition of guanidinium. Thus, RNA in plasma stored at –70°C appears to be relatively stable for 6–12 mo. Longer term studies have not been completed at this time.

As an example, prognostic data on the use of plasma HIV RNA (as an absolute RNA copy number to determine the relative risk of clinical progression) as presented by Mellors et al. *(3)*, should be interpreted with some caution. Samples from that study were analyzed by bDNA. However these samples were 5–10-yr-old, collected in heparinized plasma, and were not processed in a uniform time frame. Although the study yielded highly significant results, given all the assay and sample variables, it is unclear whether an absolute copy number derived from this study is meaningful.

With regards to HCV testing, less data has been collected. However, a recent study by Miscowski *(51)* quantitating HCV (using the Amplicor HCV Monitor RT-PCR assay) showed that samples collected in the presence of EDTA, ACD, or in serum had identical HCV RNA copy number. HCV RNA concentration was measured after storage of EDTA plasma at 4°C or room temperature for 48 h after collection. A significant reduction in HCV

RNA concentration was seen in samples held for 24 h until centrifugation was accomplished. Although a statistically significant reduction in HCV RNA was seen after 48 h, the change was relatively small (0.1 log copy/mL) and of questionable biologic significance.

Busch et al. *(52)* compared serum and fresh frozen plasma and found 62% of the sera were positive compared to all of the plasma. When fresh serum and ACD plasma were compared, there was no difference in PCR signal. Storing plasma or serum at room temperature for 7 d resulted in >$10^2$ reduction in HCV RNA in their semiquantitative RT-PCR assay. Samples held at 4°C were significantly more stable than those held at room temperature. One or more freeze-thaws resulted in a 0.5 $\log_{10}$ reduction in HCV RNA signal.

Cuyprens et al. *(53)* also confirmed that whole blood or serum held at 4°C did not result in any reduction in HCV RNA as compared to those held at room temperature, which had a 3–4 $\log_{10}$ reduction in signal. Thus if short-term storage is required prior to freezing, whole blood should be separated, and plasma stored at 4°C until freezing. Whole-blood storage, particularly at room temperature, could result in significant reduction in plasma RNA levels. Long-term freezing should always be at –80°C. Freeze-thawing of samples should be minimized.

## 5. Procedures for RNA Detection in the Presence of Inhibitors

Several procedures for whole blood and blood processing have been published and are reviewed elsewhere *(54,55)*. Although some DNA viruses can be readily detected in serum or blood just by boiling the sample prior to PCR, RNA viruses are more problematic and require a more formal extraction procedure for detection *(56)*. Inhibitors such as anticoagulant or components within blood may make viral RNA detection problematic. For prospective studies, it is best to avoid collection of whole blood in the presence of heparin. However in some clinical situations, i.e., when immune-function assays or some culture systems are also required (performed best on cells collected in heparinized blood) plasma containing heparin may be the only option. Physical separation of virions from plasma with heparin supernatant may be helpful. Ultracentrifugation or immunocapture have proven effective in allowing heparinized samples to be used. However, these methods are not standardized. Although qualitative PCR data can be produced by these methods, it is unclear whether there is a quantitative loss of template or PCR signal from any residual heparin after these separation techniques. When archived blood or plasma is collected in an unknown anticoagulant, and amplification of viral genetic material is not successful, procedures that optimize assay performance must be undertaken.

As previously stated, various techniques have been attempted to remove heparin from samples to allow efficient PCR reactions to proceed. Many assays and protocols have used some variation of an acid phenol-chloroform RNA extraction procedure prior to RT-PCR *(57)*. This process using guanidinium as a chaotrope to inhibit RNases and lyse cells was found to be unsuccessful in removing the inhibitory effects of heparin. Thus, commercially available heparinase was found by several groups to remove effectively the inhibitory effect of heparin after RNA extraction. Izraeli et al. *(27)* found that 1–3 U of heparinase I/g RNA in 5 m$M$ tris pH 7.5, 1 m$M$ $CaCl_2$ and 40 U of RNAsin incubated for 2 h at 25°C with samples containing heparin, completely removed the inhibitory effect. Heparinase I was found to be as effective as heparinase II in their study. In general, other studies used the same buffer including $CaCl_2$ and varying amounts of heparinase for 2 h at either 25° or 37°C *(24,26,28)*. Calcium removal may be important because it potentially contributes to RT-PCR inhibition. Thus, Imai et al. *(26)* recommended the use of EGTA to remove calcium prior to RT-PCR.

An additional method that can be employed to effectively remove the inhibitory effects of heparin is to adsorb viral RNA on silicon-dioxide suspensions or glass powder. Boom et al. *(46)* and others have described a capture technique, in which samples containing virus were lysed in the presence of guanidinium. Viral nucleic acid was released and bound to silica. After washing, the nucleic acid was eluted in an aqueous low-salt buffer, and the supernatant containing the nucleic acid was removed. Successful application of this technique was described for detection of hepatitis E virus from serum *(58)*. It has now been incorporated into the extraction step of the HIV RNA NASBA assay *(47)*. Several commercial preparations of this technique are widely available. As stated earlier, in HIV RNA reconstruction experiments, there was no inhibitory effect of heparin using this procedure prior to NASBA. Unpublished data indicates that clinical samples collected in heparinized plasma are acceptable for the NASBA assay. In addition, RT-PCR can also be performed on samples collected in heparin utilizing a capture technique.

## 6. Conclusions

As molecular techniques of viral nucleic acid quantification become more accepted in clinical practice, individuals involved in research, clinical-trial design, and clinical practice must be made aware of both patient- and sample-related factors that affect quantification. The optimal procedures for processing, handling, and storage of patient material has yet to be fully defined. However, a consistent procedure for handling samples, which includes the same kind of collection tube and anticoagulant, processing technique, transport, time of processing, storage procedures, and molecular assay for every sample on a given patient, is critically important to the understanding and evaluation of data from each particular patient sample.

# References


1. Holodniy, M., Katzenstein, D. A., Israelski, D. M., and Merigan, T. C. (1991) Reduction in plasma human immunodeficiency virus ribonucleic acid after dideoxynucleoside therapy as determined by the polymerase chain reaction. *J. Clin. Immunoassay* **88,** 1755–1759.
2. Holodniy, M., Katzenstein, D. A., Sengupta, S., Wang, A. M., Casipit, C., Schwartz, D. H., et al. (1991) Detection and quantificaton of human immunodeficiency virus RNA in patient serum by use of the polymerase chain reaction. *J. Infect. Dis.* **163,** 862–866.
3. Mellors, J. W., Rinaldo, C. R., Gupta P., White, R. M., Todd, J. A., and Kingsley, L. A. (1996) Prognosis in HIV-1 infection predicated by the quantity of virus in plasma. *Science* **272,** 1167–1170.
4. Katzenstein, D. A., Hammer, S. M., Hughes, M. D., Gundacker, H., Jackson, J. B., Fiscus, S., et al. (1996) The relation of virologic and immunologic markers to clinical outcomes after nucleoside therapy in HIV infected adults with 200–500 CD4 cells per cubic millimeter. *NEJM* **335,** 1091–1098.
5. Stamprans, S. I., Hamilton, B. L., Follansbee, S. E., Elbeik, T., Barbosa, P., Grant, R. M., et al. (1995) Activation of virus replication after vaccination of HIV-1-infected individuals. *J. Exp. Med.* **182,** 1727–1737.
6. Brichacek, B., Sindells, S., Janoff, E. N., Pirruccello, S., and Stevenson, M. (1996) Increased plasma human immunodeficiency virus type 1 burden following antigenic challenge with penumococcal vaccine. *J. Infect. Dis.* **174,** 1191–1199.
7. Bush, C. E., Donovan, R. M., Markowitz, N. P., Kvale, P., and Saravolatz, L. D. (1996) A study of HIV RNA viral load in AIDS patients with bacterial pneumonia. *J. Acq. Immune. Def. Syn* **13,** 23–26.
8. Donovan, R. M., Bush, Markowitz, N. P., Baxa, D. M., and Saravolatz, L. D. (1996) Changes in virus load markers during AIDS-associated opportunistic diseases in human immunodeficiency virus-infected persons. *J. Infect. Dis* **174,** 401–403.
9. Gretch, D., Corey, L., Wilson, J., dela Rosa, C., Willson, R., Carithers, R., Jr., et al. (1994) Assessment of hepatitis C virus RNA levels by quantitative competitive RNA polymerase chain reaction: high-titer viremia correlates with advanced stage of disease. *J. Infect. Dis.* **169,** 1219–1225.
10. Nousbaum, J.-B., Pol, S., Nalpas, B., Landais, P., Berthelot, P., and Brechot, C. (1995) Hepatitis C virus type 1b (II) infection in France and Italy. *Ann. Intern. Med.* **122(3),** 161–168.
11. Shindo, M., Arai, K., Yoshihiro, S., and Okuno, T. (1995) Hepatic hepatitis C virus RNA as a predictor of a long-term response to interferon-alpha therapy. *Ann. Intern. Med.* **122,** 586–591.
12. Pachl, C., Todd, J. A., Kern, D. G., Sheridan, P. J., Fong, S. J., Stempien, M., et al. (1995) Rapid and precise quantification of HIV-1 RNA in plasma using a branched DNA signal amplification assay. *J. Acq. Imm. Def. Hum. Retro.* **8,** 446–454.

13. van Gemen, B., Kievits, T., Schukkink, R., van Strijp, D., Malek, L. T., Sooknanan, R., et al. (1993) Quantification of HIV-1 RNA in plasma using NASBA™ during HIV-1 primary infection. *J. Virol. Methods* **43**, 177–188.

14. Mulder, J., McKinney, N., Christopherson, C., Sninsky, J., Greenfield, L., and Kwok, S. (1994) Rapid and simple PCR assay for quantitation of human immunodeficiency virus type 1 RNA in plasma: application to acute retroviral infection. *J. Clin. Micro.* **32**, 292–300.

15. Riddler, S. A., Holodniy, M., White, R. A., et al. (1996) Diurnal and within patient variation of HIV-viral load. In: 3rd Conference on Retroviruses and Opportunistic Infections, Washington DC. (Abstract).

16. Saag, M. S., Holodniy, M., Kuritzkes, D. R., O'Brien, W. A., Coombs, R., Poscher, M. E., et al. (1996) HIV viral load markers in clinical practice. *Nature Med.* **2(6)**, 625–629.

17. Raboud, J. M., Montaner, J. S., Conway, B., Haley, L., Sherlock, C., O'Shaughnessy, M. V., et al. (1996) Variation in plasma RNA levels: CD4 cell counts, and p24 antigen levels in clinically stable men with HIV infection. *J. Infect. Dis.* **174**, 191–194.

18. Schuurman, R., Deschamps, D., Weverling, G. J., Kaye, S., Tijnagel, J., Williams, I., et al. (1996) Multicenter comparison of three commercial methods for quantification of human immunodeficiency virus type RNA in plasma. *J. Clin. Micro.* **34**, 3016–3022.

19. Yen-Lieberman, B., Brambilla, D., Jackson, B., Bremer, J., Coombs, R., Cronin, M., et al. (1996) Evaluation of a quality assurance program for quantitation of human immunodeficiency virus type 1 RNA in plasma by the AIDS Clinical Trials Group Virology Laboratories. *J. Clin. Micro.* **34**, 2695–2701.

20. Vandamme, A. M., Schmit, J. C., Van Dooren, S., Van Laethem, K., Gobbers, E., Kok, W., et al. (1996) Quantification of HIV-1 RNA in plasma: comparable results with the NASBA HIV-1 RNA QT and the AMPLICOR HIV monitor test. *J. Acq. Immune. Def. Synd.* **13**, 127–139.

20b. Panaccio, M. and Lew, A. (1991) PCR based diagnosis in the presence of 8% (v/v) blood. *Nucleic Acids Res.* **19(5)**, 1151.

21. Gelfand, D. (1989) Taq DNA polymerase, in *PCR Technology* (Erlich, H. A., ed.), Stockton Press, New York, pp. 17–22.

22. Higuchi, R. (1989) Simple and rapid preparation of samples for PCR, in *PCR Technology* (Erlich, H. A., ed.), Stockton Press, New York, pp. 31–38.

23. Gustafson, S., Proper, J., Bowie, E., and Sommer, S. (1987) Parameters affecting the yield of DNA from human blood. *Anal. Biochem.* **165**, 294–299.

24. Beutler, E., Gelbart, T., and Kuhl, W. (1990) Interference of heparin with the polymerase chain reaction. *Biotechniques* **9**, 166.

25. Holodniy, M., Kim, S., Katzenstain, D., Konrad, M., Groves, E., and Merigan, T. (1991) Inhibition of human immunodeficiency virus gene amplification by heparin. *J. Infect. Dis.* **29(4)**, 676–679.

25a. Nakamura, S., Katamine, S., Yamamoto, T., Foung, S. K., Kurata, T., Hirabayashi, Y., et al. (1993) Amplification and detection of a single molecule of human immunodeficiency virus RNA. *Virus Genes* **4**, 325–338.

26. Imai, H., Yamada, O., Morita, S., Suehiro, S., and Kurimura, T. (1992) Detection of HIV-1 RNA in heparinized plasma of HIV-1 seropositive individuals. *J. Virologic. Methods* **36,** 181–184.

27. Izraeli, S., Pfleiderer, C., and Lion, T. (1991) Detection of gene expression by PCR amplification of RNA derived from frozen heparinized whole blood. *Nucleic Acids Res.* **19(21),** 6051.

28. Tsai, M., Miyamoto, M., Tam, S. Y., Wang, Z. S., and Galli, S. J. (1995) Detection of mouse mast cell-associated protease mRNA. Heparinase treatment greatly improves RT-PCR of tissues containing mast cell heparin. *Am. J. Path.* **146,** 335–343.

29. Panaccio, M., Georgesz, M., Lew, A. M. (1993) FoLT PCR: a simple PCR protocol for amplifying DNA directly from whole blood. *BioTechniques* **14(2),** 238–243.

30. Burckhardt J. (1994) Amplification of DNA from whole blood. *PCR Methods Appli.* **4,** 239–243.

31. Satsangi, J., Jewell, D. P., Welsh, K., Bunce, M., and Bell, J. I. (1994) Effect of heparin on polymerase chain reaction. *Lancet* **343,** 1509–1510.

32. Poli, F., Cresplanco, L., Nocco, A., and Sinclair, G. A (1993) A rapid and simple method of reversing the inhibitory effect of heparin on PCR for HLA class II typing. *PCR Methods Appl.* **2,** 356–358.

32a. Di Cioccio, R. and Strivastava, B. (1978) Inhibition of deoxynucleotide-polymerizing enzyme activities of human cells and of simian sarcoma virus by heparin. *Cancer Res.* **38,** 2401–2407.

33. Aoki-Sei, S., Yarchoan, R., Kageyama, S., Hoezema, D., Pluda, J., Wyvil,l K., Broder, S., and Mitsuya H. (1992) Plasma HIV-1 viremia in HIV-1 infected individuals assessed by polymerase chain reaction. *AIDS Res. Hum. Retrovir.* **8(7),** 1263–1270.

34. Mole, L., Margolis, D., Carroll, R., Todd, J., and Holodniy, M. (1994) Stabilities of quantitative plasma culture for human immunodeficiency virus, RNA and p24 antigen from samples collected in VACUTAINER CPT and standard VACUTAINER tubes. *J. Clin. Microbiol.* **32(9),** 2212–2215.

35. Todd, J., Pachl, C., White, R., Yeghiazarian, T., Johnson, P., Taylor, B., Holodniy, M., Kern, D., Hamren, S., Chernoff, D., and Urdea, M.(1995) Performance characteristic for the quantiation of plasma HIV-1 RNA using branched DNA signal amplification technology. *J. Acquired Immune Def. Syndromes Hum. Retrovir.* **10(Suppl. 2),** S34–S44.

36. Dickover, R., Herman, S., Saddiq, K., Wafer, D., and Byrson, Y. (1996) Stability of HIV-1 RNA in whole blood stored at room temperature and effect of anticoagulant. 36th Interscience Conference on Antimicrobial Agents and Chemotherapy, New Orleans, LA (Abstract).

37. Izopet, J., Poggi, C., Dussaix, E., Mansuy, J. M., Cubaynes, L., Profizi, N., et al. (1996) Assessment of a standardized reverse-transcriptase PCR assay for quantifying HIV-1 RNA in plasma and serum. *J Virol. Methods* **60,** 119–129.

38. Lee, T. H., Stromberg, R. R., Henrard, D., and Busch, M. P. (1993) Effect of platelet-associated virus on assays of HIV-1 in plasma [letter]. *Science* **262,** 1585.

39. Piatak, M., Shaw, G. M., Yang, L. C., et al. (1993) Effect of platelet-associated virus on assays of HIV-1 in plasma. *Science* **262,** 1585–1586.

40. Zhu, Y., Gong, Y., and Cimino, G. (1995) Quantitative analysis of HIV-1 RNA in plasma preparations. *J Virol. Methods* **52,** 287–299.

41. Holodniy, M., Margolis, D., Carroll, R., Todd, J., and Mole, L. (1996) Quantitative relationship between platelet count and plasma virion HIV RNA. *AIDS* **10(2),** 232–233.

42. Wang, J. T., Wang, T. H., Sheu, J. C., Lin, S. M., Lin, J. T., and Chen, D. S. (1992) Effects of anticoagulants and storage of blood samples on efficacy of the polymerase chain reaction assay for hepatitis C virus. *J. Clin. Microbiol.* **30,** 750–753.

42a. Willems, M., Moshage, H., Nevens, F., Fevery, J., and Yap, S. H. (1993) Plasma collected from heparinized blood is not suitable for HCV-RNA detection by conventional RT-PCR. *J. Virol. Methods* **42,** 17–30.

43. Manzin, A., Bagnarelli, P., Menzo, S., Giostra, F., Brugia, M., Francesconi, R., et al. (1994) Quantitation of hepatitis C virus genome molecules in plasma samples. *J Clin. Microbiol.* **32,** 1939–1944.

44. Conrad, A., Tong, M., Rueben, A., et al. Comparison of plasma preparation tube (PPT) and fresh frozen serum impact on stability of hepatitis C virus RNA by RT-PCR. (Abstract).

45. Van Damme, A.-M., Van Dooren, S., Kok, W., Goubau, P., Fransen, K., Kievits, T., Schmit, J.-C., De Clercq, E., Desmyter, J. (1995) Detection of HIV-1 RNA in plasma and serum samples using the NASBA amplification system compared to RNA-PCR. *J. Virolog. Methods* **52,** 121–132.

46. Boom, R., Sol, C. J. A., Alimans, M. M. M., Jansen, C. L., Weheim-van Dillen, P. M. E., and van der Noordaa, J. (1990) Rapid and simple method for purification of nucleic acids. *J. Clin. Micro.* **28,** 495–503.

47. van Gemen, B., Wiel, P. V. D., van Beuningen, R., Sillekens, P., Jurriaans, S., Dries, C., Schoones, R.. and Kievits, T. (1995) The one-tube quantitative HIV-1 RNA NASBA: precision, accuracy and application. *PCR Methods Appl.* **4,** S177–S184.

48. Holodniy, M., Mole, L., Yen-Lieberman, B., Margolis, D., Starkey, C., Carroll, R., Spahlinger, T., Todd, J., and Jackson, J. B. (1995) Comparative stabilities of quantitative human immunodeficiency virus RNA in plasma from samples collected in VACUTAINER CPT., VACUTAINER PPT., and standard VACUTAINER tubes. *J. Clinical Microbiol.* **33(6),** 1562–1566.

49. Coombs, R. W., Henrard, D. R., Mehaffey, W. F., Gibson, J., Eggert, E., Quinn, T. C., et al. (1993) Cell-free plasma human immunodeficiency virus type 1 titer assessed by culture and immunocapture-reverse transcription-polymerase chain reaction. *J. Clin. Microbiol.* **31,** 1980–1986.

50. Winters, M. A., Tan, L. B., Katzenstein, D. A., and Merigan, T. C. (1993) Biological variation and quality control of plasma human immunodeficiency virus type 1 RNA quantitation by reverse transcriptase polymerase chain reaction. *J. Clin. Micro.* **31,** 2960–2966.

51. Miskovsky, E. P., Carrella, A. V., Gutenkunst, K., Sun, C. A., Quinn, T. C., Thomas, D. L. (1996) Clinical characterization of a competitive PCR assay for quantitative testing of hepatitis C virus. *J. Clin. Microbiol.* **34,** 1975–1979.
52. Busch, M. P., Wilber, J. C., Johnson, P., Tobler, L., and Evans, C. S. (1992) Impact of specimen handling and storage on detection of hepatitis C virus RNA. *Transfusion* **32,** 420–425.
53. Cuypers, H. T., Bresters, D., Winkel, I. N., Reesink, H. W., Weiner, A. J., Houghton, M., et al. (1992) Storage conditions of blood samples and primer selection affect the yield of cDNA polymerase chain reaction products of hepatitis C virus. *J. Clin. Microbiol.* **30,** 3220–3224.
54. Greenfield, L. and White, T. (1993) Sample preparation methods, in *Diagnostic Molecular Microbiology: Principles and Applications* (Persing, D. H., Smith, T. F., Tenover, F. C., and White, T. J., eds.), American Society for Microbiology, Washington, DC, pp. 122–137.
55. Lin, L., Gong, Y., Metchette, K., et al. (1993) Simple and rapid sample preparation methods for whole blood and blood plasma, in *Diagnostic Molecular Microbiology: Principles and Applications* (Persing, D. H., Smith, T. F., Tenover, F. C., and White, T. J., eds.), American Society for Microbiology, Washington, DC, pp. 605–616.
57. Chomczynski, P. and Sacchi, N. (1987) Single step method of RNA isolation by acid quanadinium thiocyanate-phenol-chloroform extraction. *Anal. Biochem.* **162,** 156–159.
58. McCaustland, K., Bi, S., Purday, M., and Bradley, D. (1991) Application of two RNA extraction methods prior to amplification of hepatitis E virus nucleic acid by the polymerase chain reaction. *J. Virol. Methods* **35,** 331–342.

# 4

# Quantitative RT-PCR

## Paul D. Siebert

## 1. Introduction and Overview

Although reverse transcriptase polymerase chain reaction (RT-PCR) is an extremely sensitive method of mRNA analysis, obtaining quantitative information with this technique can be difficult. This is caused primarily by the fact that there are two sequential enzymatic steps involved: the synthesis of DNA from the RNA template and PCR. In practice, the exponential nature of PCR and the practical aspects of performing PCR pose the most serious obstacles to obtaining quantitative information. With some adaptations, however, RT-PCR can yield accurate quantitative results.

This chapter describes a number of methods that have been developed for using RT-PCR to determine the relative level of abundance of a particular mRNA, changes in the abundance of an mRNA over time or after induction, and the actual number of mRNA molecules in the sample. The theory and applications of each method are discussed, as well as the advantages and limitations associated with them. The chapter then expands on one method of quantitative PCR in particular, namely competitive PCR. This method, that uses nonhomologous internal standards (PCR MIMICs), is both simple and useful.

## 2. Theoretical and Practical Aspects of PCR

### 2.1. The Exponential Nature of PCR

By definition, the PCR process is a chain reaction. The twofold increase in products from one cycle of amplification serve as substrates for the next. Therefore, the amount of product increases exponentially and not linearly, as in most enzymatic processes. Under ideal or theoretical conditions, the amount of product doubles during each cycle of the PCR reaction according to **Eq. 1**. This relationship is plotted in **Fig. 1A**.

$$N = N_0 2^n \tag{1}$$

From: *Methods in Molecular Medicine, Vol 26: Quantitative PCR Protocols*
Edited by: B. Kochanowski and U. Reischl © Humana Press Inc., Totowa, NJ

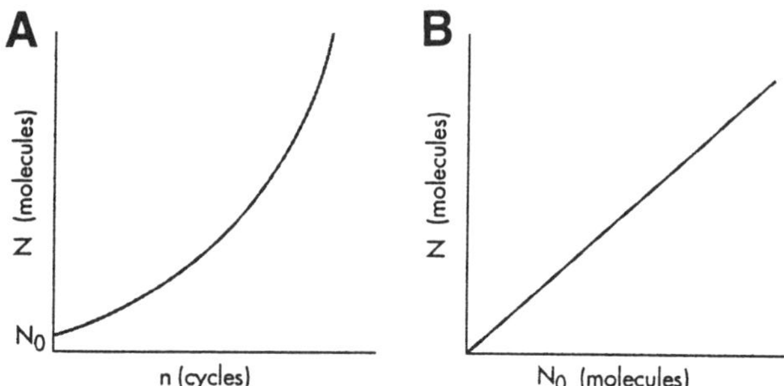

Fig. 1. Characteristics of PCR amplification in an ideal case. **(A)** Kinetics of amplification and **(B)** PCR product yield as a function of initial amount of target.

where $N$ is the number of amplified molecules, $N_0$ is the initial number of molecules, and $n$ is the number of amplification cycles. **Eq. 1** indicates a linear relationship between the number of amplified target molecules and the initial number of target molecules. This relationship is shown in Fig. 1B.

## 2.2. The Efficiency of Amplification

Amplification efficiency, that is, the fraction of the template replicated during each reaction cycle, is a crucial factor for any reliable method of quantitative PCR. Experimentally, the efficiency of amplification ($E$) is less than one, and the PCR process is thus described by **Eq. 2**.

$$N = N_0 (1 + E)n \qquad (2)$$

where $E$ is the amplification efficiency. Because of the exponential nature of PCR, a very small change in amplification efficiency, $E$, can yield dramatic differences in the amount of product, $N$, even if the initial number of target molecules, $N_0$, is the same. For example: if $E = 0.85$ and $n = 30$, then $N = N_0$ $(1 + 0.85)^{30}$ and $N = 10.4 \times 10^7 N_0$. In other words, with 85% efficiency, 30 cycles of PCR would produce a $10.4 \times 10^7$-fold increase in the amount of target molecules. However, if $E$ is reduced to 0.8, the target would only be amplified $4.6 \times 10^7$ times by PCR. Thus, a change in amplification efficiency of only 0.05 would produce a greater than twofold change in the amount of reaction product. This difference becomes even greater as the number of cycles increases.

Several experimental factors may affect the efficiency of amplification, including:

1. The sequence being amplified.
2. The sequence of the primers.
3. The length of the sequence being amplified.
4. Impurities in the sample.

The first three of these factors are important because they affect secondary structure formation and the G/C content of the target sequence—both of which may interfere with primer binding, affect the melting point of the target sequence, and reduce the processivity of the polymerase. The length of the target sequence being amplified can affect $E$ for another reason: Even with an ideal template, no polymerase exhibits 100% processivity under in vitro conditions. Because of the limited processivity of *Taq* DNA polymerase in vitro, target sequences longer than 3 kb are extremely difficult to amplify. More importantly, there is also some controversy about whether differences in target sequence lengths significantly alter the efficiency of amplification when the sequences are <1 kb. In two cases, a weak inverse correlation was observed *(1,2)*. In another case, there was no observed difference in $E$ *(3)*. Impurities in the sample can affect amplification efficiency in many ways. For example, they can degrade or inhibit the polymerase, cause conformational changes in the target DNA, or compete for primer binding sites—to name just a few of the possibilities.

There may be additional, unknown, subtle factors that affect $E$. This is illustrated by the fact that the amount of product amplified from the same target sequence after the same number of cycles and under identical experimental conditions often differs from one PCR reaction to another. This was seen even when using a master mix of reaction components *(4,5)*. Unfortunately, such tube-to-tube variation in amplification efficiency can be both significant and unpredictable. Theoretically, the efficiency of amplification, $E$, ranges from 0–1. Experimentally, the value of $E$ has been found to range from 0.46–0.99 for different genes *(3,6)*. The value of $E$ also varied, from 0.8–0.99, when the same gene was amplified in independent tubes under identical conditions *(7)*.

## 2.3. The Plateau Effect

Experimentally, the amount of product generated during PCR also deviates from the theoretical case. The amount of PCR products produced during the PCR initially increases exponentially, but then the rate of production slows and finally levels off, as shown in **Fig. 2A,B**. **Fig. 2A** is a graph of the number of amplified target molecules ($N$) plotted as a function of PCR cycles ($n$), and Fig. 2B is a graph of the number of amplified target molecules ($N$) plotted as a function of the initial number of target molecules ($N_0$). The leveling off of the rate of amplification is often referred to as the plateau effect.

The following factors can contribute for the observed plateau effect:

1. The product accumulates to a concentration at which reassociation competes with primer annealing and extension *(8)*.
2. The molar ratio of polymerase to template falls below a critical value.
3. Inhibitors of polymerase activity, such as pyrophosphates, may accumulate.
4. One or more of the components necessary for the reaction become limiting.

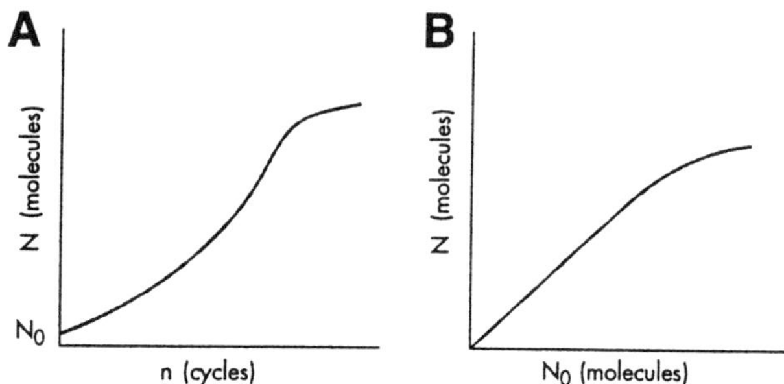



Fig. 2. Characteristics of PCR amplification in a typical case. **(A)** Kinetics of amplification and **(B)** PCR product yield as a function of initial amount of target.

The number of cycles needed to reach the plateau phase varies, depending on the sequence—and the original amount—of target mRNA. This variability makes it difficult to predict precisely the time-course of the reaction or the amount of product synthesized before plateau phase is reached. The uncertainties inherent in the plateau effect, as well as the exponential nature of PCR, contribute to the difficulty of performing quantitative PCR because they obscure the linear relationship between $N_0$ and $N$ depicted in **Eqs. 1** and **2**. Methods that employ RT-PCR to obtain quantitative information must take these factors into account.

## *2.4. Quantifying the PCR Product*

The goal of quantitative PCR is to deduce, from the final amount of PCR product, either the initial number of target molecules ($N_0$) or the relative starting levels of target molecules among several samples. Thus, the first step in this process is to measure the amount of PCR product present.

Several methods are commonly used to quantify PCR products. The most straightforward approach is to measure the incorporation of labeled nucleotides or primers into PCR products resolved by gel electrophoresis. Although direct, the use of labeled nucleotides in PCR can be problematic. High levels of unincorporated, labeled nucleotides in the PCR product mixtures result from the relatively high (to 100 $\mu M$) concentrations of nucleotides required for PCR. Consequently, trace amounts of unincorporated label often remain in the electrophoretic gel as the product bands migrate, resulting in a "trail" of label throughout the lane. Even a relatively small amount of "trailing" can make it difficult to measure the amount of incorporated label. For this reason, many researchers prefer to use labeled PCR primers rather than labeled nucleotides.

Other strategies for quantifying PCR products are based on hybridization. The most common of these methods is to probe a Southern blot of the PCR products using a radioactively labeled probe complementary to the specific, amplified sequences. To quantitate the amount of probe hybridized, the blot can either be exposed to X-ray film and the resulting autoradiogram densitometrically scanned, or the PCR product band can be excised from the blot and its radioactivity measured in a scintillation counter. Because the nucleic acid probes only hybridize to the corresponding amplified DNA sequences, this method offers the advantage of detecting only the correct PCR product. Nonspecific products do not produce a signal.

Alternative hybridization methods that avoid Southern blotting have also been utilized *(5,9,10)*. Jalava et al. *(9)* described an approach based on the capture and hybridization of biotinylated PCR products on streptavidin-coated microtiter plates. The biotin group is added to the PCR product during amplification through the use of a biotinylated primer. Biotinylated products are subsequently captured on streptavidin-coated plates, and a radioactively labeled nucleic acid probe, complementary to the biotinylated strand, is then used to measure the amount of captured product. Jalava et al. used relatively long, nick-translated DNA fragments (0.35 and 0.42 kb) as the radioactive hybridization probes; however, the results of their experiments suggest that it might also be possible to use short, nonisotopically labeled synthetic DNA probes in conjunction with an appropriate detection system.

Another hybridization method that avoids Southern blotting is solution hybridization of a radioactively labeled probe and denatured PCR products. The hybridized probes are resolved by gel electrophoresis and subsequently quantitated by scintillation counting *(5)*. Fluorescent labels also can be used instead of radioactivity. In this case, a fluorescently labeled internal primer is annealed to one strand of the PCR product and extended using *Taq* DNA polymerase. Run-off extension products are electrophoresed in an automated DNA sequencer that quantitatively detects the incorporated fluorescent label *(10)*.

Several additional methods exist for quantifying PCR products. They include measurement of the EtBr luminescence emanating from PCR products resolved by gel electrophoresis *(11)*, use of high-performance liquid chromatography *(12)*, and assays based on in vitro transcription with radioactively labeled ribonucleotide substrates *(13)*. For in vitro transcription, a transcriptional promoter is incorporated into one of the PCR primers. Following amplification, the PCR product is transcribed in vitro using radioactively labeled ribonucleotides. During transcription, the radioactive signal is amplified 100- to 200-fold, making this a very sensitive detection method. However, the additional enzymatic reaction required for in vitro transcription makes this one of the more laborious detection methods and may also increase the risk of experimental error.

## 3. Quantitative PCR Without the Use of Internal Standards

Most commonly, researchers use internal standards to control variations in amplification efficiency and to determine absolute values of mRNA (discussed

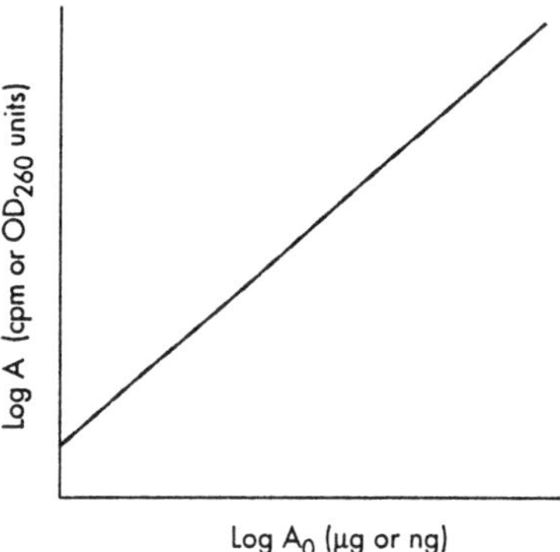



Fig. 3. Linear relationship between the log of the amount of PCR product and the log of the initial amount of sample RNA (µg) or cDNA (ng), in an ideal case.

in **Subheading 3.1.**). However, it is possible to perform quantitative PCR without internal standards if two conditions are met. First, tube-to-tube variation in the actual value of $E$ must be minimal so that a constant value can be assumed for $E$ in all related PCR reactions. Second, all data must be obtained before the reactions begin to reach the plateau phase. The methods described in this section employ mathematical models based on **Eqs. 2** and **3** to determine relative changes in mRNA levels.

$$\log N = [\log (1 + E)]\, n + \log N_0 \tag{3}$$

where $N$ is the number of amplified molecules, $N_0$ is the initial number of molecules, $n$ is the number of amplification cycles, $E$ is the amplification efficiency.

For convenience, Eqs. (2) and (3) may also be written as:

$$A = A_0 (1 + E)n \tag{2.1}$$

$$\log A = [\log (1 + E)]\, n + \log A_0 \tag{3.1}$$

where $A$ is the amount of amplified product (in cpm or $OD_{260}$ U), and $A_0$ is the starting amount of total RNA (µg) or cDNA (ng). Note: the target sequences usually comprise only a small fraction of the total. At the end of this section is a discussion of the use of linear regression analysis (also based on these equations) to estimate absolute numbers of mRNA target molecules per unit of starting RNA without using internal controls.

If the two conditions are in effect (i.e., $E$ is constant and reactions are not reaching the plateau phase), **Eq. 3.1** indicates that there is a linear relationship between the logarithm of the starting amount of target mRNA (or cDNA) (included in $A_0$) and the logarithm of the amount of amplification product generated ($A$). This relationship is illustrated in the graph of **Fig. 3**. A linear rela-

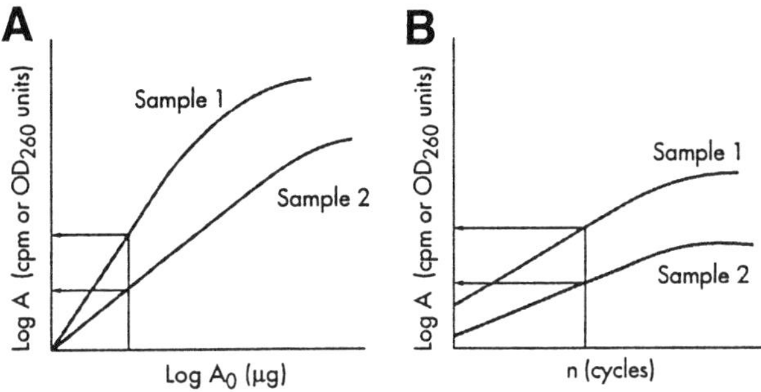


Fig. 4. Two methods for determining relative differences in the initial amount of target in two samples. **(A)** Titration method and **(B)** kinetic method.

tionship between log $A$ and log $A_0$ has also been shown to exist empirically for values of $A$ ranging over 2–3 orders of magnitude (14–16). In one case, this relationship was even found to hold for values of $A$ differing by four orders of magnitude *(11)*.

## 3.1. Determining Relative Differences in $N_0$ Between Two or More Samples

Two forms of experimental analysis, titration and kinetics, can be used to estimate the relative initial amounts of target mRNA or cDNA in two samples—when the amplification efficiencies are the same for the two samples and the data are collected before the reactions begin to reach the plateau phase.

### 3.1.1. Titration Analysis

A titration analysis is performed by making a dilution series, or titration, of RNA or cDNA, amplifying by PCR, and quantifying the signals produced (defined as $A$). **Fig. 4A** shows idealized data collected from this type of experiment, graphed as log $A$ as a function of log $A_0$. Because of the linear relationship between log $A_0$ and log $A$, and because the amount of target mRNA or cDNA is a constant proportion of the total starting material ($A_0$) for each of the various dilutions of a given sample, the relative difference in $N_0$ is proportional to the difference betwen the slopes of the two curves. Thus, a value of log $A_0$ is chosen on the X (horizonal) axis of the graph and the corresponding values of log $A$ are then extrapolated for both curves, as shown in **Fig. 4A**. The difference between the two values of log $A$ determined in this manner from the graph is equivalent to the relative difference in $N_0$ for the two samples. Singer-Sam et al. *(16)* used this method to determine the relative changes in mRNA levels for several phosphoglycerate kinases and phosphoribosyl-transferases during mouse spermatogenesis.

### 3.1.2. Kinetic Analysis

A more commonly used alternative to titration analysis is comparative kinetic analysis. To perform a kinetic analysis, values of $A$ are determined for a number of consecutive amplification cycles ($n$) for two samples. **Fig. 4B** shows idealized data from an experiment of this type, plotted as log $A$ vs $n$. The curves are consistent with **Eq. 3**.

To determine the relative difference in $N_0$ between the two samples, a value of $n$ is chosen at a point where the two curves are parallel (suggesting equal values for $E$), and the value of log $A$ is extrapolated from this value of $n$ for each curve. At this point, the difference between the two values for log $A$ is directly proportional to the difference of log $A_0$ between the two samples. Moreover, the difference of log $A_0$ between the two samples is equal to the difference of log $N_0$ between the two samples. Hence, this method can be used to determine the difference in the initial number of target molecules, but not the actual number of starting target molecules.

Comparative kinetic analyses have been used to accurately detect 2- to 10-fold changes in mRNA levels. For example, Solomon et al. *(17)* used this approach to examine differences in the levels of apolipoprotein mRNA in normal and atherosclerotic blood vessels. Dallman et al. *(18)* used a similar strategy to examine the influence of tissue transplantation on cytokine mRNA levels.

### 3.2. Using Linear Regression Analysis to Determine the Absolute Value of $N_0$

**Equation 3** describes a linear relationship in the format, $y = mx + b$, whose slope ($m$) has the value of log $(1 + E)$ and whose $y$-intercept ($b$) is $N_0$. This allows estimation of the value of $N_0$ graphically. When the value of $E$ is known, the value of $N_0$ can be determined from a linear regression analysis of the plotted data. Experimentally, a kinetic study is performed in which a constant amount of starting cDNA is amplified by PCR. During consecutive cycles, the number of product molecules, $N$, is determined. In this method it is necessary to calculate $N$, and not simply $A$. With the data graphed as log $N$ vs $n$, $E$ can be calculated from the logarithm of the slope, and $N_0$ can be derived from the $y$-intercept (**Fig. 5**).

This method was recently used by Wiesner *(7)* to estimate the number of $\alpha$- and $\beta$-myosin heavy-chain mRNA molecules per unit of total RNA extracted from rat ventricle tissue. The authors also were able to calculate the number of mRNA molecules per cell, taking into account the yield of RNA and the number of myocytes per gram of tissue.

## 4. Quantitative PCR Using Internal Standards

Thus far, a variety of methods for using quantitative PCR to determine relative initial levels of target mRNAs, and one method for estimating the absolute

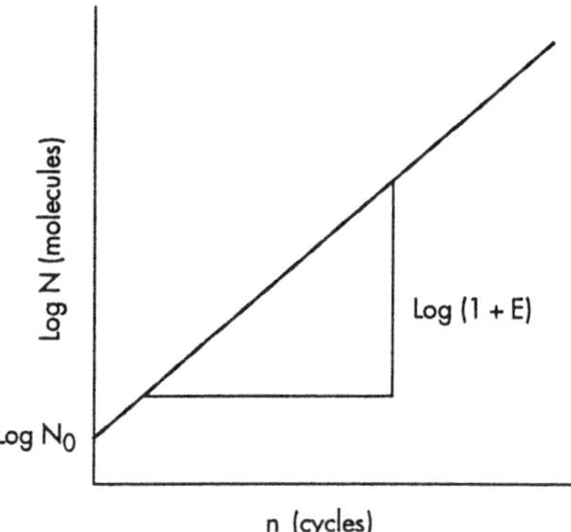


Fig. 5. Determination of initial amount of target ($N_0$) and efficiency of amplification ($E$) using linear regression. Note that the slope = log $(1 + E)$ and the $y$-intercept = $E_0$.

starting number of target molecules have been discussed. However, in all of these methods, variations in amplification efficiency ($E$) may complicate the interpretation of results. To correct for tube-to-tube variations in amplification efficiency, most investigators use internal amplification standards. Two types of internal standards can be used: an endogenous sequence or gene transcript that is normally present in the sample, or an exogenous fragment added to the amplification reaction.

## 4.1. Amplification of an Endogenous Sequence as an Internal Standard

An endogenous sequence, known to be present at constant levels throughout a series of samples to be compared, can be used as an internal standard in quantitative PCR reactions. Endogenous mRNA standards, typically for housekeeping genes or genes that are structurally or functionally related to the target mRNA *(19)*, have been used to determine relative levels of specific mRNAs *(13,20–22)*. Furthermore, endogenous single-copy gene sequences have been used as internal standards to determine relative gene copy numbers *(5,23,24)*. Finally, there is at least one case where ribosomal RNA was used as an endogenous internal standard for quantitation of mRNAs *(25)*.

In this approach, the endogenous standard sequence is amplified using a second pair of gene-specific primers, either in two separate PCR reactions, or in the same reaction as the target sequence. The ratios of the amount of PCR products generated by target and endogenous standard sequences in the different samples are then determined and compared. As with the methods described previously, the data from this type of experiment must be obtained before the amplification reactions reach the plateau phase. The data can be collected either from a titration of the sequences to be amplified, or by kinetic analysis, to ensure that signals are derived only from the exponential phase of the amplification.

The relative initial amounts of a target sequence and the endogenous standard (i.e., the ratio $N_0 t / N_0 s$) can be determined from **Eq. 4** (derived from **Eq. 2**). (The subscripts "$t$" and "$s$" refer to the target and standard sequences, respectively.) Values for the efficiency of amplification ($E$) for the target and standard may be calculated from the slope of a graph of log $N$ as a function of cycle number ($n$) (*see* **Subheading 3.2.**). Note that when the amplification efficiencies of the two reactions—target and standard—are identical, i.e., $Et = Es$, the analysis is greatly simplified *(3)*.

$$N_0 t / N_0 s = Nt(1 + Es)n / Ns(1 + Et)n \tag{4}$$

where $N_0 t$ = The initial number of target molecules, $N_0 s$ is the initial number of standard molecules, $Nt$ is the number of amplified target molecules, $Ns$ is the number of amplified standard molecules, $Et$ is amplification efficiency of the target, $Es$ is amplification efficiency of the standard, and $n$ is the number of amplification cycles.

Even without a full mathematical analysis—and even in cases where $Es$ does not equal $Et$—it has been shown empirically that endogenous mRNAs can be used to normalize target mRNA levels between samples to be compared. Thus, instead of determining the ratio of the intial absolute amounts of target and standard using linear regression, the relative amounts of PCR products generated by the target and standard templates in different samples is simply compared. Although it has not been shown theoretically, Horikoshi et al. *(13)* suggested that if the internal standard mRNA is expressed at the same level in two samples, the ratio of PCR products generated from the target and standard should indicate the relative level of expression of the target mRNA in those samples. Furthermore, it may be true that if the target and standard are amplified in the same tube, tube-to-tube variations in amplification efficiency (for example, caused by pipeting error, sample impurities, variation in the heating block, or partially degraded RNA) may be minimized as well.

This type of approach has been experimentally validated by performing PCR on mixtures of DNA. For example, Horikoshi et al. *(13)* mixed specific ratios of DNA preparations from two cell lines, one with a documented 18-fold amplification of the dihydrofolate reductase (DHFR) gene and the other carrying the gene as a single copy. In this case, two independent PCR amplifications were performed on each sample using DHFR and β-actin primers, respectively, in separate reactions; the β-actin sequence served as a single-copy standard. Amplified products were obtained under conditions in which the amount of product was still increasing linearly with increasing amounts of starting sample ($A_0$). The ratio of DHFR to β-actin PCR products obtained from the mixtures differed by only ~30% from the predicted theoretical values.

In the aforementioned experiment, the amplification of standard and target sequences was conducted in separate PCR reactions. However, a close correlation between predicted and observed target levels was similarly found by Neubauer et al. *(22)*, who performed both amplifications in a single PCR reaction in a method they referred to as differential PCR. In this case, the authors were investigating the loss of the β-interferon gene in chronic myelogeneous leukemia; the target was the β-interferon gene and the standard was the γ-interferon gene. They were able to detect changes as small as 2:1 and 3:2 in the ratio of the two genes using this method. Co-amplification also was used by Chamberlain et al. *(23)* to examine exon deletions in the Duchenne muscular dystrophy locus. In an approach they called multiplex DNA amplification, they simultaneously amplified (in one tube) six exons, each with a different set of primers. In another example, Kellogg et al. *(5)* corrected for the effects of variable amplification efficiency of an HIV-1 DNA template in several samples by using a single-copy gene from the HLA locus as a reference standard.

Many examples of the use of endogenous mRNA standards to determine relative levels of specific mRNAs (in the same tissue) can be found in the literature. The first group to use this approach was Chelly et al. *(3)*, in a study of dystrophin gene expression in different muscle tissues. Chelly et al. used aldolase A mRNA as the internal standard, and they performed the mathematical analysis, including calculation of amplification efficiencies, described at the beginning of this section. Noonan et al. *(19)* studied the relative expression of the multiple drug resistance gene (*mdr*-1) in tumor cells by normalizing PCR data to $\beta_2$-microglobulin mRNA. Horikoshi et al. *(13)* investigated expression of thymidylate synthase mRNA in tumor samples using both $\beta_2$-microglobulin and β-actin mRNA as endogenous standards. Murphy et al. *(20)* utilized both target titration and kinetic strategies to examine *mdr*-1 mRNA levels in tumor cells. Finally, Kinoshita et al. *(21)* examined levels of T-cell leukemia virus type I by performing a detailed kinetic PCR analysis that used β-actin mRNA as the endogenous standard.

Perhaps the greatest advantage of using the expression of an endogenous sequence as an internal standard is that the reference mRNA and the target mRNA are usually processed together for the entire duration of the experiment—from RNA extraction through PCR amplification. This minimizes differences in RNA yield between samples—an important advantage, particularly for analysis of small tissue samples where the quantities of RNA are too small to measure by UV spectrophotometry. In addition, if the entire population of mRNA is converted to cDNA by the use of oligo(dT) primers or random hexamers, the overall efficiency of cDNA synthesis also is somewhat normalized.

Notwithstanding the advantages to this approach, several complications may arise when amplification of endogenous mRNAs is used for semi-quantitative analysis. For this method to be reliable, the level of expression of the reference standard must be the same in each sample to be compared and must not change as a result of the experimental treatment. Unfortunately, few if any genes are expressed in a strictly constitutive manner. This is even the case for many housekeeping genes, including β-actin *(26,27)*. Therefore, the level of the mRNA used as the endogenous standard must be examined very carefully to ensure its constancy among all of the experimental conditions studied.

Another challenge of this approach is to obtain values of *At* and *As* before the amplification reactions reach the plateau phase, especially when the relative levels of expression of the standard and target sequences differ greatly. For example, if β-actin mRNA is used as the internal standard, it may be present at a much higher level than the target transcript, and amplification of the control may approach plateau phase well in advance of the target sequence. Indeed, Murphy et al. *(20)* found that their internal standard mRNA, $\beta_2$-microglobulin, entered the plateau phase before the target, *mdr*-1 mRNA, was even detectable. One solution to this problem involves simply waiting until later stages of the amplification before adding the primers for the endogenous standard *(21)*. Other researchers used gene-specific primers to synthesize cDNA from the control and target mRNAs in separate tubes and then mixed dilutions of the control and target cDNAs before performing multiplex PCR *(25)*.

Interference is a frequently observed problem when more than one set of primers is used in the same PCR reaction. For example, when Murphy et al. *(20)* added both $\beta_2$-microglobulin and *mdr*-1 primers to the same PCR reaction, they observed a premature attenuation of the exponential phase of both PCR amplifications. At Clontech (Palo Alto, CA), researchers have observed similar results; the amount of product generated (from either the target, the standard, or both) is often dramatically reduced when both sequences are amplified in a single reaction. In fact, primer pairs that function truly independently seem to be the exception rather than the rule.

## 4.2. Amplification of an Exogenous Sequence as an Internal Standard

Exogenous sequences can also be used as internal PCR standards. In this approach, an exogenous mRNA or DNA standard is added to the target sample and amplified simultaneously with the target transcript in a single PCR reaction mixture. The exogenous standard can be either a synthetic RNA added to the reverse transcription reaction or a DNA, not normally in the target sample, that is added directly to the PCR reaction.

The theory behind use of added exogenous gene sequences as internal standards is similar to that described earlier for endogenous reference sequences. With both types of internal controls, the amount of amplified standard can be

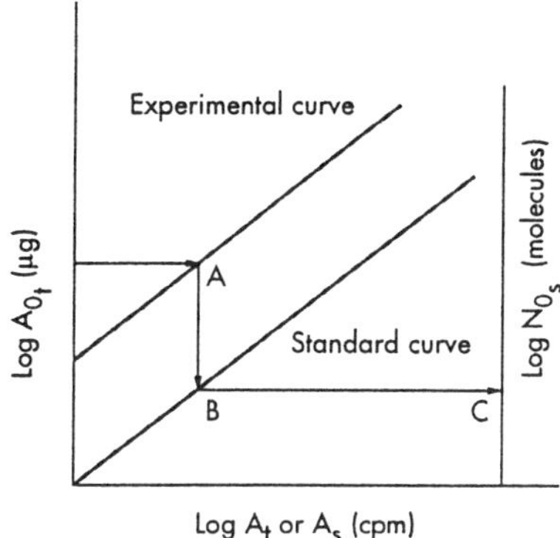



Fig. 6. Use of a standard curve, derived from an exogenously added internal standard, to quantitate initial amount of target.

quantified after the experiment, and the change in the amount of standard is proportional to the change in the amount of target. However, there is a significant advantage in using an exogenously added sequence as the internal control; namely, the initial amount of standard used in the PCR reaction is precisely known. This makes it possible to calculate the absolute level of target mRNA or cDNA present in the original sample.

A common method of obtaining quantitative results from PCR with an exogenous standard involves generating a standard curve from the data collected. This method was first described by Wang et al. *(28)*, who quantified changes in the levels of several cytokine mRNAs in stimulated macrophage cells using a synthetic internal RNA standard. In this approach, the RNA standard shares primer binding sites with the target RNA, but posseses a different "stuffer" sequence and an oligo(dT) tail. A known amount of the RNA standard is mixed with a known quantity of RNA sample (measured in micrograms, for instance) and reverse transcribed. A series of PCR reactions is then set up with dilutions of the cDNA. Because the titration is performed on a defined mixture of the target and standard mRNAs, and because the mixtures are not titrated against one another, this is not a competitive reaction (competitive PCR is discussed later). This strategy simply allows the generation of two titration curves: one for the standard RNA and one for the target RNA.

Following PCR, the amounts of PCR products obtained from the standard (*As*) and target (*At*) sequences are determined, and two curves are plotted, as illustrated in **Fig. 6**. The RNA standard curve is generated by plotting the logarithm of the starting number of RNA standard molecules (log $N_0s$) on the right vertical axis as a function of the logarithm of the amount of amplified standard product (log *As*). The target RNA curve is generated by plotting the logarithm of the initial amount of RNA sample (log $A_0 t$) on the left vertical axis as a function of the logarithm of the amount of target amplification products (log $At$). To determine the number of target mRNA molecules per unit of total RNA, a value of $A_0 t$ is chosen in the region where the curves are parallel (e.g., where values for $E$ are identical). A line is drawn from that point (labeled A in **Fig. 6**) down to the internal standard curve (point B), and from point B, a line is drawn across to the right vertical axis (point C). The value at this point is taken as the starting number of target molecules, $N_0 t$, in each microgram of total RNA. If the amount of total RNA per cell is known, the actual number of target mRNAs per cell can also be calculated. In the study by Wang et al. *(28)*, changes in mRNA levels of threefold or less were reproducibly discernible. Also, the results correlated well with data obtained from a Northern blot analysis. A similarly close correlation between this method of quantitative PCR and Northern blot analysis was found by Prendergast et al. *(29)*.

A critically important requirement of this type of experiment is that the value of $E$ be the same for both the target and standard mRNAs. This can be accomplished by designing the standard to contain the same primer binding sequences as the corresponding target mRNA. In many cases this is sufficient to make $Es$ equal to $Et$. Additional requirements for using exogenous standards are that the PCR products be of similar size and under 1 kb. The author has observed, as did Wang et al. *(28)*, that the primer sequences have the greatest effect on amplification efficiency when the sizes of the amplified sequences are similar. Wang et al. showed that the amplification efficiency of an RNA standard was the same as that of its corresponding target even though the sequence between the shared primer binding sites was completely different. Of course it is important that no regions of significant secondary structure differ between the target and standard RNA sequences. Differences in efficiency still may exist, so this parameter should always be examined before drawing firm conclusions from each study.

To calculate the absolute initial number of target molecules ($A_0 t$), the initial number of standard molecules ($A_0 s$) must be known, and a method to differentiate between the number of amplified standard and target molecules ($As$ and $At$, respectively) must be available. The most common technique used to distinguish between $As$ and $At$ is to make their sizes sufficiently different such that they can be resolved by polyacrylamide or agarose gel electrophoresis. Probe hybridization also can be used if the sequence between the two primer binding sites differs. In some cases, different restriction sites within the sequences between the primer binding sites can be used to differentiate target from standard simply by digestion with an appropriate restriction endonuclease prior to gel electrophoresis.

Since the study by Wang et al. was published, several reports have described the construction of exogenous RNA and DNA internal standards that differ from target sequences only by the presence or absence of small introns or restriction sites *(30–32)*. In these cases, there is little doubt that the amplification efficiencies of the standard and target sequences will be the same.

## 5. Competitive PCR

Competitive PCR also uses an exogenous template as an internal standard. However, the amplification takes place in a truly competitive fashion because the standard and target sequences actually compete for the same primers. In competitive PCR, a dilution series is made of either the target sequence or the standard sequence, and a constant amount of the other component is added to each of the reactions. Quantification is performed after competitive amplification of the entire series of reactions and is achieved by distinguishing the two PCR products from each tube by differences in size, hybridization properties, or restriction enzyme sites. An important advantage of competitive PCR is that, because the ratio of target to standard remains constant during the amplification, it is not necessary to obtain data before the reaction reaches the plateau phase.

In competitive PCR, the competitor fragment (usually DNA) takes the place of the standard described in the experiments discussed in the previous sections. It will still be called the standard, and the symbol "s" will be used to designate it in equations. When the amplification efficiencies of the target and standard molecules are the same, **Eq. 4** can be simplified to **Eq. 5**.

$$N_0 t \,/\, N_0 s = Nt \,/\, Ns = At \,/\, As \tag{5}$$

where $N_0 t$ is the initial number of target molecules, $N_0 s$ is the initial number of standard molecules, $Nt$ is the number of amplified target molecules, $Ns$ is the number of amplified standard molecules, $At$ is the amount of amplified target (in cpm or $OD_{260}$ U), and $As$ is the amount of amplified standard (in cpm or $OD_{260}$ U).

Thus, for any value of $n$, the initial ratio of target to standard is equal to the ratio of their amplification products (i.e., $Nt/Ns$ or $At/As$). This has been demonstrated both theoretically *(33)* and empirically *(34)*. Therefore, if the standard and target sequences amplify with the same efficiency, the absolute initial amount of target cDNA (and in turn target mRNA), can be determined by allowing known amounts of standard (DNA) molecules to compete with the target for primer binding during amplification.

In the competitive PCR method illustrated in **Fig. 7**, a dilution series of the DNA standard (referred to in the figure as the "MIMIC"*) is made, and these dilutions are added to a series of PCR reactions containing a constant amount of sample cDNA. Following PCR, the amplification products are analyzed by

---

*The use of PCR MIMICs is discussed in **Subheading 5.2.1.**

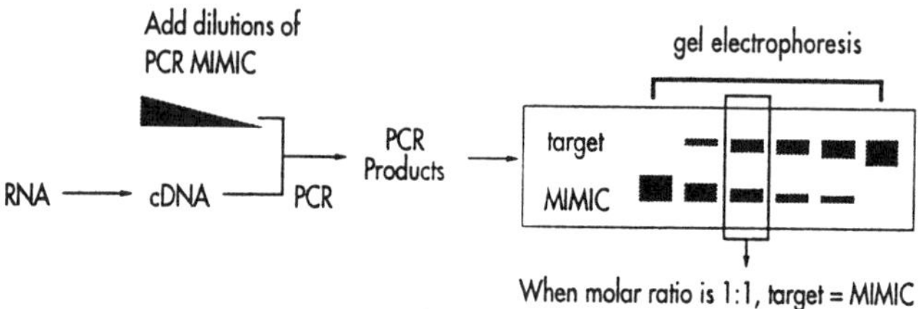



Fig. 7. Schematic diagram of competitive PCR utilizing a competitor DNA fragment (PCR MIMIC) differing in size from the target sequence. A dilution series of the competitor is added to a constant amount of cDNA. Following amplification, samples of the PCR products are resolved by gel electrophoresis, and the yields of amplified competitor and target products are quantified. The relative amounts of target product and MIMIC product in each sample are compared. The initial amounts of target and competitor are assumed to be equal in those reactions where the molar ratio of target and competitor products are judged to be equal (after correction for size differences). Because the amount of competitor added to each PCR reaction is known, the absolute initial amount of target can be determined. If the competitor is a synthetic RNA, a dilution series of the competitor is added to a constant amount of sample RNA before the reverse transcription step.

gel electrophoresis, and the amount of products generated by the standard ($As$) and the target ($At$) are determined for each individual reaction. The logarithm of the ratio of $At/As$ is graphed as a function of the logarithm of the initial molar amount of the standard ($N_0s$) (**Fig. 8**). The initial amount of target cDNA ($N_0t$) is extrapolated from the graph, assuming that $N_0t$ is equal to the amount of the standard ($N_0s$) added when an equimolar ratio of the two types of products is generated (i.e., where the log of $At/As$ = log of $1/1$ = 0). Note that if there is a difference in the size of the standard and the target sequence, $N_0t$ does not precisely equal $N_0s$ (because longer sequences incorporate more label than shorter ones). Thus, a corresponding correction must be made in the calculation of $N_0t$.

In general, when determining absolute initial amounts of mRNAs by competitive PCR using standard DNA fragments, one must take into account the fact that the efficiency of reverse transcription is <100%. The efficiency of cDNA synthesis using oligo(dT) as a primer for cDNA synthesis has been reported to be 40–50% *(34,35)*. Thus, calculations such as those described previously underestimate the number of mRNA molecules present in a given sample.

## *5.1. Homologous Competitor Fragments*

Becker-André and Hahlbrock *(31)* and Gilliand et al. *(32)* were the first to describe competitive PCR using homologous competitor fragments. Gilliand et al. used two types of internal standard: a genomic fragment corresponding to

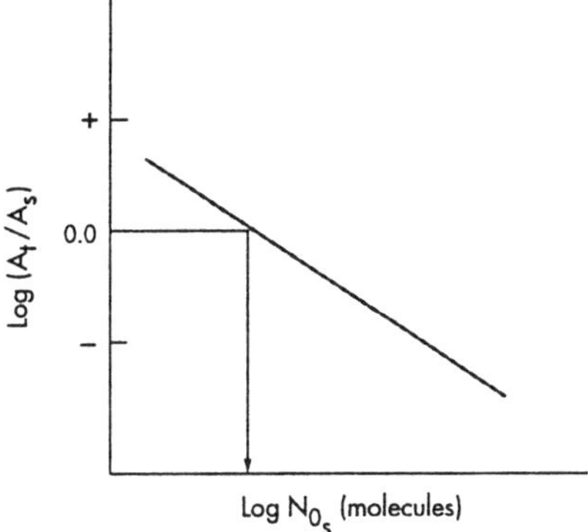

Fig. 8. Analysis of the results of a competitive PCR experiment, such as that illustrated in Fig. 7. The log of the ratio of amplified target to competitor products is graphed as a function of the log of the known amount of competitor added to the PCR reaction. Note that when the molar ratio of target and competitor is equal to 1, the log of that ratio equals 0.

the target mRNA sequence, but containing a small intron (thus yielding a PCR product slightly larger than the target mRNA); and a cDNA which was modified to contain a unique restriction site. In the latter case, PCR products were digested with the appropriate restriction enzyme before electrophoresis to differentiate between target and competitor products. To generate their internal standard, Becker-André and Hahlbrock used an in vitro transcribed mRNA designed to be identical to the target mRNA sequence except for the addition of a unique restriction site. They added different amounts of the competitor RNA into reverse transcription reactions containing a constant amount of target RNA.

Although it may be relatively easy to perform restriction endonuclease digestions to differentiate between target and competitor products, construction of such competitor fragments is often not a trivial matter. Unless there is a known small intron in the target gene, the construction of homologous competitors can require time-consuming site-directed mutagenesis and multiple cloning steps. However, recently several clever methods that use simple PCR amplification with composite primers *(36–38)* have been developed to generate homologous DNA standards. These methods can be extended to yield homologous competitor RNAs as well.

One potential problem with the use of competitor fragments that are homologous to the target is that during later stages of PCR, when the concentration of products is high, heteroduplexes can form between the standard and target sequences. This can complicate quantification of the PCR products derived specifically from the target or the standard, particularly when restriction enzyme digestion is required to distinguish between them. Therefore, heterologous DNA standards may be preferable (*see* **Subheading 5.2.**).

Although heteroduplex formation often interferes with obtaining accurate quantitative results from competitive PCR, in a novel approach described by Henco and Heibey *(39)* it is the heteroduplexes that are actually quantified. A known quantity of an internal standard, which is identical to the target except for a single nucleotide, is added to a dilution series of the target sample. Following PCR, a trace amount of radioactively labeled standard is added to the PCR products. The mixture is denatured and allowed to re-anneal; the labeled standard anneals to both target and standard sequences as a tracer. The homoduplexes and heteroduplexes are then resolved by temperature-gradient gel electrophoresis, and the amount of material in the heteroduplex (reflecting the amount of amplified target) is quantified.

## 5.2. Heterologous Competitor Fragments

DNA fragments that share the same primer template sequence but contain a completely different intervening sequence can also be used for competitive PCR. Überla et al. *(40)* prepared fragments for competitive analysis by amplifying genomic DNA fragments from another species with a low annealing stringency. Siebert and Larrick *(41)* ligated the primer template sequences to a nonhomologous DNA fragment to generate DNA standards (competitor fragments). More simply, the competitive DNA standard can be obtained by amplification of a heterologous DNA fragment using composite primers.

### 5.2.1. Generation of PCR MIMICs

One type of heterologous competitor fragment, PCR MIMIC (Clontech), is available commercially. PCR MIMICs are generated by two successive PCR amplifications as shown in **Fig. 9**. In the first PCR reaction, a heterologous DNA fragment is amplified using two composite primers. One composite primer contains the upstream primer for the target sequence linked to a 20-mer that anneals to one strand of the heterologous DNA fragment. The other composite primer contains the downstream primer for the target sequence linked to a 20-mer that anneals to the opposite strand of the heterologous DNA fragment. The two composite primers are used to amplify a small fragment of the heterologous DNA. During amplification, the target-specific primer sequences are incorporated into the PCR product. This PCR product is diluted and used to perform a second PCR amplification with primers for the target gene only. In this way the entire target primer sequences are incorporated onto the ends of the heterologous DNA fragment.

The PCR product, the newly generated PCR MIMIC, is purified by passage through a spin column that removes PCR reaction components and primers. The quantity of PCR MIMIC obtained is then determined either by measuring

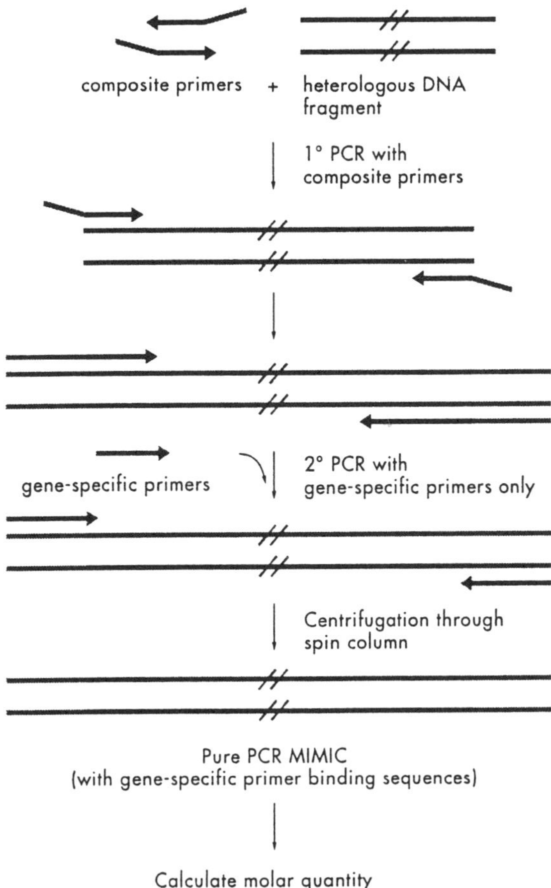



Fig. 9. Flow chart illustrating the generation of competitive PCR MIMICs. Note that the composite primers are composed of two sections; the 3' portion anneals to the heterologous DNA fragment and the 5' portion anneals to the specific target gene.

the absorbance at 260 nm or by running an aliquot of the PCR MIMIC on a gel and comparing the intensity of the band to a dilution series of DNA markers containing known quantities of DNA. The mass quantity is then converted to molar quantity using the approximation that 1 ng of a 300-bp DNA fragment is equal to $5 \times 10^3$ attomol (1 attomol = $10^{-18}$ mol).

A typical yield of PCR MIMIC, approx 200 ng, is enough to perform hundreds of competitive PCR experiments. Therefore, in practice a single determination of MIMIC yield can be used for many experiments. In this way inaccuracies in yield determination will not affect the determination of relative changes in mRNA levels (from a series of samples) by competitive PCR.

PCR MIMICs are designed so that the size of the PCR product generated from them is either slightly larger or slightly smaller than the PCR product generated from the target sequence. MIMICs of different sizes can be made simply by designing the composite primers so they anneal to different regions on the heterologous DNA fragment.

## 5.2.2. Use of RNA MIMICs

The method for generating competitive PCR MIMICs can be extended to heterologous RNA MIMICs to explicitly control for the cDNA synthesis step. To generate an RNA MIMIC, an RNA polymerase promoter and poly $(A)^+$ tail can be incorporated into the PCR product using composite primers designed for that purpose. In vitro transcription of the PCR product generates synthetic RNAs that contain the target primer sequences and a poly $(A)^+$ tail. RNA samples can then be titrated with the RNA MIMIC during reverse transcription. Transcriptional promoters have been successfully incorporated into PCR products via primer sequences *(13)*, and recently, competitive RNA fragments have been generated by this method *(42)*.

## 6. Summary of Quantitative PCR Methods

PCR is an exponential reaction in which small variations in amplification efficiency can yield large changes in the amount of products. In addition, later cycles of PCR exhibit the plateau effect, in which the rate of amplification slows and eventually levels off. These characteristics of PCR can make it difficult to obtain quantitative data. However, if specific conditions and proper controls are used, quantitative information about mRNA levels can be obtained. Of the various quantitative RT-PCR techniques currently in use, competitive PCR is often the method of choice. Competitive PCR is accurate enough to discern differences in mRNA levels of two- to threefold or smaller. This is comparable to the accuracy of quantitative methods that use either endogenous or exogenous internal standards in noncompetitive experiments.

Some investigators have observed that careful kinetic analyses can be used to determine initial concentrations of mRNAs by linear regression analysis without internal controls. At least one group, Singer-Sam et al. *(16)*, obtained satisfactory results without using either internal controls or kinetic analysis. Nonetheless, many have found it necessary to include internal controls to address the problem of tube-to-tube variation in amplification efficiency. Internal controls can be endogenous mRNA or exogenous mRNA added to the cDNA synthesis reaction. In addition, exogenous standards can be designed with the same or different primer annealing sequences as their target molecules. Each type of internal control has advantages and limitations.

One clear advantage of using endogenous internal mRNA controls is that the yield of RNA and the efficiency of the reverse transcription can vary to some extent without loss of accuracy. However, preliminary studies must be performed to ensure that the endogenous control mRNA does not change during the experiment. This must be tested because many genes, including many housekeeping genes whose expression may seem unrelated to the experimental conditions, may nevertheless vary in the experimental conditions being compared. In addition, the data must be collected before the amplification reaction reaches the plateau phase. This can be difficult if the endogenous control gene is expressed at a different level than the target gene or if their relative amplification efficiencies differ greatly.

Exogenous internal standards that share the same primer annealing sequences with the target allow calculation of the absolute amount of target mRNA, as determined by Wang et al. *(28)*. A similar method, termed competitive PCR, circumvents many of the disadvantages of the other quantitative methods. Competitive PCR can be used to measure relative changes in mRNA levels as well, for example, in gene regulation studies. However, two conditions must be met to use competitive PCR. One, the molar quantity of the competitor RNA or DNA must be known. (Usually this is not a problem because it can be measured by UV spectrophotometry.) Two, the amplification efficiency of the competitor and target must be identical. This is often true because the standard and target possess the same primer binding sequences. If the standard is a DNA fragment, the efficiency of the reverse transcription also must be considered.

Perhaps the most important advantage of competitive PCR is that useful data can be obtained during the entire course of amplification—even after the reaction has reached the plateau phase. This is not the case for methods using internal standards without competition between the standard and target molecules. Recently, however, Pannetier et al. *(43)* cautioned that competitive PCR may not provide accurate results when the sequences of the target and standard molecules are completely different (except for the primer sequences) and when the data are collected well after the plateau phase of the reaction. As stated previously, an examination of amplification efficiency is warranted.

Advantages and limitations of using homologous and heterologous competitor DNA fragments as internal standards for quantitative PCR have been discussed. In summary, homologous competitor fragments have the same amplification efficiency as their corresponding target but can form heteroduplexes that can complicate the measurement of PCR products. Heterologous competitor fragments, on the other hand, cannot form heteroduplexes, but their amplification efficiencies must be shown to be equal (or very similar to) that of the target.

## 7. Conclusions

It is possible to obtain quantitative information about specific mRNA levels using RT-PCR. The ability to accurately measure of gene expression in small amounts of tissue or in mixed cell populations will considerably expand future applications of PCR, both in research laboratories as well as in clinical settings. For example, quantitative PCR will be used increasingly in gene expression studies aimed at understanding the basic mechanisms controlling differentiation, development, immunity, and tumorigenesis. In a clinical application, competitive PCR has already been used to quantitate HIV transcripts in patient samples *(44)*. In the future, quantitative RT-PCR can be expected to aid the diagnosis and monitoring of many human diseases.

## References


1. Coker, G. T., III, Studelska, D., Harmon, S., Burke, W., and O'Malley, K. L. (1990) Analysis of tyrosine hydroxylase and insulin transcripts in human neuroendocrine tissues. *Mol. Brain Res.* **8,** 93–98.
2. Golde, T. E., Estus, S., Uslak, M., Younkin, L. H., and Younkin, S. G. (1990) Expression of β-amyloid protein precursor mRNAs: recognition of a novel alternatively spliced form and quantitation in Alzheimer's disease using PCR. *Neuron* **4,** 253–267.
3. Chelly, J., Kaplan, J. C., Maire, P., Gautron, S., and Kahn, A. (1988) Transcription of the dystrophin gene in human muscle and non-muscle tissues. *Nature* **333,** 858–860.
4. Gilliland, G., Perrin, S., Blanchard, K., and Bunn, H. F. (1990) Analysis of cytokine mRNA and DNA: detection and quantitation by competitive polymerase chain reaction. *Proc. Natl. Acad. Sci. USA* **87,** 2725–2729.
5. Kellogg, D. E., Sninsky, J. J., and Kwok, S. (1990) Quantitation of HIV-1 proviral DNA relative to cellular DNA by the polymerase chain reaction. *Anal. Biochem.* **189,** 202–208.
6. Choi, Y., Kotzin, B., Herron, L., Callahan, J., Marrack, P., and Kappler, J. (1989) Interaction of *Staphylococcus aureus* toxin "super antigens" with human T cells. *Proc. Natl. Acad. Sci. USA* **86,** 8941–8945.
7. Wiesner, R. J. (1992) Direct quantification of picomolar concentrations of mRNAs by mathematical analysis of a reverse transcription/exponential polymerase chain reaction assay. *Nucleic Acids Res.* **20,** 5863–5864.
8. Erlich, H. A., Gelfand, D., and Sninsky, J. J. (1991) Recent advances in the polymerase chain reaction. *Science* **252,** 1643–1651.
9. Jalava, T., Lehtovaara, P., Kallio, A., Ranki, M., and Söderlund, H. (1993) Quantification of Hepatitis B virus DNA by competitive amplification and hybridization on microplates. *BioTechniques* **15,** 134–137.
10. Pannetier, C., Delassus, S., Darche, S., Saucier, C., and Kourilsky, P. (1993) Quantitative titration of nucleic acids by enzymatic amplification reactions run to saturation. *Nucleic Acids Res.* **21,** 577–583.
11. Nakayama, H., Yokoi, H., and Fujita, J. (1992) Quantification of mRNA by nonradioactive RT-PCR and CCD imaging system. *Nucleic Acids Res.* **20,** 4939.
12. Katz, E. D. and Dong, M. W. (1990) Rapid analysis and purification of polymerase chain reaction products by high-performance liquid chromatography. *BioTechniques* **8,** 546–554.

13. Horikoshi, T., Danenberg, K. D., Stadlbauer, T. H. W., Volkenandt, M., Shea, L. C. C., Aigner, K., Gustavsson, B., Leichman, L., Frösing, R., Ray, M., Gibson, N. W., Spears, C. P., and Danenberg, P. V. (1992) Quantitation of thymidylate synthase, dihydrofolate reductase, and DT-diaphoroase gene expression in human tumors using the polymerase chain reaction. *Cancer Res.* **52,** 108–116.

14. Abbott, M. A., Poiesz, B.J., Byrne, B.C., Kwok, S., Sninsky, J. J., and Ehrlich, G.D. (1988) Enzymatic gene amplification: qualitative and quantitative methods for detecting proviral DNA amplified in vitro. *J. Infect. Dis.* **158,** 1158–1169.

15. Rappolee, D. A., Mark, D., Banda, M. J., and Werb, Z. (1988) Wound macrophages express TGF-α and other growth factors in vivo: analysis by mRNA phenotyping. *Science* **241,** 708–712.

16. Singer-Sam, J., Robinson, M. O., Bellve, A. R., Simon, M. I., and Riggs, A. D. (1990) Measurement by quantitative PCR of changes in HPRT, PGK-1, PGK-2, APRT, MTase, and Zfy gene transcripts during mouse spermatogenesis. *Nucleic Acids Res.* **18,** 1255–1259.

17. Solomon, R. N., Underwood, R., Doyle, M. V., Wang, A., and Libby, P. (1992) Increased apolipoprotein E and c-*fms* gene expression without elevated interleukin 1 or 6 mRNA levels indicates selective activation of macrophage functions in advanced human atheroma. *Proc. Natl. Acad. Sci. USA* **89,** 2814–2818.

18. Dallman, M. J., Montgomery, R. A., Larsen, C. P., Wanders, A., and Wells, A. F. (1991) Cytokine gene expression: analysis using northern blotting, polymerase chain reaction and in situ hybridization. *Immunol. Rev.* **119,** 163–179.

19. Sivitz, W. I. and Lee, E. C. (1991) Assessment of glucose transporter gene expression using the polymerase chain reaction. *Endocrinology* **128,** 2387–2394.

20. Noonan, K. E., Beck, C., Holzmayer, T. A., Chin, J. E., Wunder, J. S., Andrulis, I. L., Gazdar, A. F., Willman, C. L., Griffith, B., Von Hoff, D. D., and Roninson, I. B. (1990) Quantitative analysis of MDR1 (multidrug resistance) gene expression in human tumors by polymerase chain reaction. *Proc. Natl. Acad. Sci. USA* **87,** 7160–7164.

21. Murphy, L. D., Herzog, C. E., Rudick, J. B., Fojo, A. T., and Bates, S. E. (1990) Use of the polymerase chain reaction in the quantitation of *mdr*-1 gene expression. *Biochemistry* **29,** 10,351–10,356.

22. Kinoshita, T., Imamura, J., Nagai, H., and Shimotohno, K. (1992) Quantification of gene expression over a wide range by the polymerase chain reaction. *Anal. Biochem.* **206,** 231–235.

23. Neubauer, A., Neubauer, B., and Liu, E. (1991) Polymerase chain reaction based assay to detect allelic loss in human DNA: loss of β-interferon gene in chronic myelogenous leukemia. *Nucleic Acids Res.* **18,** 993–998.

24. Chamberlain, J. S., Gibbs, R. A., Ranier, J. E., Nguyen, P. N., and Caskey, C. T. (1988) Deletion screening of the Duchenne muscular dystrophy locus via multiplex DNA amplifications. *Nucleic Acids Res.* **16,** 11,141–11,156.

25. Khan, I., Tabb, T., Garfield, R. E., and Grover, A. K. (1992) Polymerase chain reaction assay of mRNA using 28S rRNA as internal standard. *Neurosci. Lett.* **136,** T18.

26. Siebert, P. and Fukuda, M. (1984) Induction of cytoskeletal vimentin and actin gene expression by a tumor-promoting phorbol ester in human leukemic cell line. *J. Biol. Chem.* **260,** 3868–3874.

27. Elder, P., French, C., Subramaniam, M., Schmidt, L., and Getz, M. (1988) Evidence that the functional β-actin gene is single copy in most mice and is associated with 5' sequences capable of conferring serum- and cycloheximide-dependent regulation. *Mol. Cell. Biol.* **8,** 480–485.

28. Wang, A. M., Doyle, M. V., and Mark, D. F. (1989) Quantitation of mRNA by the polymerase chain reaction. *Proc. Natl. Acad. Sci. USA* **86,** 9717–9721.

29. Prendergast, J. A., Helgason, C. D., and Bleackley, R. C. (1992) Quantitative polymerase chain reaction analysis of cytotoxic cell proteinase gene transcripts in T cells. *J. Biol. Chem.* **267,** 5090–5095.

30. Ballagi-Pordány, A. and Funa, K. (1991) Quantitative determination of mRNA phenotypes by the polymerase chain reaction. *Anal. Biochem.* **196,** 89–94.

31. Becker-André, M. and Hahlbrock, K. (1989) Absolute mRNA quantification using the polymerase chain reaction (PCR): a novel approach by a PCR aided transcript titration assay (PATTY). *Nucleic Acids Res.* **17,** 9437–9446.

32. Gilliland, G., Perrin, S., Blanchard, K., and Bunn, H. F. (1990) Analysis of cytokine mRNA and DNA  detection and quantitation by competitive polymerase chain reaction. *Proc. Natl. Acad. Sci. USA* **87,** 2725–2729.

33. Nedelman, J., Heagerty, P., and Lawrence, F. (1992) Quantitative PCR with internal controls. *CABIOS* **8,** 65–70.

34. Bouaboula, M., Legoux, P., Pességué, B., Delpech, B., Dumont, X., Piechaczyk, M., Casellas, P., and Shire, D. (1992) Standardization of mRNA titration using a polymerase chain reaction method involving co-amplification with a multispecific internal control. *J. Biol. Chem.* **267,** 21,830–21,838.

35. Berger, S. L., Wallace, D. M., Puskas, R. S., and Eschenfeldt, V. H. (1983) Reverse transcriptase and its associated ribonuclease H: interplay of two enzymes activity controls the yield of single-stranded complementary deoxyribonucleic acid. *Biochemistry* **22,** 2365–2372.

36. Diviacco, S., Norio, P., Sentilin, L., Menzo, S., Clementi, M., Biamonti, G., Riva, S., Falaschi, A., and Giacca, M. (1992) A novel procedure for quantitative polymerase chain reaction by coamplification of competitive templates. *Gene* **122,** 313–320.

37. Celi, F., Zenilman, M., and Shuldiner, A. (1993) A rapid and versatile method to synthesize internal standards for competitive PCR. *Nucleic Acids Res.* **21,** 1047.

38. Vanden Heuval, J., Tyson, F., and Bell, D. (1993) Construction of recombinant RNA templates for use as internal standards in quantitative RT-PCR. *BioTechniques* **14,** 395–398.

39. Henco, K. and Heibey, M. (1990) Quantitative PCR: the determination of template copy numbers by temperature gradient gel electrophoresis (TGGE). *Nucleic Acids Res.* **18,** 6733,6734.

40. Überla, K., Platzer, C., Diamantstein, T., and Blankenstein, T. (1991) Generation of competitor DNA fragments for quantitative PCR. *PCR Methods Appl.* **1,** 136–139.

41. Siebert, P. D. and Larrick, J. W. (1993) PCR MIMICs: competitive DNA fragments for use as internal standards in quantitative PCR. *BioTechniques* **14,** 244–249.
42. Henvel, J.P.V., Tyson, F.L., and Bell, D.A. (1993) Construction of recombinant RNA templates for use as internal standards in quantitative RT-PCR. *BioTechniques* **14,** 395–398.
43. Pannetier, C., Delassus, S., Darche, S., Sancier, C., and Kourilsky, P. (1993) Quantitative titration of nucleic acids by enzymatic amplification reactions run to saturation. *Nucleic Acids Res.* **21,** 577–583.
44. Piatak, M., Luk, K-C., Williams, B., and Lifson, J. (1993) Quantitative competitive polymerase chain reaction for accurate quantitation of HIV DNA and RNA species. *BioTechniques* **14,** 70–80.

# 5

# Kinetic Quantitative PCR vs End-Point Quantitative PCR with Internal Standard

Olivier Lantz, Elizabeth Bonney, Franck Griscelli, and Yassine Taoufik

## 1. Introduction

Quantitative PCR can be done either by measuring the amount of PCR products at a given number of cycle (end-point quantitative PCR) *(1–4)* or by following the amount of products during the PCR at several cycles (kinetic quantitative PCR) *(5,6)*. In this chapter, we define these two quantitative PCR methods, give their main characteristics, and compare their advantages and drawbacks. We then give a few examples of applications of the kinetic PCR method we have been using during the past few years.

### 1.1. End-Point Quantitative PCR with Internal Standard

When doing end-point quantitative PCR, most workers agree that an internal standard must be spiked in every reaction to check the efficiency of the PCR *(7)*. In our opinion, the term "internal standard" should be reserved for synthetic construct amplifiable with the same primers as the ones used for the target. This internal standard can be either homologous to the target (differing by only a few nucleotides) or nonhomologous (only the two extremities match the target). Some authors have proposed to use another gene in the same reaction tube amplified with another pair of primers, but this is more multiplex PCR than PCR with an internal standard. This is a procedure to control for the RNA recovery and the cDNA yield, but not really for the efficiency of the target amplification with the primers of interest. It is also very difficult to control for primer artefacts and for competition phenomena, especially if the cDNA used for the control amplification is a housekeeping gene, which has a much higher copy number than the experimental gene.

From: *Methods in Molecular Medicine, Vol 26: Quantitative PCR Protocols*
Edited by: B. Kochanowski and U. Reischl © Humana Press Inc., Totowa, NJ

There are two formats when using an internal standard: in the first format, few doses *(1–4)* of internal standard may be used and the results are computed according to the ratio of unknown vs standard amplicons (noncompetitive PCR). In the other, several *(6–9)* doses of internal standard are used, and one looks for the equivalence between the standard and the unknowns (competitive PCR). In the noncompetitive format, the amount of both amplicons is measured either during the exponential phase or at saturation. In the competitive format, the PCR is run to saturation. In the past, there have been some controversies about the characteristics of the internal standard: should it be or should it not be homologous to the target sequence? It is now clear that if one uses a format where the amount of amplicons is measured during the exponential phase, nonhomologous standard may be used. On the other hand, at saturation, the amount of primers becomes limiting, and reannealing without new DNA synthesis becomes frequent. If the standard is nonhomologous to the target, heteroduplexes cannot be formed between the standard and the target, and therefore the least abundant molecular species will be amplified preferentially, as stressed by Pannetier et al. *(8)* and Grandchamps *(9)*. This phenomenon is especially important if there is a large difference between the amounts of the two molecular species. Thus, at saturation, homologous standard must be used if the ratio of the standard and target differs by a factor of more than 5- or 10-folds.

How different from the target may be a homologous standard? A few nucleotide differences allowing new restriction site or differentiation with an automatic sequencer are not harmful. The number of nucleotide differences that can be tolerated has not been studied, and will probably depend on many parameters, such as the nucleotide sequences, PCR conditions, and ratio of standard vs target sequences.

## 1.2. Kinetic Quantitative PCR

In the kinetic method, first described by Dalmann et al. *(5)* in 1991, an internal standard is not mandatory and an external scale is sufficient if the reproducibility of the PCR is good. The PCR efficiency of every reaction is checked by looking at the slopes of the increase in PCR products for the experimental samples and for the external scale.

One of the advantages of a kinetic PCR method is the speed to quantitate a new gene because it is not necessary to construct a standard. It is then possible to semiquantitatively compare the expression of any known gene in different samples in a matter of days. If one wants to do true quantitation, an external scale should be devised: this is most easily done by using serial dilutions of purified PCR products of the gene studied or by amplifying dilutions of the sample containing the highest quantity of the cDNA studied in the same

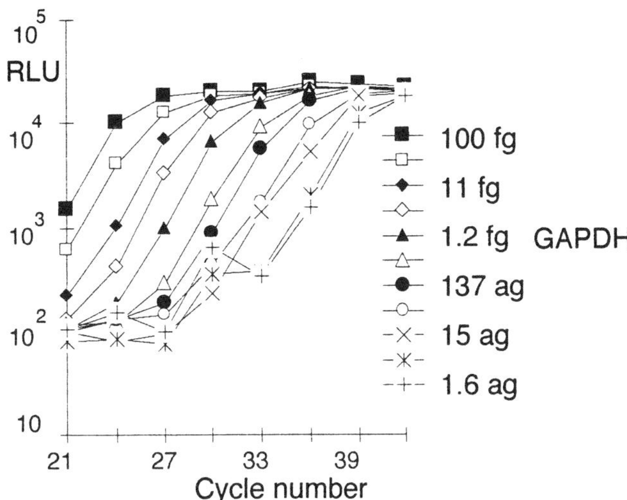



Fig. 1. Example of scale. Serial dilutions of purified IL4 PCR products were amplified and the reactions were sampled at the indicated cycle. The amount of amplicon was quantified by ELISA with luminometry readings.

experiment. **Figure 1** shows the results obtained with such a scale. By choosing early or late cycle curves, one can measure samples containing very different amounts of targets. As shown in **Fig. 2**, it is then possible to obtain meaningful results without previous knowledge of the amount of the target in the samples studied.

This method requires sampling every reaction at regular intervals during the PCR. At first, this may seem cumbersome, but with multichannel pipettors and 96-well format thermocyclers, this can be easily accomplished. Moreover, the whole process can be automated with a robotic work station.

## 1.3. Kinetic vs End-Point Quantitative PCR

The advantage of using internal standards is that the PCR reactions can be run to saturation without attendance. One of the drawbacks is that an internal standard needs to be constructed for each gene studied.

The quantitation and the storage of the standards are not always trivial procedures. Moreover, the sensitivity and the dynamic range of the method used for quantifying the PCR products, as well the relative abundance of standard and experimental targets, will determine the number of PCRs to be done for each sample. **Figure 3** shows that if the assay is not sensitive enough (method A in **Fig. 3**), the product in lowest amount will not be detected because of the inhibition of its amplification by the most abundant. This would preclude any

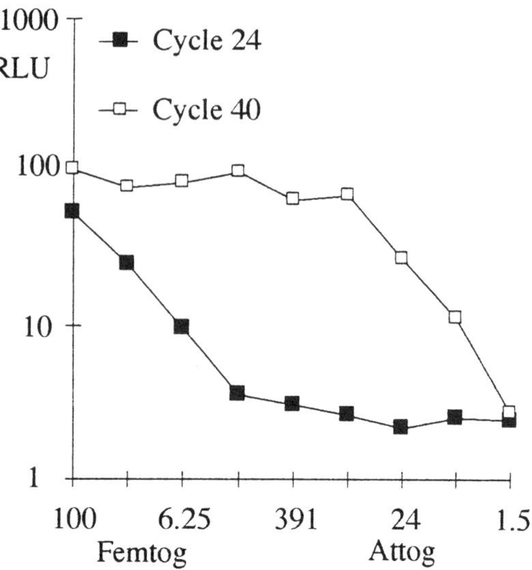


Fig. 2. Dynamic range of kinetic quantitative ELISA-PCR. According to the cycle (24 or 40) used for the quantitation, one can study high (1.62–100 fg) or low (1.5–100 ag) copy number samples. 1 ag corresponds approximately to four copies of IL2. Serial dilutions of gel-purified PCR products were amplified and sampled at regular intervals during the PCR. The amount of amplicon was measured by ELISA and luminometry.

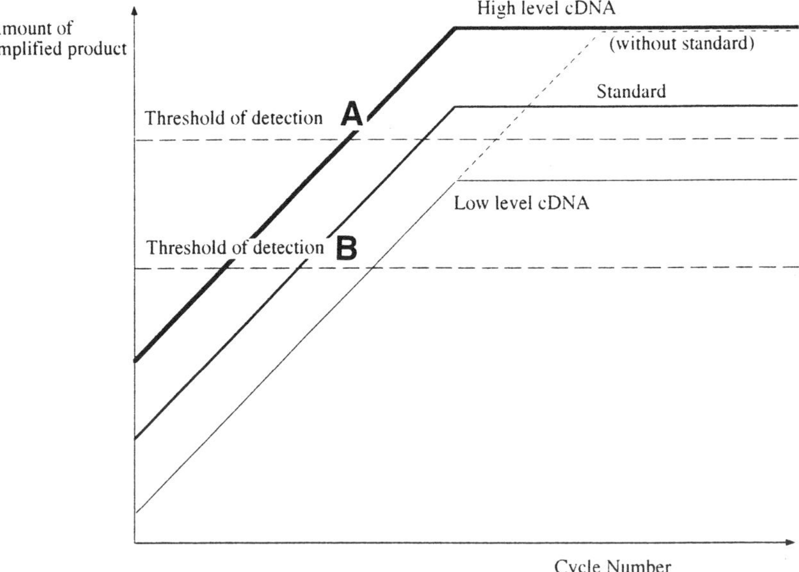


Fig. 3. End-point quantitative PCR with internal standard requires a very sensitive method with a wide dynamic range to quantify the amount of amplicons (*see* text for explanations).

quantitation. Thus, if samples containing very different amounts of target are studied, a very sensitive assay (method B) and/or several doses of standard should be used to quantify the amplicons when doing end-point quantitative PCR with internal standard.

If the dynamic range of the assay is not broad enough, the most abundant sample will be at saturation and will be underestimated; the assay of the amount of amplicons should have the widest dynamic range possible. Thus, most methods of quantitative PCR using internal standards require assaying several dilutions of the samples to be tested. This increases the cost of testing each sample and decreases the rate at which samples can be tested.

In kinetic PCR, the sensitivity of the assay to quantitate the amount of amplicon is not as important as in endpoint PCR: with method A, the samples will become positive later during the PCR than with method B, but that will not change the possibility of quantitating an unknown even in low amounts. However, the more sensitive the assay, the less likely are the artefacts, because samples are quantified earlier in the PCR. The main problem in kinetic PCR is the need to sample every reaction every 3–4 cycles: 2 min work every 12–15 min for 1 h 15 min.

Kinetic quantitative PCR requires a fast, easy, and low-cost method to quantify the amount of amplicons. Enzyme-linked immunosorbent assay (ELISA) is ideal to process the high number of samples obtained by kinetic quantitative PCR *(6)*. Moreover, although it may not be as sensitive as ELISA, the recently available PERKIN-ELMER 7700 apparatus (PE Applied Biosystems, Foster City, CA) is very appealing because it monitors the amount of PCR products on line during the PCR.

In conclusion, for the time being, kinetic quantitative PCR with ELISA and luminometry readings is probably the most versatile and inexpensive method available; there is only one PCR per sample, and the cost of the ELISA assay is small compared to the cost of the PCR itself. This method is well-adapted to research settings wherein one wants to quantitate the level of several mRNA in a limited number of samples, and be able to rapidly change the targets. As for hospital laboratories where the number of samples is much higher and the number of genes studied less numerous and less varied, the use of internal standards may be more convenient despite its higher cost and lower throughput. Furthermore, when very small differences are studied, as, for instance, in the case of the loss of alleles in oncology (twofold at most), an endogenous internal standard is necessary.

## 2. Examples of Application

We have been using a kinetic quantitative PCR method with ELISA measurement of the amount of amplicons *(6)* in the last three years to measure the

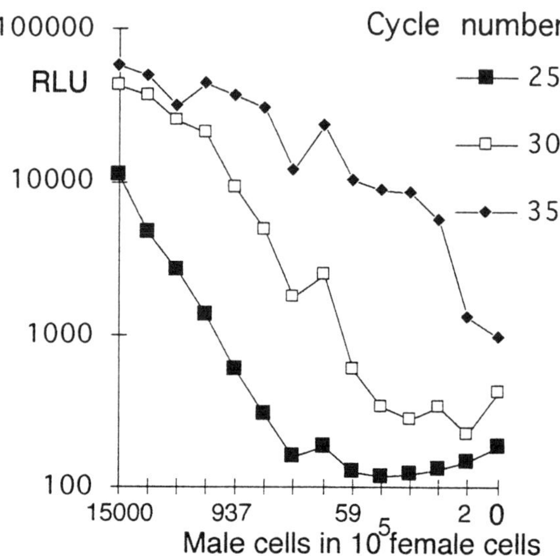



Fig. 4. Detection of male cells diluted in female cells. Serial dilutions of male cells in 1 $10^5$ female cells were lysed and the amount of a Y-specific sequence (sry) was quantified. The results obtained at three different samplings (cycles 30, 35, and 40) are shown.

expression of a large variety of cellular genes (e.g., interleukins, growth factors, T-cell receptors). Without automation, one technician is able to quantify 200 duplicate samples (400 PCR total)/wk. To follow are a few examples of applications.

## 2.1. Quantitation of DNA Targets

One should note that the complexity of DNA preps is about 10-fold higher than that of RNA preps. To get a few copy sensitivity in $10^6$ cells required special procedures during the processing of the samples and high stringency PCR (high annealing temperature and true hot start).

### 2.1.1. Quantitation of Microchimerism

In the course of studying the role of the persistence of antigen in different models of tolerance, we have devised a method to detect a few male cells in a large number of female cells. To do so, we amplify a Y chromosome-specific locus in genomic DNA *(11)*. An example of results is displayed in **Fig. 4**.

### 2.1.2. Detection of HIV DNA in Cellular DNA

Quantitation of integrated HIV DNA is of unknown clinical significance and the number of copies is rather low in a high-complexity DNA sample. We

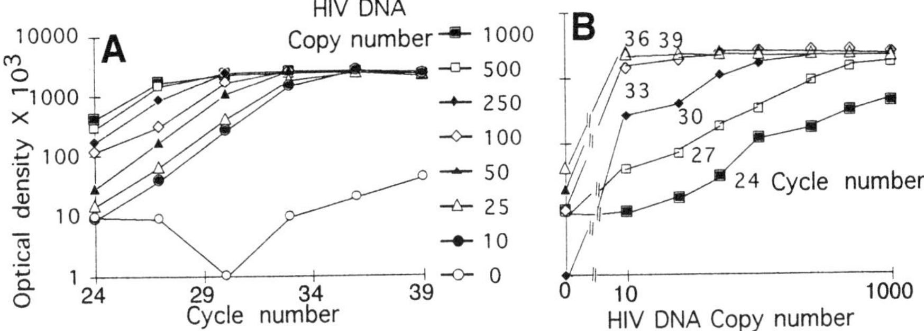



Fig. 5. Tat amplification of the standard scale: 8E5 cells were diluted in uninfected PBL in order to obtain a scale containing known number of HIV copies. This scale was amplified with tat primers and 5 µL were sampled through oil from cycles 21–42. The amount of tat products was quantified by ELISA. Results are plotted in **(A)** as OD signal versus cycle number, in **(B)** as OD signal versus initial copy number.

have studied the amount of tat and gag sequences in 32 patients with CD4 counts between 200 and 400/µL. After crude cell lysis with proteinase K and Tween, $10^5$ cell equivalent are amplified in parallel with an external scale made of serial dilutions of the line 8E5 (containing one HIV genome/cell) in $10^5$ noninfected cells. Examples of such scales are displayed in **Fig. 5**. It is shown that a linear relationship is obtained between the copy input and the signal. As shown in the appendices, by comparing the curves obtained with the unknowns to those obtained with the scale, one can compute the copy numbers in the unknowns for both gag and tat. The two independent measurements were compared and, as shown in **Fig. 6**, a very good correlation was obtained between the two measurements, which further validates the technique.

This assay has been applied to samples harvested longitudinally in patients undergoing low dose (AZT) treatment. Samples from 10 patients obtained at several time intervals before and during (AZT) therapy (500 mg/d) were studied. The number of HIV DNA copies at baseline varied widely among the patients, ranging from $<10$–$810/1.5$ $10^5$ (PBL), despite the fact that at the initiation of AZT treatment, all these AZT-naive patients had CD4 cells between 200 and 300 µL. As shown in **Fig. 7**, the 10 patients could be divided in three groups according to the increase or decrease of HIV DNA copy numbers in their PBL. Two patients displayed an increase in the HIV DNA load, whereas two other had a decrease. The six others showed no change in their HIV DNA load. Thus, on this limited number of samples studied, no general modifications of the number of integrated HIV DNA genome numbers was observed under AZT treatment.

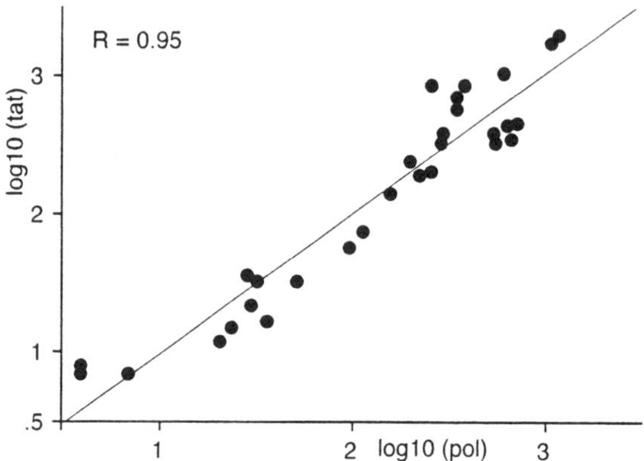

Fig. 6. Correlation between pol and tat measurements in 30 samples from randomly selected patients (with CD4 between 40 and 900/µL). Tat and pol copy numbers were measured by kinetic quantitative PCR by comparison with the scale described in **Fig. 5**.

## *2.2. mRNA Quantitation with Kinetic ELISA-PCR*

### *2.2.1. Method*

In our experience, if a true hot start is used (most easily done with the anti-Taq antibody from Clontech and now with the thermo-activable Gold-Taq from Perkin-Elmer or the Taq Platinum from GIBCO-BRL) the main source of variability for quantitative RT-PCR is not the reverse-transcriptase (RT) or the PCR steps, but the amount of starting material and the yield of the RNA extraction. Therefore, all samples are tested in duplicates beginning at the RNA extraction step. For cellular genes, the amount of starting material and the efficiency of RNA extraction can be checked by quantifying a housekeeping gene such as GAPDH, actin, or HPRT. However, for extracellular targets, such as viruses, one has to use an internal RNA standard spiked in the samples at the time of the RNA extraction to verify the extraction efficiency and RT yield.

Because the kinetic ELISA-PCR is very reproducible *(6)*, quantitation can be done by comparing the signal of the unknown samples obtained at successive sampling to that of an external scale containing known amounts of the gene studied. Semiquantitative results can be obtained in a matter of days by using serial dilutions of the cDNA of the most concentrated sample as an external scale. If one wants absolute quantitation, serial dilutions of purified PCR products can be used. This will not assess the RT yield. Because in our experience the most variability comes from the number of cells and the efficiency of RNA extraction, all results are normalized to the amount of GAPDH or to some other housekeeping gene.

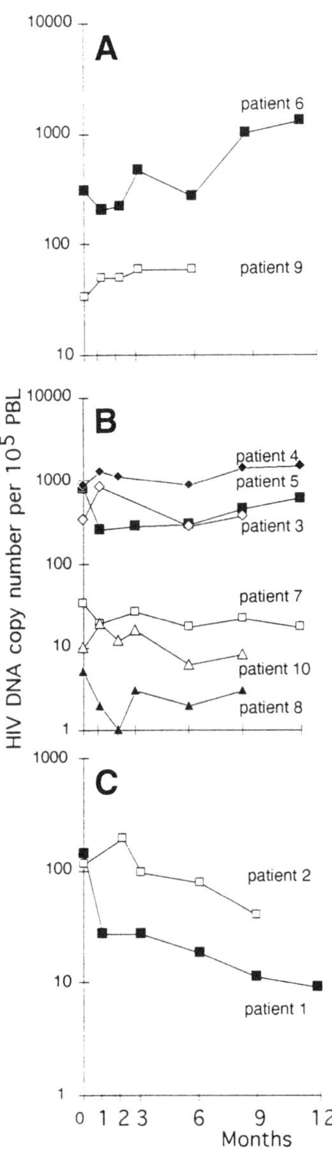



Fig. 7. Evolution of tat copy numbers in PBL after beginning low dose (500 mg) AZT treatment in 10 patients with CD4 between 200 and 300/µL. The 10 patients were divided in three groups according to the variation of tat copy numbers during the follow-up. Tat copy number was measured by comparing the kinetic PCR curves of the unknown samples to those of the scale described in **Fig. 5**.

Most of the time, only comparative results are needed. Therefore, equivalent numbers of cells are extracted and one has just to check that the amount of cDNA is the same in the different preparations by verifying that the curves obtained after amplifying a housekeeping gene are identical. Thereafter, any shift of the curves obtained after amplifying the gene of interest will be meaningful. In other cases, absolute values or normalization (because of unequal amount of starting material, for instance) are required, and absolute values must be computed. An external scale made with serial dilutions of purified PCR products is amplified in parallel to the experimental samples.

The comparison of the signal obtained from the different samples to the external scale is not a straightforward procedure. Two methods for processing the data numerically are included in the appendices. The first method is a simple linear regression method in which, for each cycle, the unknown values are compared to the external scale. If signal values are in the linear range of the assay (colorimetry or luminometry) for two or more successive samplings, the values are individually computed and averaged. It can easily be done using a spreadsheet. The second method is a logistic regression analysis, where a four-parameter curve is fitted to the experimental curves for every sample, allowing one to compute a "corrected cycle at half maximum." The value of the unknowns is then compared to that of the external scale by linear regression analysis. This second method is well-suited for data obtained with luminometry reading and takes into account all available information. However, a statistical package equipped for curve fitting with iterative procedures is required.

### 2.2.2. Examples of RNA Quantitation

#### 2.2.2.1. T-Cell Repertoire

We have studied the expression of an invariant alpha chain of the T-cell receptor (TCR) in a peculiar murin T-cell subpopulations (the NK1$^+$ T-cells). This chain uses a peculiar V$\alpha$ and J$\alpha$ segment (V$\alpha$14-J$\alpha$281) with a conserved VJ jonction. In this work, we have extensively used our kinetic ELISA-PCR assay: cells from different mouse strains are sorted using FACS according to certain membrane markers and the amount of C$\alpha$ (reflecting the number of T-cells), V$\alpha$14-C$\alpha$, V$\alpha$14-J$\alpha$281 (hybridized either with a probe on the V$\alpha$14 or on the VJ jonction) are compared in the different cell fractions. An example is displayed in **Fig. 8**.

#### 2.2.2.2. Quantitation of Lymphokine mRNA After Mitogenic Activation

We have been studying the effect of antiviral treatment decreasing the viral load on interleukin secretion in HIV patient. Because this work was carried out in a routine hospital laboratory, we have adapted the kinetic PCR technique and devised a method to construct homologous internal standards suitable for ELISA (*see* Chapters 10 and 11) allowing end-point quantitative PCR. We

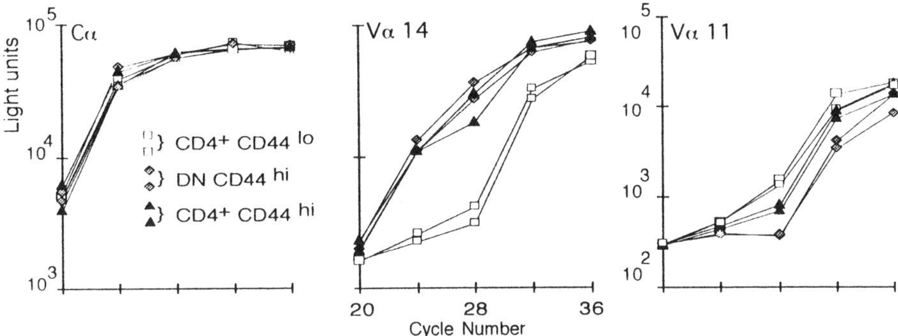


Fig. 8. T-cell repertoire analysis of mature thymocytes sorted according to CD44 and CD4/CD8 expression. Mature thymocytes were FACS sorted according to the indicated markers and Vα14, Vα11, and Cα mRNAs were quantified.

have constructed such internal standards for human GAPDH, IL2, IL4, IL10, IL12 p40, IL13, IFNγ, TNFα and TGFβ1. In the example reported here, a patient was given two new antireverse transcriptase drugs, and PBL were harvested and frozen at d 0, 3, 7, 14, and 28. Cells were thawed simultaneously and stimulated in vitro with Ionomycine and PMA in duplicate cultures. After a 4-h culture, RNA was extracted and quantitation of several genes carried out by quantitative RT-PCR. **Figure 9** displays the kinetic curves obtained for GAPDH and IL2 amplification. **Figure 9C** shows the results normalized to GAPDH. **Figure 10** shows the correlation between the results obtained with the two methods, kinetic or end-point with internal standard in 28 samples stimulated with various mitogens (anti-CD3 or ionomycin + PMA). It should be stressed that, if the use of internal standard decreases the labor and the cost of quantitative PCR with ELISA readings (the PCR volume reaction is lower and the number of ELISA plates smaller) compared with kinetic quantitative PCR, it requires the construction of an internal standard for every gene studied and a fine tune-up of the system to avoid competition phenomena. This whole process can require many preliminary experiments, and thus the whole procedure becomes quite labor-intensive if the number of samples is not high enough.

## 3. Conclusion

The homogenous phase Taq-Man assay on cycler (Perkin-Elmer 7700 apparatus) is now available and allows one to follow the amount of PCR products during the PCR on the thermocycler itself. Thus, kinetic PCR with or without internal standard will be done easily. In research settings, kinetic ELISA-PCR remains the most versatile and inexpensive method available for quantifying specific sequences in small samples. For clinical applications, even though the quantitative PCR step can be carried out reliably, as previously seen, in our

                                                                    *Lantz et al.*

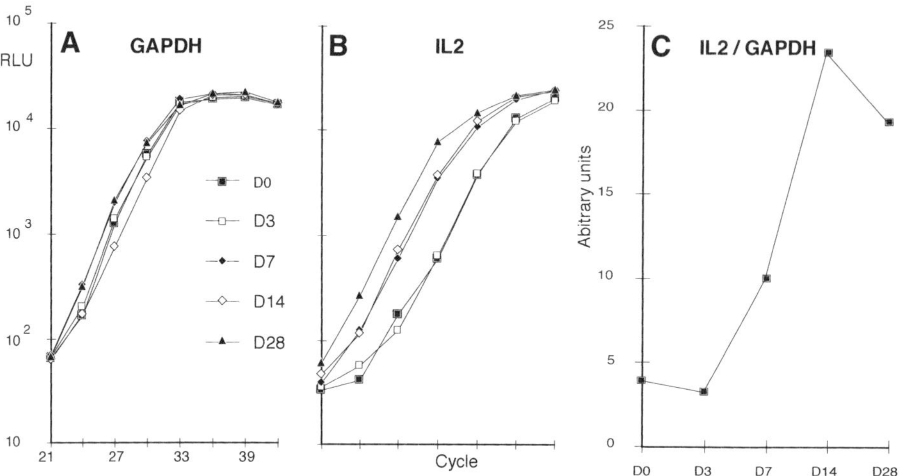



Fig. 9. Cytokine mRNA expression after mitogen stimulation of PBL from a HIV patient undergoing antiviral treatment. PBL were harvested and frozen at the indicated time after beginning a bitherapy with reverse transcriptase inhibitors. Cells were thawed and $2 \times 10^5$ PBL were stimulated with ionomycin and PMA for 4 h before extracting the RNA. The kinetic curves obtained with GAPDH (**A**) and IL2 (**B**) are shown. An external scale was amplified for both genes in the same experiments allowing to compute the amount of cDNA of the two genes in every samples. Normalized results are shown in (**C**).

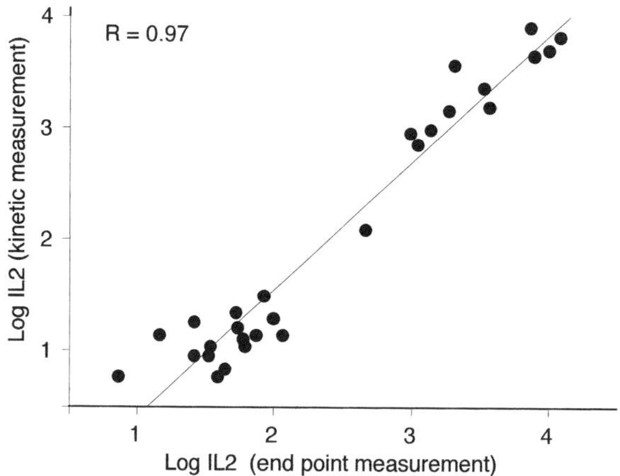



Fig. 10. Correlation between kinetic quantitative PCR and end-point quantitative PCR with internal standard for IL2 cDNA quantitation. IL2 cDNA was measured in 28 samples with both kinetic quantitative PCR or with end-point quantitative PCR with internal standard.

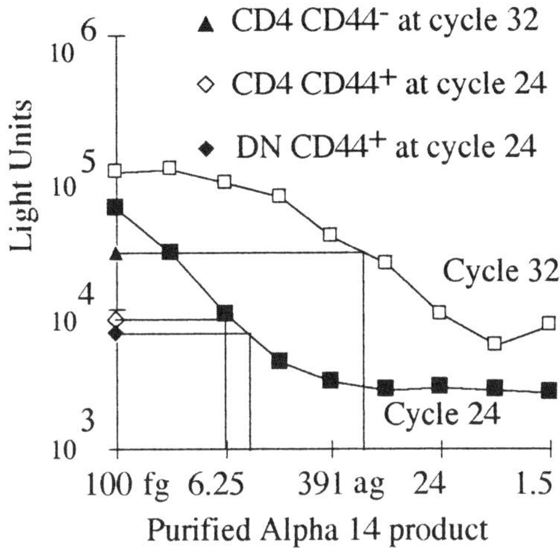


Fig. 11. Analysis of ELISA-PCR data by linear regaression analysis. IN this example, the signal obtained after Va14 sequence amplification in cDNA from the indicated subpopulations of mature thymocytes purified by FACS was compared tot he signal obtained after amplification of serial dilutions of purified Va14 amplicon.

opinion there is still a very strong obstacle to wider use of PCR for studying gene expression in tissue biopsies or even m cell suspensions: there is no RNA extraction method allowing processing many samples easily and reliably. Indeed, the mostly used "-zol" kits are all very operator-dependent and quite time-consuming. Some other kits based on beads and spin column give RNA highly contaminated with genomic DNA. Thus, for wider clinical applications, a more reproducible (less dependent on the operator skills) and easier RNA extraction method first needs to be devised.

## 4. Procession the Results of Kinetic ELISA-PCR

In parallel to the unknown samples, an external scale made of serial dilutions of purified amplified products is amplified. Every reaction is sampled at regular intervals during the PCR, and PCR products are quantified by ELISA. All the OD or RLU values are exported to a spreadsheet. For every reaction, one can draw a curve S = f(cycle number) where S is the signal. We have used two methods to process the data to get absolute values.

### 4.1. Linear regression analysis (Fig. 11)

This method is just a regular linear regression of ELISA data. Using signal (S) values above background and below saturation, for a given cycle one can draw a straight line:

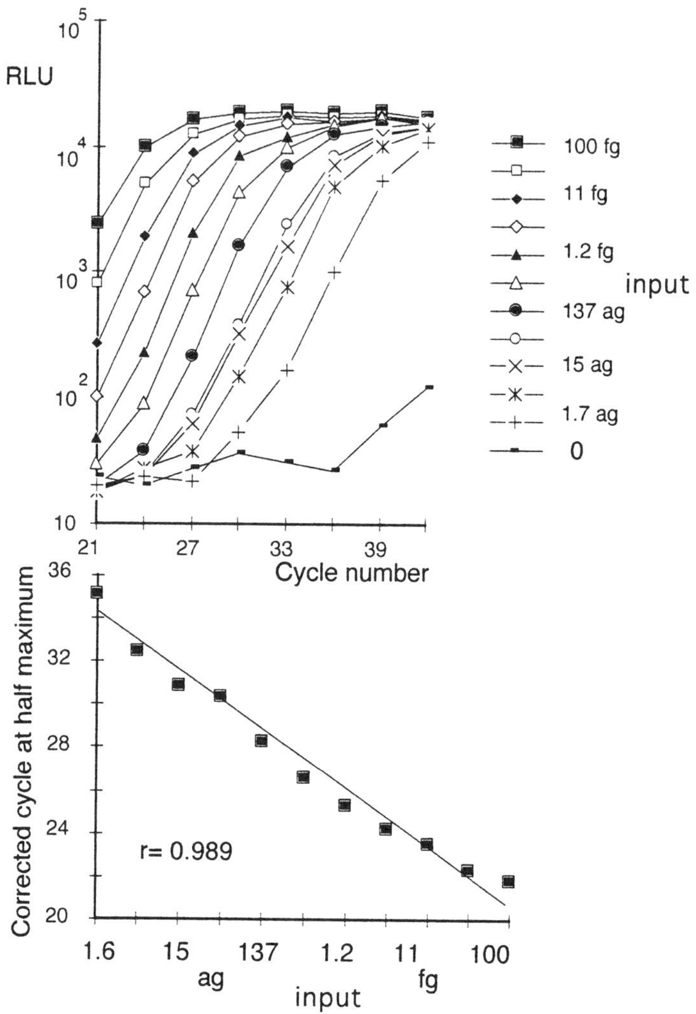


Fig. 12. Logistic regression for analysis of ELISA-PCR data (*see* text for explanations). Serial dilutions of IL4 product were amplified and amplicons were quantified by ELISA at the indicated cycle.

$$\log_{10}(S) = a \log_{10}(C) + y_0 \qquad (1)$$

where C is the concentration of the standard, a and $y_0$ are the parameters (slope and ordinate intercept) of the straight line estimated by least square regression.

One can then compute the concentration of the unknown samples as

$$C = 10^{([\log_{10}\{S\}-y0]/a)} \qquad (2)$$

The estimation can be done at several samplings (cycle number) and the values are averaged.

## *4.2. Logistic Regression*

The principle of the method is to compute the "corrected cycle at half maximum" for every sample and for the scale and to compare its value for the unknown to that of the scale. An example is shown in **Fig. 12**.

One can compute the four-parameter curve that fit the experimental values:

$$S = C + \{(D - C) / [1 + Exp(-A - B X)]\} \tag{3}$$

A and B are curve parameters describing the intercept and slope of the line. C and D are lower and upper plateau of the logistic curve. The cycle (X) at the half-maximum occurs for a value $X^* = -B/A$. In **Fig. 11B**, values of $X^*$ are plotted against the corresponding input of the scale and the parameters of the straight lines are calculated by least squares regression. By linear regression analysis, the values of the unknown can then be calculated.

## References


1. Becker-Andre, M. and Hahlbrock, K. (1989) Absolute mRNA quantification using the polymerase chain reaction (PCR). A novel approach by a PCR aided transcript titration assay (PATTY). *Nucleic Acids Res.* **17,** 9437–9446.
2. Bouaboula, M., Legoud, P., Pessegué, B. Delpech, B., Dumont, X., Piechaczyk, M., Cassellas, P., and D. Shire. (1992) Standardization of mRNA titration using a polymerase chain reaction method involving coamplification with multispecific internal control. *J. Biol. Chem.* **267,** 21,830–21,838.
3. Gilliland, G., Perrin, S., Blanchard, K., and Bunn, H. F. (1990) Analysis of cytokine mRNA and DNA: Detection and quantitation by competitive polymerase chain reaction. *Proc. Natl. Acad. Sci. USA* **87,** 2725–2729.
4. Wang, A. M., Doyle, M. V., and Mark, D. F. (1989) Quantification of mRNA by the polymerase chain reaction. *Proc. Natl. Acad. Sci. USA* **86,** 9717–9721.
5. Dallman, M. J., Larsen, C. P., and Morris, P. J. (1991) Cytokine gene transcription in vascularised organ grafts: analysis using semiquantitative polymerase chain reaction. *J. Exp. Med.* **174,** 493–496.
6. Alard, P., Lantz, O., Sebagh, M., Calvo, C. F., Weill, D., Chavanel, G., Senik, A., and Charpentier, B. (1993) A versatile ELISA-PCR assay for mRNA quantitation from a few cells. *Biotechniques* **15,** 730–737.
7. Ferré, F., Marchese, A., Pezzoli, S., Griffin, S., Buxton, E., and Boyer, V. (1994) Quantitative PCR: an overview, in *PCR the Polymerase Chain Reaction* (Mullis, K. B., Ferré, F., and Gibbs, R. A., eds.), Birkauser, Boston.
8. Pannetier, C., Delassus, S., Darche, S., Saucier, C., and Kourilsky, P. (1993) Quantitative titration of nucleic acids by enzymatic amplification reactions run to saturation. *Nucleic Acids Res.* **21,** 577–583.

9. Grandchamp, B. (1995) Quantitative PCR with internal standard, in *Quantitative PCR Workshop.* Atelier INSERM, Paris, October.

10. Lantz, O. and Bendelac, A. (1994) An invariant T cell receptor a chain is used by a unique subset of MHC class I-specific $CD^{4+}$ and $CD^{4-8-}$ T cells in mice and humans. *J. Exp. Med.* **180,** 1047–1106.

11. Bonney, E. A. and Matzinger, P. (1997) The maternal immune system's interaction with circulating fetal cells. *J. Immunol.* **158,** 40–47.

# 6

# Comparison of Competitive PCR and Positive Control-Based PCR

François Mallet

## 1. Introduction

The exponential amplification of small amounts of nucleic acids makes polymerase chain reaction (PCR) not only powerful but also challenging as a quantitative method. Variations in nucleic acid preparation, thermal cyclic performance, the choice of the polymerase, and the amplification procedure can cause large differences in final product yield. To address the challenges of quantitative PCR, the procedure has been critically examined, leading to an understanding of the critical parameters involved in quantitative amplification. Accepted parameters can be summarized as a series of choices: external vs internal standard, exogenous vs endogenous standard, competitive vs noncompetitive amplification, and exponential vs plateau amplification *(1–3)*.

The first quantitative approach in PCR was semiquantitative, based on the amplification of sample in limiting dilutions *(4)*. Another approach to quantitation is based on amplification with known amounts of an external standard, such as a cell line carrying a defined wild-type gene copy number *(5,6)*. These two approaches, however, failed to control tube to tube variation. Thus, alternative strategies including an internal standard have been developed. The main difficulty using internal standards is competition resulting from coamplification of the gene of interest and a sequence of reference *(3)*. Multiple sets of primers in the same reaction generally interfere with amplification of either the target sequence and/or the reference sequence. Furthermore, differences in the composition and quantity of the gene of interest and of the reference sequence may influence amplification efficiency. In quantitative competitive-PCR (QC-PCR), the gene of interest is coamplified with different concentrations of an added

*From: Methods in Molecular Medicine, Vol. 26: Quantitative PCR Protocols*
*Edited by: B. Kochanowski and U. Reischl © Humana Press Inc., Totowa, NJ*

standard; generally four different PCR reactions are done. The internal standard is defined to be as closely related as possible to the target, with differentiation occuring in the detection method (e.g., by difference in size, hybridization sequence, or changes in restriction pattern *[7–9]*). Because the sequences of target and control are very close, this situation approaches the "equivalency of replication efficiencies" as defined by Nedelman et al. *(10)*. The process of coamplification is truly competitive *(3)*; therefore, PCR is performed to the plateau. The yield of generated target product can be directly correlated to the internal standard. In an alternate approach, the gene of interest is coamplified, either with an endogenous standard, such as the cellular HLA-DQ-α gene *(11,12)*, or with a fixed amount of a DNA fragment or plasmid that carries a heterologous sequence that is flanked by sequences homologous to the amplification primers *(13,14)*. Internal standards act as a control for amplification efficiency. However, because the sequences and the amount of target and control are different, the "equivalency of replication efficiencies" *(10)* is rarely approached; therefore, the amplification process must be stopped during the exponential phase. The yield of generated target product can be directly correlated to standard curves.

The detection method must also be taken into consideration as a part of the overall quantitative process. Standardization of routine procedures, such as microtiter plate-based DNA and RNA hybridization assays *(15–17)*, has considerably simplified this step of the quantitation procedure.

To our knowledge, no comparison of the different internal standards-based methods has been published so far. To address this question, we defined and compared two quantitative PCR methods that use internal standards *(18)* and that were linked to our nonisotopic enzyme-linked oligosorbent assay (ELOSA) *(17)*, based on sandwich hybridization in a microtiter plate format. We will now present an overview of both methodologies and address the advantages as well as the limitations of individual protocols.

## 2. Strategies to Obtain
## a Quantitative Amplification Using Internal Standards
### 2.1 Overview of the Methods

The main difficulty with using internal standards is competition that results from coamplification of the gene of interest and a sequence of reference, independent of the degree of homology of both *(3)*. Thus, the definition of quantitative PCR procedures using internal standards goes through the achievement of two major characteristics: to find conditions approaching "equivalency of replication efficiency" and to define criteria of acceptance validating a result. Internally controlled- (IC-) and quantitative competitive- (QC-) PCR-ELOSA were defined as a positive control-based process and as a true competitive process, respectively, and applied to human immunodeficiency virus (HIV)-1 quantitation *(18)*.

We based IC-PCR approach on the coamplification of the HIV-1 *nef* gene, with an internal endogenous standard, the *ras* gene, as a positive control of amplification (IC-PCR-ELOSA is schematized in **Fig. 1**). This cellular target shares neither the primer binding sites nor the region in between with the target of interest. Acceptable OD intervals for RAS and NEF amplification products were defined to validate IC-PCR amplification and to quantitate the HIV-1 copy number using external standard of known amounts of HIV-1 DNA. The sensitivity we observed (10 copies of HIV-1 in 1 µg of DNA) and the range of quantitation (10–2000 copies) correlate well with results described by others: 1–1000 copies *(12)*; 12–400 copies *(11)*; 10–1000 copies *(19)*; and 100–10,000 copies *(20)*.

We based QC-PCR on the coamplification of the HIV-1 *nef* gene with different amounts of a pNEFmut plasmid that contains the *nef* region but with mutations in the capture probe recognition region (QC-PCR-ELOSA is schematized in **Fig. 2**). The NEF wild-type (NEF) and the NEF mimic (NEFmut) amplification products were differentiated in ELOSA by different capture probes and a common detection probe. NEFmut OD to NEF OD ratios were plotted against the number of mimic copies. The deduced linear curve was characterized by the equation and the coefficient of correlation $R^2$. The $R^2$ cut-off was defined to validate QC-PCR amplification and to permit quantitation of HIV-1 copy number by the equation. Curves obtained by QC-PCR-ELOSA were similar to those previously described using ethidium bromide fluorescence and computer video imaging *(9)*, radiolabeled PCR products and scintillation counting *(9)*, dye-labeled oligonucleotide fluorescence and an automated sequencer *(7)*, or digoxigenin-labeled probe and a microtiter plate assay *(16)*.

The ELOSA detection procedure *(17,21)* used in both techniques can be summarized in the following four steps.

1. Denaturation of the PCR product and hybridization to capture probe preadsorbed to the microtiter plate.
2. Hybridization of the captured PCR product to the detection probe.
3. Washing and removing excess of detection probe.
4. Colorimetric detection by addition of O-phenylenediamine substrate and reading absorbance at 492 nm.

The ELOSA procedure is suitable for automation: human papillomarius (HPV) typing *(22)* and *Mycobacterium tuberculosis (23)* have been defined on VIDAS (BioMérieux/VITEK) multiparametric immunoanalysis automated analyzer. The ELOSA system has been also used for nonisotopic and single-temperature detection of points mutations in the HLA DR phenotyping *(24)*, using the microtiter plate format.

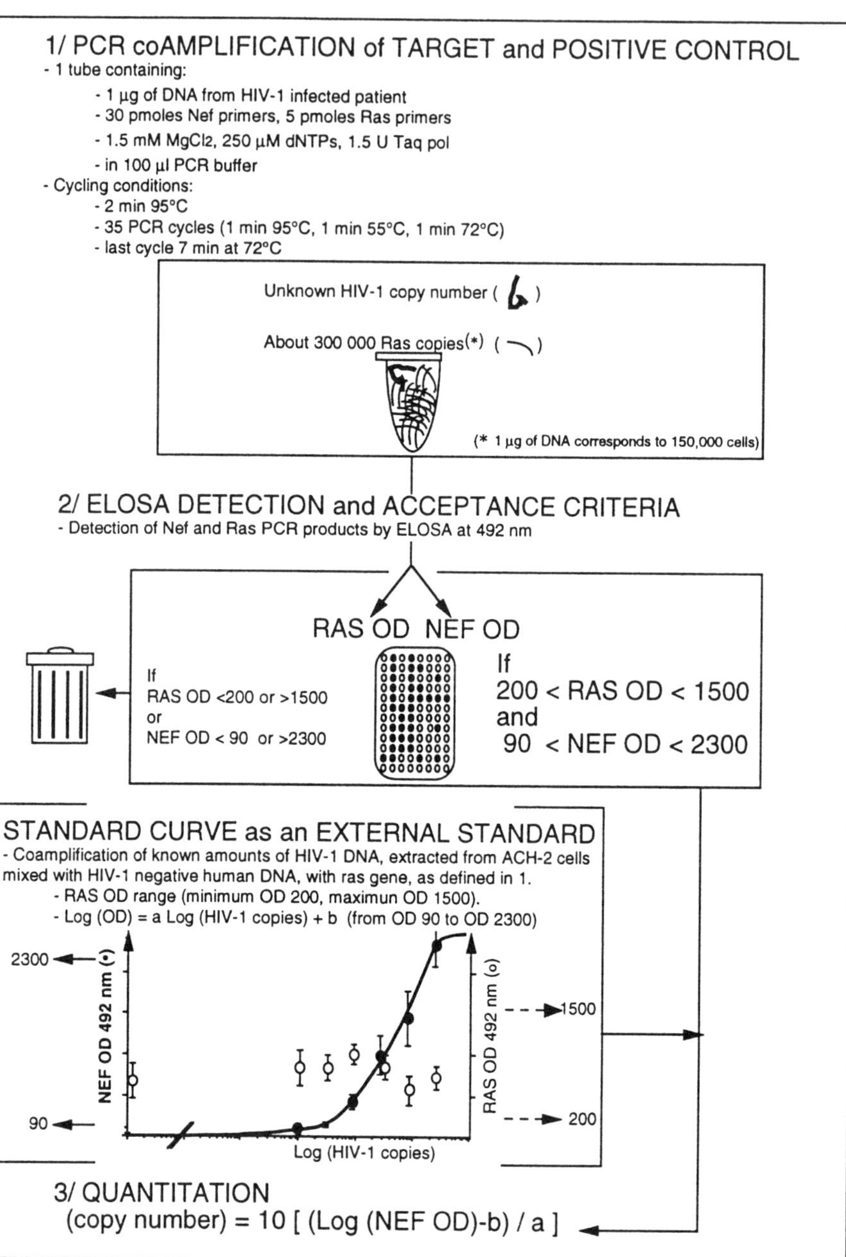


Fig. 1. IC-PCR-ELOSA procedure.

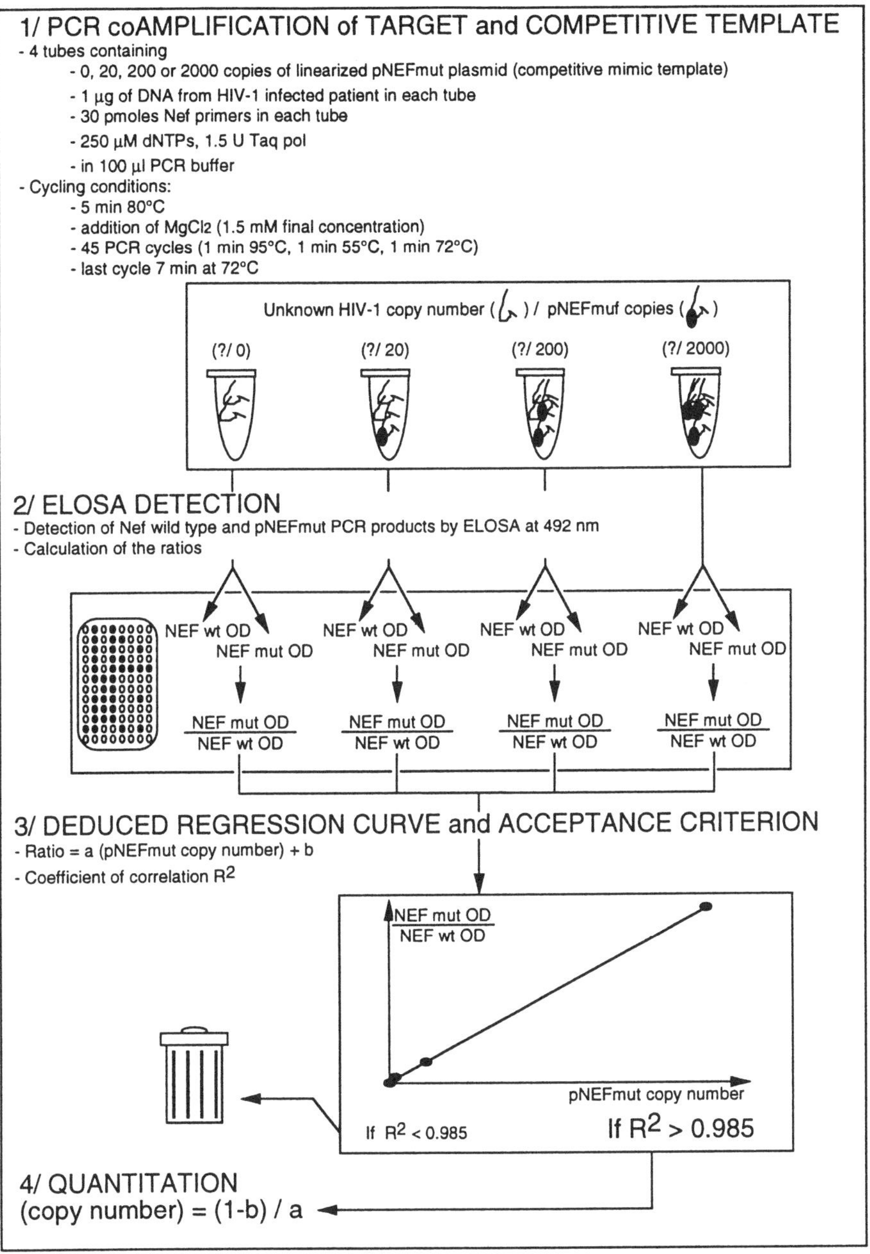


Fig. 2. QC-PCR-ELOSA procedure.

The reproducibility of QC-PCR-ELOSA and IC-PCR-ELOSA varied, with coefficient of variations (CVs) from 11–20% and 15–24%, respectively. These CVs are comparable to those described by others: from 2.2–39.9% *(11)* and 9.6% *(20)* for IC-PCR–like formats linked to isotopic detection, and from 8.2–28.9% *(3)* and 25% *(7)* for QC-PCR formats linked to nonisotopic detection.

## 2.2. Advantages and Limitations of Individuals Protocols

### 2.2.1. IC-PCR-ELOSA Procedure

The main advantages of IC-PCR-ELOSA procedure are its simplicity and the fact that no profound molecular biology is needed. First, only a single tube is required. Thus, duplication can be eventually achieved without tedious experiments. Second, as far as the acceptance criteria are carefully defined, the use of an endogenous internal control rule out tube-to-tube and sample-to-sample variations, including spectrophotometric dosage errors, pipetting errors, and presence of inhibitors. Third, this method is particularly suitable for automation, using predefined cut-off (acceptance criteria) and calibration curves (external standard) as in routine immunoassays.

However, the IC-PCR method has several intrinsic constraints, and attention must be paid to precise definition of each step. The nature of sample and sample treatment may have an influence on definition of acceptance criteria, on the equation, and the range of utilization of the standard curve. As discussed by Ferre *(2)*, the established limits of a quantitative PCR method are reliable only when using the same batch of standard within a given study. Using IC-PCR-ELOSA, this seems particularly relevant to the nature of the sample. We observed that RAS OD ranges and standard curves differed for whole blood and cell-line samples. For example, RAS ODs ranged from 197–933 and 254–1742 in the presence of exogenous DNA from whole blood and H9 cells, respectively. Standard curves of Log(OD) as a linear function of Log (HIV-1 copy number) were different, depending on the exogenous DNA source, Log(OD) = 0.6242 [Log(HIV-1 copies)] + 1.3417 and Log(OD) = 0.6365 [Log(HIV-1 copies)] + 1.4290 for whole blood and H9 cells, within the ranges of 10–1762 copies and 7–848 copies of HIV-1, respectively. These observations imply that calibration curves should be designed using a source of sample and a nucleic acids preparation as close as possible to those used for patients. In our experiments, DNAs of patients were extracted from whole blood using an Applied Biosystems 340A nucleic acid extractor. Thus, the standard curve of HIV-1 DNA was established by using known amounts of HIV-1 DNA, extracted from ACH-2 cells (an HIV-LAV infected cell line that has been shown to contain one HIV proviral DNA copy per cell *[25]*) and mixed with DNA extracted from whole blood obtained from a pool of three HIV-negative volunteers as a source of exogenous DNA.

Conditions of coamplification are obviously very critical. Owing to the large amount of positive control vs target copy number, coamplification of both genes would induce detrimental competition for target amplification by pumping out dNTP and polymerase. Consequently, one has to determine which positive control primer concentration permits efficient amplification of the positive control gene without affecting amplification of the target gene—that is an attempt to reach "equivalency of replication efficiencies". In our model system, tens to thousands of HIV-1 targets are coamplified with large amount of *ras* gene, about 300,000 copies per 1 µg of DNA. DNA was extracted from mixtures of HIV-infected cells, and HIV noninfected cells, and *nef* and *ras* genes were coamplified using 30 pmol of NEF primers and 0, 2.5, 5, 10, and 20 pmol of RAS primers. The lowest RAS primer concentration, 2.5 pmol, did not permit significant amplification of *ras,* whereas the highest primer concentration, 20 pmol, inhibited *nef* amplification. Efficient amplification of *ras* gene, which did not affect amplification of *nef* gene, was determined at 5 pmol of RAS primers in coamplification. One must keep in mind that conditions of coamplification are only valuable for a determined number of cycles (35 cycles in our experiments), in the exponential phase.

The RAS endogenous standard cannot be used as an absolute quantitative marker of DNA input, as previously hypothesized by us *(17)* and suggested by others *(11,12)*, because the determined ratio of NEF and RAS primers (30 pmol/5 pmol) was valid only within a short range of DNA concentration, roughly 1 µg ± 50%. Interestingly, the range of coamplification application in IC-PCR coincides with the range of true "equivalency of replication" between wild-type and mimic template in competitive amplification, as described elsewhere *(26)*. RAS conditions for coamplification were considered suitable for use of RAS endogenous standard as a positive control of DNA integrity and amplification efficiency, as used by others with different genes *(14,20,27)*, as no significant difference was observed in the detection level of the control cellular gene among 12 seronegative controls and 53 HIV-1 seropositive patients.

Finally, for reverse transcriptase-PCR, the task becomes more difficult. A cellular mRNA has to be selected that has an even level of transcription and is independent of different degrees of cellular activation. First attempts had been performed with several mRNA, such as aldolase *(28)*, β-actin *(29)*, and GAPDH *(30)*, but it is still unknown if all of them fulfill the criteria of an even and undisturbed transcription. In the laboratory, the strategy we retained for quantifying specific target mRNA in well-defined cellular systems consists of normalizing mRNA content and checking for the presence of inhibitor prior competitive RT-PCR-ELOSA quantitation. This is achieved using RT-PCR-ELOSA detection of the aldolase mRNA *(31)* and validated using an OD range similar to RAS OD acceptance criteria in IC-PCR-ELOSA.

In conclusion, positive control-based PCR, as defined above, seems to be a method of choice for quantifying a target DNA in a background of nontarget DNA, as is the case for HIV-1 provirus. The method is extremely simple and informative, including DNA integrity and amplification efficiency, but it is not at all flexible. Therefore, each new set up of the assay requires a complete reevaluation of the parameters, with regard to pertinent definition of acceptance criteria. Depending on the level of the pathogen, coamplification conditions should be defined, including primers ratio and number of cycles. Furthermore, standard curves should be defined accordingly to the source of the sample, the nature of the sample, and sample treatment.

### 2.2.2. QC-PCR-ELOSA Procedure

Advantages (and disadvantages) of QC-PCR-ELOSA procedure are generally the opposite of positive control-based method. First, as the amplification is truly competitive, the amplification can reach the plateau, reducing the risk of errors and potentially the variability of an assay. In fact, recent studies have shown that both wild-type and competitive mimic templates proceeded during the exponential phase up to the plateau phase with equal efficicency *(32,33)*. Second, for quantifying target in absence of nontarget DNA or RNA, (for example HIV or HCV viral RNA), competitive-based PCR is the only amplification procedure available that uses internal standard. This was recently demonstrated for HIV using continuous RT-PCR *(34)* coupled with ELOSA, the competitive template consisting of in vitro synthesized mimic RNA *(35)*. Lastly, QC-PCR was neither affected by the concentration of exogenous DNA in the 0.25–2 µg range nor by the nature of the sample. Consequently, regarding total DNA content, QC-PCR was less informative but more flexible than IC-PCR. Thus, the influence of sample treatment, the presence of inhibitors and pipetting errors as well are not taken into account.

In fact, the main difficulty of the method relies on understanding what is an appropriate competitive template and what are the limitations in coamplifying to "false twins" templates. The internal standard that competes with the primary target for enzyme, nucleotides, and primer molecules has to be designed carefully. The competitors bear the same primer binding region, but the sequence in between is modified in such a way that amplification products derived from the competitor and the target of interest can be differentiated in ELOSA. Although the ELOSA system has been shown to discriminate between point mutations *(24)*, because of the HIV-1 genome variability, it is more suitable to define a capture probe including several mutations. For example, the mutated NEF capture probe used to detect competitive internal standard in

QC-PCR-ELOSA was designed by changing residues in the central core of the NEF wild-type capture probe, swapping G and C tracks: the length and Tm of both probes remained identical. Consequently, the pNEFmut mimic template used in coamplification is equal in size and composition to the *nef* wild-type gene, except for the nine point mutations in the capture region.

The competitive template shape and dosage as well as the amplification procedure may greatly influence the overall sensivity and precision. We observed that the sensitivity of QC-PCR could be improved using linearized pNEFmut plasmid rather than supercoiled plasmid and a hot-start step followed by addition of $MgCl_2$ rather than classical PCR. This was illustrated by the difference observed between the expected copy numbers of 10, 100, 1000, and 10,000, and the measured copy numbers of 65, 200, 1078, and 8362, repectively, using supercoiled plasmid and nonhot-start PCR. This difference was reduced using the optimized protocol, 10 vs 15 and 90 vs 75, for the expected and measured copy numbers, respectively. However, we cannot ignore that the difference between the expected and determined copy number at low copy level may have reflected the difficulty in obtaining reproducible batches of standard at low concentrations.

Finally, the precise quantitation relies on obtaining a precise experimental curve that we believe should not include less than four experimental points. For example, coamplification of 1000 copies of pNEF with 10, 100, 1000, and 10,000 of pNEFmut allowed to generate the regression curve defined by the equation NEF mut OD/NEF wt OD = 0.7242 $10^{-3}$ (pNEFmut copies) + 0.2210 and the coefficient of correlation $R^2$ = 0.999, leading to a calculated copy number of 1078. In those conditions, it is obvious that it is not reasonable to consider duplicates.

In conclusion, quantitative competitive PCR seems to be the method of choice for quantifying a target nucleic acid in the absence of nontarget nucleic acids. In fact, the method is extremely flexible and can be theorically applied to any context of quantitation, although it is less simple and informative than positive-control based PCR in a target DNA-nontarget DNA context. However, each new set up of the assay requires a complete reevaluation of the parameters discussed above, including mainly the design of the competitor (sequence and shape), careful titration of the competitor to obtain reproducible batches of standard at low concentrations, and the range of applicability, depending on the level of the pathogen.

## 2.3. Comparison Between PCR-ELOSA Procedures: Sample Analysis

QC-PCR-ELOSA and IC-PCR-ELOSA and a limiting dilution method (LD-PCR-ELOSA) were tested on a sample panel consisting of 53 seropositive

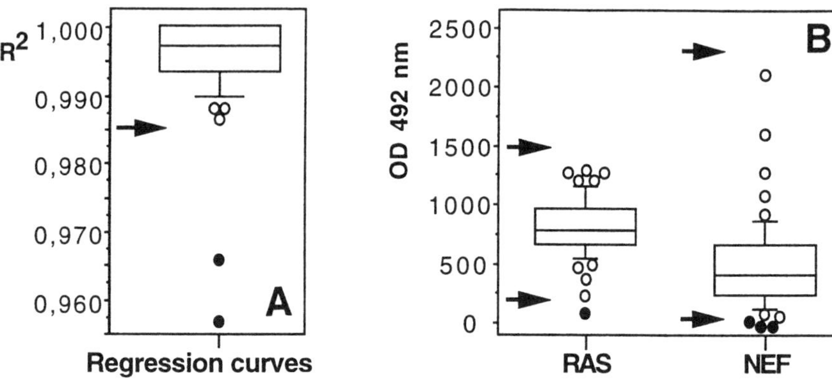


Fig. 3. Distribution of the acceptance criteria determined for PCR-ELOSA methods. Box plots indicate the distribution, the median OD, the 10th, 25th, 50th, 75th, and 90th percentiles. Data fitting or not fitting acceptance criteria are plotted. **(A)** Distribution of the coefficients of correlation ($R^2$) of the regression curves defined as NEF OD/NEFmut OD ratio in relation to pNEFmut mimic plasmid copy number is indicated for QC-PCR-ELOSA. The $R^2$ cut-off of 0. 985 is shown by an arrow. **(B)** Distribution of the ODs determined by IC-PCR-ELOSA, in relation to RAS and NEF oligonucleotide quartets. 200 < RAS OD < 1500 and 90 < NEF OD < 2300 acceptance criteria are shown by arrows.

patients at different CDC stages and 12 seronegative controls *(18)*. Quantitation was possible within acceptance criteria for 96% of samples for QC-PCR-ELOSA, and for 92% of samples for IC-PCR-ELOSA **(Fig. 3)**. The LD-PCR-ELOSA method correlated poorly with these two methods, 42% and 40% correlation for QC- and IC-PCR-ELOSA, respectively. The discrepancy between LD-PCR-ELOSA and internal standard-based methods may have been attributed to the absence of a control for tube-to-tube variation in the former. It may also have reflected that the fixed NEF OD cut-off *(17)* used in LD-PCR-ELOSA was not suitable, as linear relationship between the initial amount of target and the amplification product was only maintained for a limited range of starting DNA, as previously described *(36)*. In contrast, quantitations by QC-PCR-ELOSA and IC-PCR-ELOSA were identical for 77% of patient samples with a 95% confidence level, based on the ratio method *(37)*. Discrepancies were not the result of a difference in the sensitivity of these methods, since discrepancies were not localized at low copy level, but spread throughout the observed concentration range. Discrepancies between these methods were also not caused by the chosen acceptance criteria, as $R^2$ and RAS and NEF ODs were distributed randomly within their respective acceptability ranges. On the other hand, discrepancy could be a result of the method of analysis: an underestimated

CV in QC-PCR-ELOSA could result in a too stringent test, as was possible, because CV was defined using only a short range of target concentration.

## 3. Future Prospects

Nucleic acids quantitation is, in essence, a three-step process: the extraction of the nucleic acids target from a complex mixture, the amplification of a segment of the nucleic acids target, and finally, the detection of the amplified target fragment. Although PCR-ELOSA- (and RT-PCR-ELOSA-) based quantitative methods are simple and reliable, improvements are necessary to define routine diagnostic tests wherein each step of the process is controlled. The two main points to resolve are the preamplification step, consisting of sample preparation and calibration, and the quantitative-amplification step, which is dependent on obtaining reliable batches of internal and external standards. Advantages of positive control based method rely on simplicity, as it uses a single tube, and control of sample preparation. Advantage of competitive-based method relies on versatility. Future improvements of quantitative amplification processes should include features of the two methods, i.e., an added positive control in sample, in order to control all the steps from sample processing to the detection step, and several competitive templates added in a single reaction tube. Consistant with this hypothesis, addition of known amount of a virus mutant in plasma has been used to control extraction efficiency of RNA virus *(38)* and addition of a positive control to the amplification vial has been used to control the absence of PCR inhibitors *(14)*. Addition of two or more known concentrations of different-sized competitive standards to the same vial has been achieved using nucleic sequence-based amplification (NASBA) *(39)* and, more recently, using PCR *(40,41)*. Together, these are important steps to global automation required in routine diagnostic and clinical applications.

## Acknowledgments

I thank Christine Hebrard for excellent technical assistance, Christelle Brun and Nathalie Ferraton for oligonucleotide synthesis and enzyme coupling. The ACH-2 cell line was obtained through the AIDS Research and Reference Reagent Program, Division of AIDS, NIAID, NIH: **(25)**.

## References

1. Clementi, M., Menzo, S., Bagnarelli, P., Manzin, A., Valenza, A., and Varaldo, P. E. (1993) Quantitative PCR and RT-PCR in virology. *PCR Methods Applic.* **2,** 191–196.
2. Ferre, F. (1992) Quantitative or semi-quantitative PCR: reality versus myth. *PCR Methods Applic.* **2,** 1–9.

3. Sninsky, J. J. and Kwok, S. (1993) The application of quantitative polymerase chain reaction to therapeutic monitoring. *AIDS* **7(Suppl.),** S29–S34.

4. Simmonds, P., Balfe, P., Peutherer, J. F., Ludlam, C. A., Bishop, J. O., and Leigh Brown, A. J. (1990) Human immunodeficiency virus-infected individuals contain provirus in small numbers of peripheral mononuclear cells and at low copy numbers. *J. Virol.* **64,** 864–872.

5. Dickover, R. E., Donovan, R. M., Goldstein, E., Dandekar, S., Bush, C. E., and Carlson, J. R. (1990) Quantitation of human immunodeficiency virus DNA by using the polymerase chain reaction. *J. Clin. Microbiol.* **28(9),** 2130–2133.

6. Innocenti, P., Ottmann, M., Morand, P., Leclercq, P., and Seigneurin, J. M. (1992) HIV-1 in blood monocytes: frequency of detection of proviral DNA using PCR and comparison with the total CD4 count. *AIDS Res. Hum. Retroviruses* **8,** 261–268.

7. Pannetier, C., Delassus, S., Darche, S., Saucier, C., and Kourilsky, P. (1993) Quantitative titration of nucleic acids by enzymatic amplification reactions run to saturation. *Nuc. Acids Res.* **21,** 577–583.

8. Piatak, M., Jr., Luk, K. -C., Williams, B., and Lifson, J. D. (1993) Quantitative competitive polymerase chain reaction for accurate quantitation of HIV DNA and RNA species. *BioTechniques* **14,** 70–81.

9. Siebert, P. D. and Larrick, J. W. (1993) PCR MIMICS: competitive DNA fragments for use as internal standards in quantitative PCR. *BioTechniques* **14,** 244–249.

10. Nedelman, J., Heagerty, P., and Lawrence, C. (1992) Quantitative PCR: procedures and precisions. *Bul. Math. Biol.* **54,** 477–502.

11. Kellog, D. E., Sninsky, J. J., and Kwok, S. (1990) Quantitation of HIV-1 proviral DNA relative to cellular DNA by polymerase chain reaction. *Anal. Biochem.* **189,** 202–208.

12. Lee, T. H., Sunzeri, F. J., Tobler, L. H., Williams, B. G., and Busch, M. P. (1991) Quantitative assessment of HIV-1 DNA load by coamplification of HIV-1 gag and HLA-DQ-a genes. *AIDS* **5(6),** 683–691.

13. Arnold, B. L., Itakura, K., and Rossi, J. J. (1992) PCR-based quantitation of low levels of HIV-1 DNA by using an external standard. *Genet. Anal. Tech. Appli.* **9,** 113–116.

14 Cone, R. W., Hobson, A. C., and Huang, M. L. W. (1992) Coamplified positive control detects inhibition of polymerase chain reactions. *J. Clin. Microbiol.* **30(12),** 3185–3189.

15 Dyster, L. M., Abbott, L., Bryz-Gornia, V., Poiesz, B. J., and Papsidero, L. D. (1994) Microplate-Based DNA hybridization assays for detection of human retroviral gene sequences. *J. Clin. Microbiol.* **32(2),** 547–550.

16. Kohsaka, H., Taniguchi, A., Richman, D. D., and Carson, D. A. (1993) Microtiter format gene quantification by covalent capture of competitive PCR products: application to HIV-1 detection. *Nuc. Acids Res.* **21(15),** 3469–3472.

17. Mallet, F., Hebrard, C., Brand, D., Chapuis, E., Cros, P., Allibert, P., Besnier, J. -M., Barin, F., and Mandrand, B. (1993) Enzyme-linked oligosorbent assay for detection of polymerase chain reaction-amplified human immunodeficiency virus type 1. *J. Clin. Microbiol.* **31,** 1444–1449.

18. Mallet, F., Hebrard, C., Livrozet, J. M., Lees, O., Tron, F., Touraine, J. L., and Mandrand, B. (1995) Quantitation of human immunodeficiency virus type 1 DNA by two PCR procedures coupled with enzyme-linked oligosorbent assay. *J. Clin. Microbiol.* **33(12),** 3201–3208.

19. Yerly, S., Chamot, E., Hirschel, B., and Perrin, L. H. (1992) Quantitation of human immunodeficiency virus provirus and circulating virus: Relationship with immunologic parameters. *J. Infect. Dis.* **166,** 269–276.

20. Aoki, S. A., Yarchoan, R., Thomas, R. V., Pluda, J. M., Marczyk, K., Broder, S., and Mitsuya, H. (1990) Quantitative analysis of HIV-1 proviral DNA in pheripheral blood mononuclear cells from patients with AIDS or ARC: decrease of proviral DNA content following treatment with 2', 3'-dideoxyinosine (ddI). *AIDS Res. Hum. Retroviruses* **6(11),** 1331–1339.

21. Mallet, F., Cros, P., and Mandrand, B. (1995) Enzyme-linked oligosorbent assay for detection of PCR-amplified HIV-1, in *PCR: Protocol for Diagnosis of Human and Animals Virus Diseases* (Becker, Y. and Darai, G., eds.), Springer-Verlag, Heidelberg, pp. 19–28.

22. Allibert, P., Cros, P., and Mandrand, B. (1992) Automated detection of nucleic acid sequences of HPV 16, 18 and 6/11. *Eur. J. Biomed. Tech.* **3(14),** 152–155.

23. Mabilat, C., Desvarenne, S., Panteix, G., Machabert, N., Bernillon, M.H., Guardiola, G., and Cros, P. (1994) Routine automated identification of *Mycobacterium tuberculosis* complex isolates with a DNA probe. *J. Clin. Microbiol.* **32(11),** 2702–2705.

24. Cros, P., Allibert, P., Mandrand, B., Tiercy, J.M., and Mach, B. (1992) Oligonucleotide genotyping of HLA polymorphism on microtitre plates. *Lancet* **340,** 870–873.

25. Clouse, K. A., Powell, D., Washington, I., Poli, G., Strebel, K., Farrar, W., Barstad, P., Kovacs, J., Fauci, A. S., and Folks, T. M. (1989) Monokine regulation of human immunodeficiency virus-1 expression in a chronically infected human T cell clone. *J. Immunol.* **142,** 431–438.

26. Souazé, F., Ntodou-Thomé, A., Tran, C. Y., Rostène, W., and Forgez, P. (1996) Quantitative RT-PCR: Limits and Accuracy. *BioTechniques* **21,** 280–285.

27. Coutlée, F., He, Y., Saint-Antoine, P., Olivier, C., and Kessous, A. (1995) Coamplification of HIV type 1 and b-globin gene DNA sequences in a nonisotopic polymerase chain reaction assay to control for amplification efficiency. *AIDS Res. Hum. Retroviruses* **11,** 363–371.

28. Chelly, J., Kaplan, J. C., Maire, P., Gautron, S., and Kahn, A. (1988) Transcription of the dystrophin gene in muscle and non-muscle tissues. *Nature* **333,** 858–860.

29. Yamamura, M., Uyemura, K., Deans, R. J., Weinberg, K., Rea, T. H., Bloom, B. R., and Modlin, R. L. (1991) Defining protective responses to pathogens: cytokine profiles in leprosy lesions. *Science* **254,** 277–279.

30. Alard, P., Lantz, O., Sebagh, M., Calvo, C. F., Weill, D., Chavanel, G., Senik, A., and Charpentier, B. (1993) A versatile ELISA-PCR assay for mRNA quantitation from a few cells. *BioTechniques* **15,** 730–737.

31. Mallet, F., Oriol, G., and Mandrand, B. (1998) Characterization of RNA using Continuous RT-PCR coupled with ELOSA in *Methods in Molecular Biology, Vol. 86: RNA Isolation and Characterization Protocols* (Rapley, R. and Manning, D. L., eds.), Humana Press, Totowa, NJ, pp. 161–172.

32. Morrison, F. and Gannon, F. (1994) The impact of the PCR plateau phase on quantitative PCR. *Biochim. Biophys. Acta* **1219,** 493–498.

33. Siebert, P. D. and Larrick, J. W. (1992) Competitive PCR. *Nature* **359,** 557–558.

34. Mallet, F., Oriol, G., Mary, C., Verrier, B., and Mandrand, B. (1995) RT-PCR using AMV-RT and Taq DNA polymerase: characterization and comparaison to uncoupled procedures. *BioTechniques* **18,** 678–687.

35. Mallet, F. (1996) Continuous RT-PCR using AMV-RT and Taq in *PCR: Essential Techniques* (Burke, J. F., ed.), BIOS Scientific Publishers Ltd., Oxford, pp. 82–85.

36. Diviacco, S., Norio, P., Zentilin, L., Menzo, S., Clementi, M., Biamonti, G., Riva, S., Falaschi, A., and Giacca, M. (1992) A novel procedure for quantitative polymerase chain reaction by coamplification of competitive templates. *Gene* **122,** 3013–3020.

37. Sholler, R., Gervasi, G., Avigdor, R., and Castanier, M. (1987) Comparaison de deux méthodes de dosage in *Colloque sur les actualités en immunoanalyse,* CORATA, Université de Bordeaux II, Bergeret, Bordeaux, 2–4 Dec 1987, pp. 225–233.

38. Natarajan, V., Plishka, R. J., Scott, E. W., Lane, H. C., and Salzman, N. P. (1994) An internally controlled virion PCR for the measurement of HIV-1 RNA in plasma. *PCR Methods Appl.* **3,** 346–350.

39. Van Gemen, B., Van Beuningen, R., Nabbe, A., Van Strijp, D., Jurriaans, S., Lens, P., and Kievits, T. (1994) A one-tube quantitative HIV-1 RNA NASBA nucleic acid amplification assay using electrochemiluminescent (ECL) labelled probes. *J. Virol. Methods* **49,** 157–168.

40. Vener, T., Axelsson, M., Albert, J., Uhlén, M., and Lundeberg, J. (1996) Quantitation of HIV-1 using multiple competitiors in a single-tube assay. *BioTechniques* **21,** 248–255.

41. Zimmermann, K., Schögl, D., Plaimauer, B., and Mannhalter, J. W. (1996) Quantitative multiple competitive PCR of HIV-1 DNA in a single reaction tube. *BioTechniques* **21,** 480–484.

# II

## Protocols

# 7

# End-Point Titration-PCR for Quantitation of Cytomegalovirus DNA

Jerzy K. Kulski

## 1. Introduction

Polymerase chain reaction (PCR) is an important qualitative procedure in the routine microbiology laboratory for detecting the presence or absence of potentially harmful microorganisms in clinical specimens *(1,2)*. The use of PCR to quantify an infectious agent in a clinical specimen (e.g., viral or bacterial load) is advantageous for monitoring disease progression and efficacy of treatment, for differentiating between asymptomatic and symptomatic infection, or for quality control of false positive samples. End-point titration-PCR (ET-PCR) is a simple method for differentiating between the presence of low, medium, or high amounts of viral, fungal, or bacterial DNA in a test sample. Basically, the qualitative PCR method *(3)* is used in an ET-PCR to amplify a specific target sequence in serial dilutions of a DNA sample *(4)*. The limit of detection of the amplified product, which is the end-point dilution or titer, is the quantitative index for the DNA target in the sample. End-point titers obtained by ET-PCR have been shown to increase proportionally with increasing amounts of standard DNA *(4)*. The result of an ET-PCR can be presented as a titer, dilution, DNA copy number, or amount of a specific DNA sequence relative to an external standard or as relative differences between samples. On this basis, ET-PCR has been used to quantitate the presence of viral and bacterial DNA in clinical specimens *(4–10)*. The ET-PCR method described here is for the quantitation of cytomegalovirus (CMV) DNA in leukocytes *(4)*. However, the general principles of quantitation by ET-PCR are applicable to any other etiologic agent that can be detected by a qualitative PCR method.

From: *Methods in Molecular Medicine, Vol 26: Quantitative PCR Protocols*
Edited by B. Kochanowski and U. Reischl © Humana Press Inc., Totowa, NJ

## 2. Materials

1. Oligonucleotides for PCR or DNA hybridization can be ordered as custom-made products from GIBCO BRL (Life Technologies, Gaithersburg, MD) or from one of many other commercial companies. Reconstitute the newly synthesized oligonucleotides in sterile distilled water at a stock concentration of 500 µg/mL, and store at −20°C in small aliquots to prevent frequent thawing and freezing (*see* **Note 1**).

2. CMV oligonucleotide primers and probes: The oligonucleotide primers chosen for amplification of CMV DNA fragments of the major immediate early (MIE) gene and/or the late antigen (LA) gene are those described by Demmler et al. *(11)*. The MIE4 sense primer (5'-CCAAG CGGCC TCTGA TAACC AAGCC-3') and the MIE5 antisense primer (5'-CAGCA CCATC CTCCT CTTCC TCTGG-3') are used to amplify a 435 bp MIE fragment from CMV DNA in an assay designated as MIE-PCR *(4)*. The oligonucleotide MIEPR (5'-GAGGC TATTG TAGCC TACAC TTTGG-3') is an intervening sequence between the MIE4 and MIE5 primer sequences of the amplified MIE fragment, which can be used as a specific hybridization probe to specifically detect the MIE-DNA amplified product. The LAPR sense primer (5'-GTCGC CTGCA CTGCC AGGTG CTTCG-3') and the LA6 antisense primer (5'-CACCA CGCAG CGGCC CTTGA TGTTT-3') are used to amplify a 200 bp LA fragment of CMV DNA in an assay designated as LA-PCR *(4)*. The oligonucleotide LA9 (5'-GACCT GCGTA CCAAC ATAGA GGTGA GC-3') is an intervening sequence between LAPR and LA6 sequences of the amplified LA fragment that can be used as a specific hybridization probe to detect the LA DNA amplified product. The primers for LA-PCR and MIE-PCR may be used in a single PCR or combined in a duplex PCR with little loss in sensitivity.

3. Primers for amplication of human cellular DNA: One of two different sets of oligonucleotide primers can be used to amplify human cellular DNA as external controls. A 536 bp fragment of the human β-globin gene can be amplified by GLBN-PCR with GLBN primers RS42 (5'-GCTCA CTCAG TGTGG CAAAG-3') and KM29 (5'-GGTTG GCCAA TCTAC TCCCA GG-3'), which have been described by Greer et al. *(12)*. Alternatively, a 282 bp DNA fragment from the human androgen receptor gene can be amplified by HARE-PCR *(4,7)* using the HARE primers E1 (5'-CAACC CGTCA GTACC CAGAC TGACC-3') and E2 (5'-AGCTT CACTG TCACC CCATC ACCAT C-3'), which have been described by Lubahn et al. *(13)* (*see* **Notes 2** and **3**).

4. CMV and cellular DNA standards: *Hin*dIII restriction fragment E or *Hin*dIII restriction fragment L of CMV recombinant DNA harbor the MIE gene and LA gene, respectively *(14)*. These *Hin*dIII restriction fragments of CMV (strain AD 169) in vector pAT153 were obtained from Dr. Helena Browne (Virology Division, Department of Pathology, Cambridge University, UK). Recombinant DNA plasmids from *Escherichia coli* were prepared by standard methods *(15)*. A preparation of 1 µg of recombinant plasmid is sufficient for about one million PCR assays if 1 pg or less is used as a positive control per assay. If recombinant plasmids are not available, CMV DNA and human placental DNA can be purchased from Sigma (St. Louis, MO) for use as positive controls or standards (*see* **Note 3**).

5. Aerosol resistant tips: Aerosol resistant tips (ART, Molecular BioProducts, San Diego, CA) for use with pipets (Gilson Medical Electronics, Villiers-le-Bel, France) ranging from 1–20, 20–200, and 200–1000 μL.

6. Clinical specimens: Collect whole blood (in tubes with EDTA or citrate as anticoagulants), serum, plasma, urine, or cerebrospinal fluid (CSF) and store frozen at –20°C to –70°C (*see* **Note 4**).

7. Preparation of DNA from clinical specimens: Two different methods will be described for the preparation of DNA from blood and bronchial washings. A commercial QIAamp Blood Kit (QIAGEN, Chatsworth, CA) can be used routinely to prepare DNA for PCR from whole blood, bronchial washings, plasma, serum, urine, and CSF by following the manufacturer's protocol. The QIAamp Blood Kit contains silica spin columns, collecting microtubes (2 mL), two reagents (AL1 and AL2)—one of which contains guanidine hydrochloride, a wash buffer that is reconstituted with 70% ethanol (buffer AW), and proteinase K stock solution (19 mg/mL). Alternatively, a simple and an inexpensive cell lysis and proteinase K digestion method that works well for the preparation of DNA from whole blood or bronchial washings will be described in Methods *(16)* (*see* **Note 5**).

8. Red cell lysis (RCL) buffer: 0.34 $M$ sucrose, 0.01 $M$ Tris-HCl, pH 8.0, 0.05 $M$ MgCl$_2$, 1% Triton X-100.

9. White cell lysis (WCL) buffer: 0.01 $M$ Tris-HCl, pH 8.0, 0.5% Nonidet P40, 0.5% Tween-20.

10. Proteinase K: Prepare proteinase K (Sigma) at 20 mg/mL and store at –20°C in small aliquots to avoid repeated freezing and thawing.

11. Dri-bath: One or more dri-baths (Thermolyne, Dubuque, IA) for 1.5-mL Eppendorf tubes. The availability of two or three dri-baths is more convenient with temperatures set at 37°C, 65°C, and 95°C.

12. Aliquot mixer: The Hema-Tek aliquot mixer (Miles Inc., Elkhart, IN) is useful for mixing solutions or samples by inversion.

13. Electrophoresis running buffer: Prepare a stock of 5x TBE buffer as follows: 54 g of Tris base (Sigma), 27.5 g of boric acid (Sigma), and 20 mL of 0.5 $M$ disodium EDTA.2H$_2$O (pH 8.0) per liter. Alternatively, purchase TBE buffer as 5X concentrate or working solution from Sigma.

14. Ethidium bromide: 10 mg/mL from Sigma.

15. Thermostable DNA polymerase: A total of 250 units of *Tth* Plus DNA polymerase at a concentration of 5.5 units/μL was purchased from Biotech International (Bentley, Australia) and stored at –20°C in small aliquots to avoid repeated freezing and thawing.

16. Centrifuge: Biofuge 13 bench-top centrifuge (Heraeus, Sepatech, Germany) for 1.5-mL Eppendorf tubes.

17. Eppendorf tubes, 1.5 mL, hinged cap: Obtain from your local source (BioRad, Hercules, CA or Australian Biosearch, Perth, Western Australia) and sterilize by autoclaving.

18. 5X PCR polymerization buffer can be purchased from Biotech International as 5X PCR polymerization buffer in lots of 1 mL consisting of 335 m$M$ Tris-HCl,

pH 8.8, 83 m$M$ (NH$_4$)$_2$SO$_4$, 2.25% Triton X-100, 1 mg/mL gelatin), dNTPs (1 m$M$ each). Dispense into small aliquots, and store at –20°C to avoid repeated freezing and thawing.

19. 25 m$M$ MgCl$_2$· 5.08 g MgCl$_2$· 6H$_2$O/L: Stock 25 m$M$ MgCl$_2$ in 1 mL lots (Biotech International).

20. Mineral oil: Purchase 500 mL from Sigma and aliquot into 5 or 10 mL lots.

21. PCR final mixture: Each final PCR mixture (50 µL) for amplification consists of a buffer (67 m$M$ Tris-HCl, pH 8.8, 16.6 m$M$ (NH$_4$)$_2$SO$_4$, 0.45% Triton X-100, 200 µg/mL gelatin), dNTPs (200 µ$M$ each), primers (100 ng each of MIE4 and MIE5 or 50 ng each of LAPR and LA6; 50 ng each of RS42 and KM 29 or E1 and E2), 1.5 m$M$ MgCl$_2$, 1.6 U of *Tth* Plus DNA polymerase and template DNA (less than 1 µg) overlaid with a few drops of mineral oil.

22. Master mix for LA-GLBN duplex PCR: This formula will make up a master mix of PCR reagents for 20 reaction tubes, where 40 µL of the mix will be added to each PCR tube. Add the following to a sterile 1.5-mL Eppendorf tube labeled "Master Mix:" 200 µL of 5X PCR polymerization buffer, 60 µL of 25 m$M$ MgCl$_2$, 20 µL of each primer (50 ng/µL) LAPR, LA6, RS42, and KM29, 6 µL of *Tth* Plus DNA polymerase (5.5 units/µL), and 454 µL of sterile distilled water.

23. PCR tubes: 0.5 mL thin-walled GeneAmp reaction tubes with flat caps for use with DNA thermal Cycler 480 (Perkin Elmer, Norwalk, CT).

24. Electrophoresis gel loading solution: Consists of 0.05% bromophenol blue, 40% sucrose, 0.1 $M$ EDTA, pH 8.0, 0.5% sodium lauryl sulfate (Sigma).

25. DNA molecular size marker: Low-molecular-mass-marker representing DNA fragments of 1,746, 1,434, 800, 634, 303, 279, 249, and 222 bp (BioRad).

26. Horizontal electrophoresis tank: Mini-Sub DNA Cell, gel tray and two 15-well combs (BioRad).

27. Agarose: MetaPhor Agarose (FMC BioProducts, Rockland, ME).

## 3. Methods

### 3.1. Extraction of DNA From Clinical Samples

#### 3.1.1. DNA Extraction from Whole Blood or Bronchial Washings by Proteinase K Cell Lysis

1. Remove a specimen of whole blood or bronchial washings from the freezer and thaw at 37°C for 10 min and mix well (*see* **Note 6**).

2. Add 0.2 mL of the blood or bronchial washing to 1.0 mL of RCL buffer in a sterile 1.5 mL-Eppendorf tube. Mix by inversion and centrifuge at 13,000$g$ for 1 min.

3. Wash the white cell pellet twice in 1.0 mL of RCL buffer or until the pellet is free of visible red cells. Centrifuge after each wash at 13,000$g$ for 1 min.

4. When the white cell pellet is free of red cells, resuspend in 0.2 mL of WCL buffer containing 50 µg of proteinase K. Incubate the proteinase K digestion at 65°C for at least 2 h in a dri-bath.

5. Heat the tubes at 95°C for 10 min to inactivate the proteinase K reaction.

6. Store the digested samples frozen at –20°C until required for PCR.

### 3.1.2. Extraction of DNA from Whole Blood, Bronchial Washings, Urine, Plasma, Serum or CSF Using QIAmp Blood Extraction Kit

1. Add 0.2 mL of whole blood or other biological fluids to an equal volume of a guanidine hydrochloride lysis buffer (reagent AL) containing 25 μL of proteinase K. Mix and incubate at 70°C for 10 min and then at 95°C for 10 min to inactivate the proteinase K (*see* **Note 6**).
2. After cooling to room temperature, add 0.21 mL of absolute isopropanol to the lysed and digested sample, mix and load into a QIAamp silica-spin column. Centrifuge three times at 6000*g* for 1 min, and wash the contents of silica spin column twice with 0.5 mL of 70% isopropanol-salt wash solution (buffer AW).
3. Elute the DNA from the spin column with 0.2 mL of distilled water (preheated to 70°C) by centrifugation at 6000*g* for 1 min.
4. If required, determine the concentration of the eluted DNA by UV absorbance at 260 nm, spectrofluorometrically with Hoeschst dye, or visually by comparing the intensity of ethidium-stained samples and known amounts of placental DNA in 1% agarose gels *(15)*.

## 3.2. Serial Dilutions of Samples and Standards

1. Prepare six serial 10-fold dilutions (neat, 1 in 10, 1 in 100, 1 in 1000, 1 in 10,000, and 1 in 100,000) of a DNA sample that was extracted from a blood specimen and that was positive for CMV-DNA by LA-GLBN duplex PCR in the initial screening assay (*see* **Notes 6** and **7**).
2. Prepare Eppendorf tubes with six serial 10-fold dilutions of standard CMV DNA starting with a concentration of 0.1 pg/10 μL (*see* **Note 8**).
3. Prepare three Eppendorf tubes with serial 10-fold dilutions (neat, 1 in 10, 1 in 100) of CMV-negative sample DNA.

## 3.3. PCR

1. Dispense 40 μL of PCR master mix into each of the 16 labeled -PCR tubes. To these tubes, add 10 μL from each dilution of DNA sample, standard and negative control that were prepared in **Subheading 3.2.** Keep each tube covered between pipeting steps to prevent contamination with aerosols.
2. Include at least two reagent blanks by adding 10 μL of sterile distilled water to two different reaction tubes each containing 40 μL of the PCR reagents (Master mix).
3. To prevent evaporation during thermal cycling, add two or more drops of mineral oil with a sterile plastic disposable pipet to cover the aqueous phase in each reaction tube. Close the lids tightly and transfer the tubes to a cycle-room containing a thermal cycler for PCR.
4. Program the Thermal Cycler for the following conditions: an initial denaturation step for 5 min at 95°C, followed by 41 cycles of an annealing step at 60°C for 30 s, an extension step at 72°C for 2 min, and a denaturation step at 95°C for 1 min per cycle. Use a final temperature of 72°C for 7 min to ensure complete extension of the PCR product. After completion of PCR cycles maintain the tubes at 4°C or prepare to detect products immediately.

## *3.4. Detection of PCR Products*

1. Prepare a 3% MetaPhor agarose gel (0.75 g/25 mL) in 1X TBE buffer and dissolve in a steam bath or boiling water bath for 15–20 min. Cool the liquid gel to 70°C, and then pour into a taped gel tray. Place one of the 15-well combs at one end of the gel tray and place the other 15-well comb in the middle of the gel tray. Allow the agarose to cool and gel at room temperature for 30 min, and then at 4°C for a further 30 min to obtain optimal gel handling characteristics and optimal resolution. Overlay the gel with a thin layer of buffer (2–3 mm), and remove the combs and tape from the tray. Place the gel in the electrophoresis tank submerged under a thin layer (2–3 mm) of 1X TBE running buffer.

2. Add 5 μL of each PCR sample to 1 μL of gel loading solution that has been aliquoted onto the surface of a plastic petri dish or a square of paraffin film. Prepare a separate aliquot of gel loading solution with DNA size standard (*see* **Note 9**).

3. Mix PCR samples or DNA size standard with gel loading solution by pipeting up and down a few times. Transfer 5 μL of each sample and DNA size standard to wells of the agarose gel.

4. Electrophoresis in 1X TBE buffer at 80 V for 1 h or until the bromophenol blue dye in the gel loading buffer has migrated 3–4 cm down the gel.

5. Stain the gel in 100 mL of freshly prepared ethidium bromide (0. 5 μg/mL) for 15–20 min in a plastic container protected from light.

6. Place the stained gel on a UV transilluminator to visualize the ethidium bromide-stained DNA fragments.

7. Photograph the gel (*see* **Note 10**).

8. If desired, perform a DNA hybridization on the PCR products with a labeled probe (*see* **Note 11**).

## *3.5. Quantitation (see* **Note 12***)*

1. The end-point titer for MIE-PCR or LA-PCR, performed as either a uniplex or a duplex PCR, should be seen clearly in the photograph of the gel as shown in **Fig. 1** for blood samples obtained from two renal transplant patients infected with CMV (*see* **Note 13**).

2. The end-point titer of 10-fold dilutions of sample or standard DNA is the highest dilution that produces a specific DNA band detected by ethidium bromide staining on an agarose gel. This highest dilution is the one which precedes the dilution at which the amplified DNA band can no longer be clearly detected on the agarose gel (*see* **Note 14**).

3. The end-point titer can be expressed either as the highest dilution (or titer), as presented in **Fig. 2**, or as a concentration of DNA (*see* **Notes 15** and **16**).

4. To calculate the concentration of CMV DNA, the lowest amount of standard DNA that can be detected by the PCR assay is multiplied by the dilution factor that is needed to determine the limit of detection of DNA in a positive test sample (*see* **Note 16**).

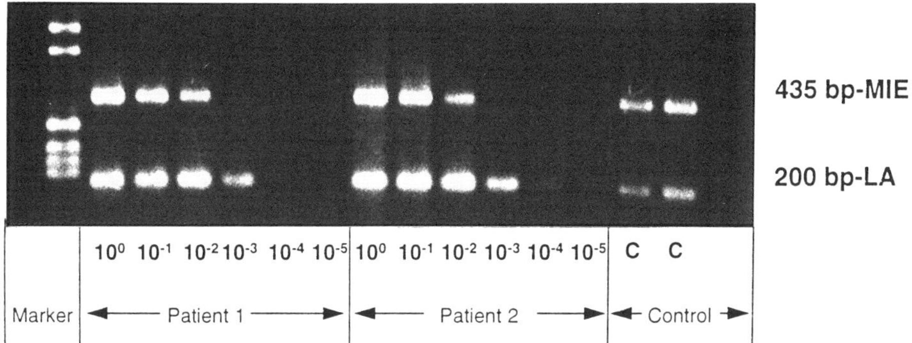



Fig. 1. Electrophoretic detection of amplified CMV DNA after MIE-LA duplex PCR of serial 10-fold dilutions of DNA extracted from blood samples of two patients with CMV infection (*see* **Notes 13** and **14**).

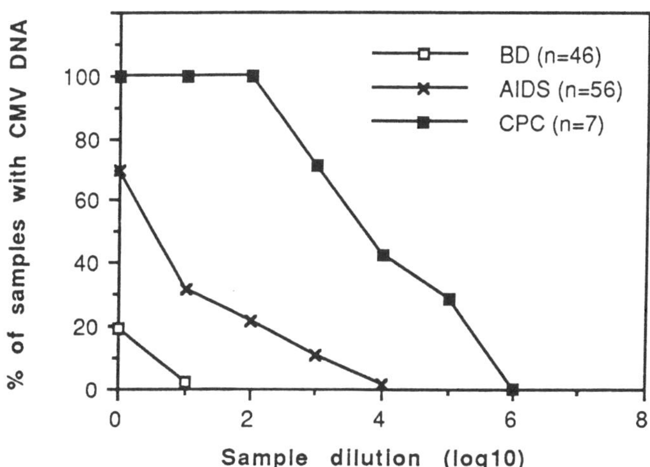



Fig. 2. The frequency of detection of CMV DNA (% of samples with CMV DNA) by LA-E PCR in 10-fold dilutions (log10) of extracted DNA from blood samples obtained from 46 blood donors (BD), 56 patients with AIDS, and 7 patients diagnosed with symptomatic CMV infection (CPC) (*see* **Notes 15** and **16**).

5. The concentration of CMV can be expressed either as the number of CMV copies (log10) per mL of blood or the amount of CMV DNA (pg) per 10 µg of leukocytes in blood assuming that the number of CMV genome copies is equiv to approx either 100 copies per fg of 11.7 kb recombinant LA-CMV DNA or five copies per fg of 240 kb of genomic CMV DNA, and 1 mL of blood contains 33 µg of leukocytic DNA or $5 \times 10^5$ leukocytes (*see* **Note 17**).

## 4. Notes

1. A series of ET-PCR optimization assays can be performed using a known amount of standard template DNA to confirm that both the new and old set of primers are comparable in regard to specificity, sensitivity, and reproducibility.
2. A set of cellular DNA primers can be combined with a set of CMV primers to amplify cellular DNA and CMV DNA in a single duplex PCR assay *(4)*. The HARE-primers are combined with the MIE-primers in a duplex PCR to give two distinctly separate PCR products of 282 and 435 bp, respectively. The GLBN-primers are combined with the LA-primers in a duplex PCR to give two distinctly separate PCR products of 536 and 200 bp, respectively, for the routine screening or quantitation of CMV DNA in clinical specimens.
3. The primers for amplification of cellular DNA by HARE- or GLBN-PCR can be used as a positive internal control to account for false negative results caused by inhibitors in the sample, poor specimen preparation, or DNA loss and suboptimal PCR reagents. If there is no amplification of CMV DNA, but there is amplification of globin DNA in a sample, detection of the positive internal control shows that the sample is a true negative rather than a PCR failure as a result of other reasons. The internal cellular DNA control is also useful for ET-PCR as an indicator of reproducibility between samples and assays and for providing a reference point for quantitating CMV DNA levels *(4)*.
4. Whole blood should be collected into tubes with EDTA or citrate as anticoagulants rather than heparin as the later is a strong inhibitor of PCR *(17)*.
5. The cost of extracting DNA from leukocytes with the QIAamp DNA extraction method is approx $3 US per extraction, which is 30 times more than the cost of extracting with the proteinase K cell lysis method. Although more expensive, the DNA purified by the QIAamp is free of protein and other contaminants that may influence the PCR. There was no statistically significant difference between the proteinase K cell lysis method and the Qiamp DNA extraction method for the frequency of detection of CMV DNA by LA-PCR in 43 blood samples (McNemar's test for symmetry was $X^2 = 0$). The DNA extraction methods were concordant for 35 of 43 samples (81.4%). The eight discordant samples (18.6%) were quantitated for CMV DNA by ET-PCR and found to be close to the limit of sensitivity of the LA-PCR assay. Thus, there are no marked differences between the performance of PCR on DNA prepared by proteinase K cell lysis or QIAamp method.
6. For improved economy and efficiency, ET-PCR is used only on those samples that were positive by LA-GLBN PCR during the initial screening.
7. The range of dilutions from neat (equiv to 330 ng of cellular DNA from whole blood using either one or other of the extraction procedures in **Subheading 3.1.**) to 1 in 100,000 should detect the end-point titer for CMV DNA in most samples (*see* **Fig. 2**). The range and increment of dilutions can be extended or reduced depending on the required accuracy and precision of quantitation results. In this example, the serial dilutions are designed to detect the end-point titers for CMV DNA in leukocytes. The amount of CMV DNA in plasma, urine, or CSF is usually less than that found in leukocytes and therefore may require fewer serial 10-fold

dilutions than that used for leukocytes. On the other hand, a smaller increment between dilutions (e.g., twofold or fivefold) may be preferred for end-point titration of plasma, urine, and CSF or other biological fluids where CMV DNA is frequently present at lower copy number than in leukocytes. The number of serial dilutions required to obtain a convenient end-point titer can be estimated by comparing the intensity of the ethidium bromide-stained product of the unknown sample to the intensity obtained with a known amount of standard DNA when the samples are first screened for the presence of CMV DNA by PCR *(18)*.

8. Only one standard amount (0.1 pg/10 µL) of CMV DNA is used in this experiment. Previous experiments have shown that the end-point titers correlated with increasing amounts of CMV DNA standard ranging between 10 fg and 1000 pg *(4)*. Nevertheless, the relationship between end-point titers and different amounts of standard DNA should be determined empirically for each new PCR assay that will be used for quantitation. The relationship between end-point titers and different amounts of DNA target may be linear or curvilinear over a quantitative range depending on the efficiency and rate of PCR. The line of best fit for end-point titers vs amount of DNA can be obtained by optimizing the number of cycles and/or primer concentrations.

9. If preferred, higher volumes of PCR products and gel loading solution can be mixed in 0.5-mL Eppendorf tubes.

10. To photograph gels stained with ethidium bromide, a Polaroid Land camera with Polaroid 667 instant film can be used.

11. A Southern blot hybridization of the gel by standard methods *(15)* during the screening step when it is used initially to detect the presence of CMV DNA in clinical specimens can be performed. Other hybridization procedures, such as dot blots or ELISA microtiter plate formats using nonradiolabeled probes, can be used to detect PCR products *(19,20)*.

12. Traditionally, quantitative PCR involves an end-point detection method to examine the final amount of PCR product that has accumulated after a fixed number of cycles. Studies on the kinetics of PCR have shown that amplification remains exponential for a limited number of cycles after which the rate of amplification will plateau *(21,22)*. The rate or the level at which the plateau is reached is dependent on the amount of starting DNA target, number of PCR cycles and PCR assay conditions. The amount of PCR product amplified from a standard amount of DNA target in repeat assays is generally more variable in the plateau phase than in the initial exponential phase of amplification caused by variation in reaction conditions or in sampling or presence of inhibitors. Consequently, real time monitoring of DNA amplification reactions over a range of cycle numbers is considered to be more accurate and reproducible for quantitation because measurements are obtained when reagents are not limited and are less subject to variation *(21,22)*. However, the cost of instrumentation and reagents to perform real time PCR may be too expensive for many laboratories. In this regard, LA-GLBN duplex PCR of serially diluted samples to determine end-point titers remains an useful option for quantitation of CMV DNA, particularly to differentiate between greater than 100-fold differences in clinical samples.

13. In **Fig. 1**, lane 1 is a molecular-mass marker with molecular sizes of 1,746, 1,434, 800, 634, 303, 279, 249, and 222 bp. Six serial 10-fold dilutions of DNA extracted from a blood sample obtained from one patient (P1) are in lane 2 (undiluted) to lane 7 (1 in 100,000), and from another patient (P2) are in lane 8 (undiluted) to lane 13 (1 in 100,000). A standard amount of recombinant CMV-DNA L and E fragments (100 fg each) are in lane 14 and 15. The 200-bp PCR product is LA-DNA and the 435-bp PCR product is MIE DNA of CMV.

14. The end-point titers in **Fig. 1** for MIE-PCR and LA-PCR are $10^2$ and $10^3$, respectively for both P1 and P2. Both end-point titers are equiv to 1 pg of CMV DNA because the limit of sensitivity for the recombinant CMV standards was 10 fg by MIE-PCR, and 1 fg by LA-PCR.

15. In **Fig. 2**, the products amplified from serial dilutions of extracted DNA were detected by electrophoresis and ethidium bromide staining, as described in the Methods section. For this series of experiments, PCR was performed using a LA-E duplex PCR. However, the results were essentially the same with the LA-GLBN duplex PCR.

16. Quantitation of relative differences between samples in **Fig. 2** is sufficient to show the differences among the three population groups. However, the normalization of relative differences to absolute values or concentration of CMV DNA in samples or specimens is helpful for comparison between different assays and studies. As expected, the frequency of detection of CMV DNA is significantly lower in the group of blood donors than AIDS patients or the group of symptomatic patients with culture-proven CMV viremia (CPC group). In addition, the end-point titers are generally less in the blood donor group than in the AIDS group or CPC group. The amount of CMV DNA at each dilution is equivalent to the dilution factor multiplied by 100 copies of CMV genome (or 1 fg of recombinant CMV DNA standard), which is the limit of detection of CMV DNA by LA-E PCR.

17. The use of ET-PCR for quantitation of DNA target has limitations concerning application, economy, sensitivity, accuracy, and reliability. ET-PCR is labor intensive and therefore best suited to a small number of samples per assay run. For a large number of clinical samples, other quantitative methods that allow the use of a single sample may be preferred *(8,9,22)*. However, the use of serial dilutions for PCR is reasonably economical *(9)* given the reduction in costs of *Taq* DNA polymerase and reagents in recent times. The accuracy of ET-PCR is affected by reliability, reproducibility and precision. The inclusion of three or four different amounts of external or internal standard DNA will help to assess the reliability, reproducibility, and linear range of the method. However, if you use cellular DNA as an internal reference for quantitation of CMV DNA in blood leukocytes by LA-GLBN PCR, the amount of amplified β-globin DNA may be less in samples obtained from patients with neutropenia. The precision and accuracy of ET-PCR also can be influenced in different ways by the increment used between dilutions. The accuracy of ET-PCR using 5- or 10-fold increments between dilutions to differentiate between low and high viral load of a magnitude of at least 100-fold is sufficient for most purposes in a routine microbiology laboratory. Furthermore, the accuracy of ET-PCR to differentiate between small

differences in serial two-fold dilutions may be poor because of an increased sensitivity in the measurement of errors and variations introduced by inhibitors, pipeting, and sample preparation. ET-PCR using 5- or 10-fold increments can provide useful information on the relative quantity of a microorganism which may increase from a few fold to a million-fold in some infections *(4,7,9)*. For further information and discussion on the validation of PCR for quantitative applications by determining the limits of sensitivity, linear range, reproducibility, precision, accuracy and specificity see the review by Farre et al. *(23)*.

## Acknowledgments

I thank Julie Pearson for her expert help in the laboratory and Dr. John Pearman and members of the Management Committee of the Microbiology Department for supporting this work.

## References


1. Persing, D. H., Smith, T. F., Tenover, F. C., and White, T. J. (eds.) (1993) *Diagnostic Molecular Microbiology. Principles and Applications.* Mayo Foundation, Rochester, MN.
2. Innis, M. A., Gelfand, D. H., Sninsky, J. J., and White, T. J. (eds.) (1990) *PCR Protocols: A Guide to Methods and Applications.* Academic, San Diego, CA.
3. Saiki, R. K., Scharf, S., Faloona, F., Mullis, K. B., Horn, G. T., Erlich, H. A., and Arnheim, N. (1985) Enzyme amplification of beta-globin genomic sequences and restriction site analysis for diagnosis of sickle cell anemia. *Science* **230,** 1350–1354.
4. Kulski, J. K. (1994) Quantitation of human cytomegalovirus DNA in leukocytes by end-point titration and duplex polymerase chain reaction. *J. Virol. Methods* **49,** 195–208.
5. Simmonds, P., Balfe, P., Peutherer, J. F., Ludlam, C. A., Bishop, J. O., and Leigh Brown, A. J. (1990) Human immunodeficiency virus-infected individuals contain provirus in small numbers in peripheral mononuclear cells at low copy numbers. *J. Virol.* **64,** 864–872.
6. Brillanti, S., Garson, J. A., Tuke, P. W., Ring, C., Briggs, M., Masci, C., Miglioli, M., Barbara, L., and Tedder, R. S. (1991) Effect of alpha-interferon therapy on hepatitis C viraemia in community acquired chronic non-A, non-B hepatitis: a quantitative polymerase chain reaction study. *J. Med. Virol.* **34,** 136–141.
7. Kulski, J. K. and Pryce, T. (1996) Preparation of mycobacterial DNA from blood culture fluids by simple alkali wash and heat lysis method for PCR detection. *J. Clin. Microbiol.* **34,** 1985–1991.
8. Erhardt, A., Schaefer, S., Athanassiou, N., Kann, M., and Gerlich, W. H. (1996) Quantitative assay of PCR-amplified hepatitis B virus DNA using a peroxidase-labelled DNA probe and enhanced chemiluminescence. *J. Clin. Microbiol.* **34,** 1885–1891.
9. Rawal, B. K., Booth, J. C., Fernando, S., Butcher, P. B., and Powles, R. L. (1994) Quantification of cytomegalovirus DNA in blood specimens from bone

marrow transplant recipients by the polymerase chain reaction. *J. Virol. Methods* **47,** 189–202.

10. Cagle, P. T., Buffone, G., Holland, V. A., Samo, T., Demmler, G. J., Noon, G. P., and Lawrence, F. C. (1992) Semiquantitative measurement of cytomegalovirus DNA in lung and heart-lung transplant patients by in vitro DNA amplification. *Chest* **101,** 93–96.

11. Demmler, G. J., Buffone, G. J., Schimber, C. M., and May, R. A. (1988) Detection of cytomegalovirus in urine from newborns by using polymerase chain reaction DNA amplification. *J. Infect. Dis.* **158,** 1177–1184.

12. Greer, C. E., Peterson, S. L., Kiviat, N. B., and Manos, M. M. (1991) PCR amplification from paraffin-embedded tissues. Effects of fixative and fixation time. *Am. J. Clin. Pathol.* **95,** 117–124.

13. Lubahn, D. B., Brown, T. R., Simental, J. A., Higgs, H. N., Migeon, C. J., Wilson, E. M., and French, F. S. (1989) Sequence of the intron/exon junctions of the coding region of the human androgen receptor gene and identification of a point mutation in a family with complete androgen insensitivity. *Proc. Natl. Acad. Sci. USA.* **86,** 9534–9538.

14. Fleckenstein, B., Muller, I., and Collins, J. (1982) Cloning of the complete human cytomegalovirus genome in cosmids. *Gene,* **18,** 39–46.

15. Sambrook, J., Fritsch, E. F., and Maniatis, T. (1989) *Molecular Cloning: A Laboratory Manual,* 2nd ed. Cold Spring Harbor Laboratory, Cold Spring Harbor, NY.

16. Higuchi, R. (1989) Rapid, efficient DNA extraction for PCR from cells or blood. *Amplifications* **2,** 1–3.

17. Poli, F., Cattaneo, R., Crespiatico, L., Nocco, A., and Sirchia, G. (1993) A rapid and simple method for reversing the inhibitory effect of heparin on PCR for HLA class II typing. *PCR Methods Appl.* **2,** 356–358.

18. Lee, W. T., Antoszewska, H., Powell, K. F., Collins, J., Doak, P. B., Williams, L. C., Munn, S., Verran, D., and Croxson, M. C. (1992) Polymerase chain reaction in detection of CMV DNA in renal allograft recipients. *Aust. N.Z. J. Med.* **22,** 249–255.

19. Inouye, S. and Hondo, R. (1990) Microplate hybridization of amplified viral DNA segment. *J. Clin. Microbiol.* **28,** 1469–1472.

20. Tanaka, M., Onoe, S., Matsuba, T., Katayama, S., Yamanaka, M., Yonemichi, H., Hiramatsu, K., Baek, B-K., Sugimoto, C., and Onuma, M. (1993) Detection of Theileria sergenti infection in cattle by polymerase chain reaction amplification of parasite-specific DNA. *J. Clin. Microbiol.* **31,** 2565–2569.

21. Higuchi, R., Fockler, C., Dollinger, G., and Watson, R. (1993) Kinetic PCR analysis: Real-time monitoring of DNA amplification reactions. *BioTechnology* **11,** 1026–1030.

22. Bassam, B. J., Allen, T., Flood, S., Stevens, J., Wyatt, P., and Livak, K. J. (1996) Nucleic acid sequence detection systems: revolutionary automation for monitoring and reporting PCR products. *Aust. Biotech.* **6,** 285–294.

23. Ferre, F. (1992) Quantitative or semiquantitative PCR: reality versus myth. *PCR Methods Appl.* **2,** 1–9.

# 8

# Analysis of Amplified DNA Molecules by Capillary Electrophoresis and Laser Induced Fluorescence

## Michael J. Fasco

## 1. Introduction

The polymerase chain reaction (PCR) has revolutionized molecular biology. Portions of single-copy per cell genes (and cDNAs) prepared from very small tissue or cell samples can be specifically amplified for use in sequence determination, gene identification, and quantitation. Improvements to the method, such as the introduction of genetically engineered, thermostable polymerases, more precise thermocyclers and more efficient reverse transcriptases for mRNA conversion to cDNA, have combined to make RNA-PCR (also called reverse transcriptase, or RT-PCR) and PCR more reproducible and specific. Coupled with the high sensitivity of the reactions, RT-PCR and PCR are increasingly used as quantitative bio-analytical techniques.

PCR amplification of a particular target sequence becomes exponential after the first two or three cycles and remains so for several additional cycles. Following the exponential phase, the amplification process slows appreciably. Unfortunately, the number of cycles in the exponential phase is variable for each target. The variability is dependent on many factors, including target nucleotide composition, the nucleotide sequences of the primer set, and the size of the product. For relative abundance measurements, accurate comparison between a target gene, and some "housekeeping" internal reference gene, such as $\beta2$-microglobulin *(1)*, must be derived from the linear, exponential region of the amplification reaction. Thus, reaction parameters have to be customized for each target and primer set to obtain linearity for as many cycles as possible. However, for some templates, the exponential portion of amplification is too short to allow easy detection of the products. An additional compli-

*From: Methods in Molecular Medicine, Vol 26: Quantitative PCR Protocols*
*Edited by: B. Kochanowski and U. Reischl © Humana Press Inc., Totowa, NJ*

cation, unique to RNA-PCR, is that the extent of target mRNA that is reverse-transcribed to cDNA is often incomplete, introducing another variable.

Early studies of "quantitative PCR" were labor-intensive and subject to error because of tube-to-tube variations. Understandably, quantitative PCR and quantitative RT-PCR were used sparingly as analytical methods during the initial stages of PCR development. A significant advance in quantitative RT-PCR was reported by Wang et al. *(2)*, who used an internal standard to quantitate specific cellular mRNAs present in low copy numbers. A premeasured quantity of generic cRNA internal standard, constructed from a phasmid insert that contained promoter and polyadenylation sites, and primer sites shared by the mRNA target sequence, were reverse transcribed and amplified with the sample mRNA in the presence of $^{32}$P labeled 5'-primer. At distinct cycle numbers or total RNA concentrations, the two different-size products of the reaction were separated by agarose-gel electrophoresis, visualized by staining with ethidium bromide, and their quantity determined by scintillation counting of the excised bands. The range of cycles undergoing exponential amplification was determined for each product and the target mRNA concentration calculated from its proportion to the internal standard.

Other quantitative RT-PCR methods were subsequently devised, each with one or more advantageous features. Gilliland et al. *(3)* showed that an internal standard, containing complementary sequences for the same primer pair as the target sequence, was amplified identically to the target sequence during both the exponential and postexponential phases of the PCR. Co-amplification of the target and internal-standard templates also occurred independently of the cDNA or DNA concentration. In a normal assay, various concentrations of the internal standard are added to a fixed amount of cDNA or DNA, and following amplification, the internal standard-to-target ratio is determined. The concentration of the target is determined from the internal-standard concentration when the ratio of the two is unity. Alternatively, the internal-standard concentration can be fixed and the cDNA or DNA concentration varied. Competitive PCR, as the method is known, greatly increased the use of RT-PCR and PCR as bio-analytical methods. Apostolakos et al. *(4)* combined competitive RT-PCR with the use of a "housekeeping" mRNA in multiplex competitive RT-PCR. In their method, the mRNA of glyceraldehyde-3-phosphate dehydrogenase (GAPDH) and a target mRNA were simultaneously reverse-transcribed and dilutions of the cDNA amplified in the presence of the appropriate competitor sequences and primer sets. The advantage of this method is that tube-to-tube variations in the amplification process are reduced.

Despite these advances, all competitive and relative abundance methods of quantitative RT-PCR and PCR remain labor-intensive and require the careful preparation and analysis of many samples. With respect to the latter, the analytical methods used are often cumbersome and require multiple steps to quantitate the DNA products formed. Most quantitative RT-PCR and PCR methods now employ a "competitor" DNA or cRNA that differs from the corresponding target sequence by the presence or absence of a restriction-enzyme site or part of the internal sequence, both of which confer eventual size-separation capability. In the case where a DNA competitor is used, an internal reference gene, such as GAPDH, is usually also measured either in the same tube or in a separate reaction. Although several analytical methods for determining competitor-to-target ratios have been described, most involve some combination of an agarose gel separation step, coupled to either scintillation counting, radioimaging, or densitometric scanning of autoradiographs or photographs of ethidium bromide-stained gels. Distinct disadvantages of these methods include: the use of radioisotopes; the need to frequently relabel $^{32}$P containing primers or purchase $^{32}$P labeled nucleotide triphosphates; a narrow range of linear detection response; and multiple steps such as restriction-enzyme cleavage, gel electrophoresis and detection.

Capillary electrophoresis (CE) in the presence of liquid or solid polymers is highly efficient in separating DNA fragments in the size range frequently used in quantitative RT-PCR and PCR of 200–800 bp *(5–7)*. Advantages of CE over the methods previously cited are that: the separation times are short (25–35 min); baseline separation of DNA products that differ by as few as 10 bp is achieved between 200 and 300 bp; the peaks are very sharp, allowing accurate area integration with sophisticated integration software; the detection systems have a wide linear range; the need for radioisotopes is eliminated; and many samples can be run unattended overnight. The principal disadvantage of early CE-based methods was that the PCR products were detected from their ultraviolet (UV) absorption at 254 or 260 nm *(5,8)*. Although suitable for the detection of products produced in some PCR applications, UV absorption was not sensitive enough for the detection of RT-PCR products formed from low copy numbers without employing large reaction volumes that required desalting and concentration before assay. The introduction of laser-induced fluorescence (LIF) and the discovery of DNA intercalators that exhibited high fluorescence only when bound to DNA *(9)*, and that were compatible with the excitation and fluorescent-emission wavelengths of the argon ion laser, greatly enhanced the potential for CE use in PCR product quantitation. Thiazole orange coupled with LIF detection provided up to a 400-fold increase in detection over UV *(5)*. Srinivasan et al. *(10)* achieved femtogram detection by LIF of PCR products, separated by CE, using the highly fluorescent, dimeric intercalating agents of oxazole yellow (YOYO-1) and thiazole orange (TOTO).

Although these highly fluorescent intercalators made possible the detection of extremely low levels of PCR products, their usefulness in accurately reflecting the concentrations of PCR products separated by CE had not been shown. The fluorescent intercalators have a very high affinity for DNA, particularly the dimeric dyes, and bind to both the interior and exterior regions of the double-stranded helix in a concentration-dependent manner. Unless strict experimental conditions are followed, broad peaks and inconsistent amount/area values result. The DNA to TOTO ratio affects peak shape and fluorescent intensity of DNAs separated by CE *(5)*. The quality of DNA-YOYO-1 complex separations in agarose and their fluorescent intensity measured by laser-excited confocal fluorescence is sensitive to the order of component addition, the DNA concentration, and molar ratio of DNA to fluorescent dye *(9)*. Srinivasan et al. *(10)* observed similar dependencies in their separation of DNA-YOYO-1 complexes by CE; optimum separation and peak shape being dependent on premixing of the DNA with YOYO-1 or TOTO at a molar ratio of 5 base pairs to 1 of the intercalator. Because both the agarose and CE methods of DNA-YOYO-1 complex analysis require fairly accurate, prior estimates of the DNA concentrations in the sample, their use was not well-suited for the analysis of the many samples normally generated during competitive RNA-PCR or competitive PCR studies.

A CE analytical method, using LIF detection of DNA-YOYO-1 complexes, that reproducibly binds the intercalator independently of the initial DNA concentration and its size, was developed in this laboratory *(11)*. Low levels of multiple DNA species generated during competitive or multiplex-competitive RT-PCR can be measured without the use of special primers or additional treatment of the amplified samples. Besides the advantages of CE cited above, the method requires only 10 μL of sample that can be recovered except for the few nanoliters used for the analysis. Details of the method, including its limitations, are presented in this chapter. Also included are some methods we have found best for the reverse transcription and competitive amplification of a variety of templates. Competitive RT-PCR applications using CE that have been published by this laboratory can be found in the following references: P-450 1A1 and 1A2 *(11,12)*, optimized conditions for DNase use in quantitative RNA-PCR *(13)*; and competitive PCR of the human estrogen receptor mRNA and its exon deletion mRNA forms in breast tumors and cell lines *(14)*.

## 2. Materials

### 2.1. Instruments

1. The capillary electrophoresis unit is a P/ACE 2200 equipped with an argon ion laser and System Gold system and integration software (Beckman Instruments, Fullerton, CA).

2. RNA reverse transcription reactions and PCR amplifications are done in a Perkin Elmer 9600 thermal cycler (Applied Biosystems, Foster City, CA).
3. The high performance liquid chromatograph (HPLC) unit is from Waters Associates and consists of a 600E Pump/System controller, WISP, 996 detector, fraction collector, and Millennium software for visualization and integration of the separated peaks.

## 2.2. Chemicals and Reagents

4. MilliQ (Millipore Corporation, Milford, MA) water is used throughout unless otherwise stated.
5. Hydroxypropylmethylcellulose (HPMC) was purchased from Sigma Chemical Co. (St. Louis, MO).
6. 5-carboxyfluorescein (Sigma).
7. YOYO-1 is obtained from Molecular Probes (Eugene, OR).
8. PhiX 174 RF DNA *Hae*III (Beckman).
9. LiFluor dsDNA 1000 Kits (Beckman).
10. RNA-PCR and PCR Core Kits and reaction tubes are obtained from Applied Biosystems.
11. Superscript II is purchased from Gibco BRL (Grand Island, NY).
12. RNase (Gibco BRL).
13. Taq extender from Strategene (La Jolla, CA).
14. Taq-start antibody from Clontech (Palo Alto, CA).
15. TRI reagent is purchased from Molecular Research Center, Inc. (Cincinnati, OH).
16. MicroCon 30 and Centricon 30 spin-filter units are from Amicon (Beverely, MA).
17. Oligonucleotides are synthesized at our Molecular Genetics Core Facility (Wadsworth Center, Albany, NY) using a Milligen 8750 DNA synthesizer (Millipore). The forward primer for actin is (2104–2125): 5'-GCGGGAAATCGTGCGTGACATT, and the reverse primer is (2409–2432): 5'-GATGGAGTTGAAGGTAGTTTCGTG. The size of the amplified actin fragment is 328 bp.

    The forward primer for GAPDH (296-315) is 5'-TCTTCACCACCATGGAGAAG; the reverse primer (917–936) is 5'-GTCATACCAGG<u>AAAATGA</u><u>GCT</u>; and the sequence specific (nonbolded portion 767–786) linker reverse primer is 5'-**GAAATGAGCT**CTGCTTCACCACCTTCTTGA:
18. Confluent HepG2 or MCF-7 cells cultured in DMEM with phenol red and 10% FBS under a 5% $CO_2$ atmosphere at 37°C are obtained from our tissue culture facility (Wadsworth Center, Albany, NY).
19. Diethylpyrocarbonate (DEPC; Sigma). Autoclaved distilled water is incubated with 0.1% DEPC for 1 h and than autoclaved.
20. The CE separation and wash buffer is TBE containing 0.5% HPMC and 100 n*M* YOYO-1 (*see* **Note 5**). A 5X solution of TBE is made by dissolving 54 g of Tris base and 27.5 g of boric acid in 700 mL of water. Ethylenediaminetetraacetic acid (EDTA) (20 mL of a 0.5 *M* solution, pH 8.0) is added and the volume brought to 1 L with water. The solution is stored in a plastic container. The HPMC solu-

tion (0.5% in 1X TBE) is prepared by stirring the polymer and buffer overnight at room temperature. The solution is filtered first through paper and then through a 0.45 μ sterilization filter (optional). It is stable for a minimum of 6 months at room temperature. YOYO-1 (1 m*M* in DMSO; 2 μL) is dissolved in 0.2 mL of TBE and a 0.15 mL aliquot dissolved in 15 mL of the HPMC-TBE solution by gentle mixing. This solution is divided into three buffer vials supplied with the instrument. The remaining TBE solution of YOYO-1 (0.05 mL) is diluted with an equal volume of TBE in a 0.5-mL microfuge tube (5 μ*M* final concentration). The tube is cut slightly above the fluid line and placed in a buffer vial equipped with a spring so that it can be contacted by the capillary. YOYO-1 solutions in DMSO are not repeatedly frozen and thawed. A new vial of YOYO-1 is aliquoted into 2 μL portions in 0.5-mL Eppendorf tubes and stored at –20°C until use.

A polyacrylamide polymer buffer can be purchased from Beckman that is also compatible with the conditions described. The separation is very similar to that obtained with HPMC except that the elution times are somewhat longer.

## 2.3. Capillaries and Columns

21. The capillaries used are either the dsDNA 1000 coated capillary (Beckman) or the DB-1 or DB-17 coated capillaries from J&W Scientific (Folsum, CA).
22. Purification of DNA internal standards is done by high performance liquid chromatography (HPLC) using a Gen-Pak Fax column (Waters Associates, Millipore).

## 3. Methods
## 3.1. Molecular Biology
### 3.1.1. RNA Isolation

The isolation of total RNA from untreated or 2,3,7,8-tetra-chloro-dibenzo-β-dioxin (TCDD)-treated HepG2 or MCF-7 cells using TRI reagent is as recommended by the manufacturer. The method also works very well for RNA isolation from normal or tumor tissues.

1. Solubilize tissue-culture cells ($5–10 \times 10^6$) or tissues (50–100 mg) in 1 mL volume of TRI reagent. Cultured cells are dissolved by incubation in the reagent for 5–10 min at room temperature. Tissue samples are dispersed by homogenization or sonication in the TRI reagent.
2. Precipitate DNA and protein by the addition of 0.2 mL chloroform, mix thoroughly, and after 2 min, centrifugate at 12,000*g* for 15 min at 4°C for isolation of RNA.
3. Remove the RNA in the clear upper phase and precipitate with 0.5 mL isopropanol.
4. After 5 min at room temperature pellet RNA by centrifugation at 12,000*g* for 10 min at 4°C.
5. Suspend the pellet by light vortex mixing in 1 mL 70% ethanol.

6. Pellet again by centrifugation at 7500*g* for 5 min at 4°C.
7. Decant the ethanol supernatant and dry the pellet partially under a stream of air flowing under a chemical fume hood.
8. Dissolve the RNA in DEPC-treated-water at a concentration of 1–7 mg/mL and stored at –85°C.
9. Determine spectometrically the absorbance at 260 and 280 nm. If the ratio is greater than 1.6, then the amount of DNA is negligible. The absorbtion at 260 nm multiplied with the extinction coefficient of 40 mg/mL gives the RNA concentration.

## 3.1.2. RNA Reverse Transcription (see **Note 1**)

Reverse transcription of mRNA is accomplished with either of two enzymes.

### 3.1.2.1. REVERSE TRANSCRIPTION WITH MURINE LEUKEMIA VIRUS (MuLV)

The conditions for MuLV's use are exactly as described in the instructions included with the Applied Biosystems RNA-PCR Core Kit.

1. mRNA is reverse-transcribed with Oligo(dT)$_{16}$ priming at a concentration of 1 µg RNA/20 µL of reaction for 30 min at 42°C.
2. Inactivate the reverse transcriptase by heating at 95°C for 5 min. Samples are either used immediately after cooling or stored at –20°C. (For cDNA amplification, *see* **Subheading 3.1.3.1.**)

### 3.1.2.2. REVERSE TRANSCRIPTION WITH SUPERSCRIPT II

1. Reaction mixtures of 20 µL contain: 4 µL of 5X buffer (included with the enzyme); 2 µL of 0.1 *M* dithiothreitol (included with the enzyme); 8 µL of dNTP mixture (2.5 m*M* each base; RNA-PCR Core Kit); 0.5 µL of a gene specific primer (50 pmol/µL) or (1 µL of Oligo(dT)$_{16}$ [50 pmol/µL–RNA PCR Kit]); 1 µL of Superscript II; 5 µg of total RNA; and DEPC-treated water up to 20 µL.
2. The reverse transcription reaction mixture is incubated at 25°C for 5 min followed by: incubation for 50 min at 42°C, enzyme inactivation is at 95°C for 5 min, and cooling to 4°C.
3. mRNA – cDNA heteroduplexes are destroyed by incubation with 2 units of RNase H at 37°C for 20 min.
4. The RNase H is then inactivated by heating at 95°C for 5 min and the reaction mixture cooled to 4°C. cDNA samples are stored at –20°C. (For cDNA amplification, *see* **Subheading 3.1.3.2.**)

## 3.1.3. cDNA Amplification

PCR primers used in our investigations are designed with the aid of a primer program such as DNAsis, and span an intron whenever possible. The primer concentrations are 50 pmol each/0.1 mL of reaction. Most PCR reactions are scaled down to 25 µL and run without oil. Master mixes are used whenever possible to reduce tube-to-tube variations.

### 3.1.3.1. cDNAs Produced by Reverse Transcription with MuLV

For cDNAs prepared by MuLV reverse transcription, the PCR reactions are prepared exactly as described in the directions accompanying the RNA-PCR Core kit (Applied Biosystems).

### 3.1.3.2. cDNAs Produced by Reverse Transcription with Superscript II

1. Reaction mixtures of 0.1 mL contain (in order of addition): 7.2 µL of $MgCl_2$ (from RNA-PCR Core Kit); 9.6 µL of 10 X buffer (from RNA-PCR Core Kit); 69.3 µL of DEPC-treated water; 6.4 µL of dNTP mixture (2.5 m$M$ each base; from RNA-PCR Core Kit); 0.5 µL of "Taq extender"; 1 µL of "Taq-Taq start antibody complex" (*see* **Note 2**); 4 µL of cDNA; and 1 µL of each primer (*see* **Note 3**).

   The reaction is initiated by heat-denaturing the sample at 94°C for 1 min. This is followed by 30–35 cycles of denaturation at 94°C for 10 s, primer annealing at 60°C or 65°C (dependent on the $T_m$ of the primer set) for 15 s, and amplification at 72°C for 15 to 60 s (dependent on the length of the DNA amplificon). Reaction is terminated by cooling to 4°C.
2. An aliquot (9 µL) is mixed with 1 µL of a suitable external DNA standard (normally actin, 25 ng), and the mixture transferred to a micro-vial for analysis by CE.

### 3.1.4. Preparation of the Actin External Standard (see Note 4)

The human actin external standard is prepared from the total RNA of HepG2 cells. Reverse transcription of actin mRNA and PCR is done using the protocol described by Applied Biosystems for their RNA-PCR Core Kit.

1. Prepare actin external standard in 10 repetitive 0.1 mL amplification reactions for 35 cycles. The cycling conditions are as previously described using the 65°C annealing temperature and extension for 30 s.
2. Combine the contents of the reactions.
3. Desalt the sample with a Centricon 30 spin filter.
4. Purify it by HPLC using the solvent and wash solutions for nucleic acid separation provided with the Gen-Pak Fax column. The gradient is 30% solvent B to 80% solvent B over 30 min at a flow rate of 0.75 mL/min.
5. Collect the eluted actin fragment in 0.2-mL fractions and analyze each fraction for purity by CE.
6. Combine those fractions containing the pure fragment and concentrate again in a Centricon 30 spin filter.
7. Determine the concentration of the actin external standard solution by mixing an aliquot with a known quantity of PhiX 174 RF DNA *Hae*III and, following separation by CE, comparing its integrated area with those of the PhiX 174 RF DNA components (*see* **Note 5**).

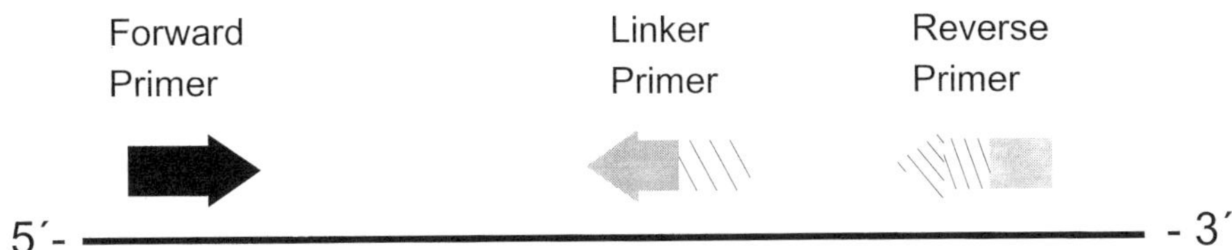


Fig. 1. Schematic order of the primers used for the construction of the competitor.

### 3.1.5. Preparation of DNA Competitors

Competitors are made from the cDNA of TCDD-treated HepG2, or untreated HepG2 or MCF-7 cells using a method modified from that described by Förster *(15)*. A schematic diagram of the strategy is shown in **Fig. 1**. A summary of the method using primers to synthesize a GAPDH competitor that can be used with either rat or human cDNAs follows (*see* **Note 6**).

1. Prepare cDNA by one of the reverse transcription methods described in **Subheading 3.1.2.**
2. First-round PCR is done in 25 µL as previously described using the forward primer and the linker reverse primer. The nonbolded portion of the linker reverse primer (hatched in **Fig. 1**) is complementary to a region of the GAPDH sequence that is upstream from the reverse primer. The bolded 10 nucleotides of the linker primer are the same as those underlined in the 3' end of the reverse primer (gray in **Fig. 1**). During the amplification, these nucleotides become part of the DNA product.
3. Following completion of the reaction, dilute the template 1/1000 with water and amplify again in a preparative PCR reaction (one or more tubes of 0.1 mL) using the forward and reverse primers. During this amplification reaction, the remaining nucleotides (diagonal marking in **Fig. 1**) that are present in the reverse primer, but absent in the reverse linker primer, are added, thus producing a DNA product that has the reverse primer built in, but is 131 bp shorter than the original 641 bp cDNA template.
4. Purify the template by HPLC and quantitate it as described for the actin external standard.

### 3.2. Capillary Electrophoresis

### 3.2.1. Preparation and Conditioning of the Capillary

The total length of the capillary is 47 cm. It is very important that the ends be cut squarely, otherwise the eluted peaks will be broad and irregularly shaped. Newly installed capillaries are normally not as sensitive to detection by LIF as are capillaries conditioned with buffer overnight. If the capillary must be used immediately, the PMT100 value on the instrument can be adjusted to compensate for the initial loss in sensitivity.

### 3.2.2. Separation Conditions

Do not subject the single vial containing the wash buffer to current.

1. Reverse the polarity of the CE from its normal setting so that loading of the samples occurs at the negative electrode.
2. Before separation of the sample, wash the column with high pressure in either direction with wash buffer for 4 min.
3. Submerge the tip of the capillary into a vial filled with just enough water so that it is used as the waste receptacle for the wash solution.
4. Rinse into a water reservoir in order to reduces the build-up of the HPMC on the lift units that house the capillary and electrodes and seals the vial caps during loading and separation.
5. Load forward a solution of YOYO-1 (5 $\mu M$) in TBE with high pressure onto the column for 12 s (*see* **Note 7**). This is followed by forward, low pressure loading of the sample for 10–18 s and separation for 35 min at 200 V/cm (*see* **Note 8**).
6. If enhanced sensitivity is required, dilute part of the PCR sample with 100 volumes of water, desalted, and concentrated in a Microcon 30 or Centricon 30 spin-filter unit (*see* **Note 9**). The sample is then loaded electrokinetically on the column at 200 V/cm for 5–10 s.

## 4. Notes

1. We now routinely use Superscript II in our reverse-transcription reactions. It has the advantage that larger quantities of RNA (5 $\mu g$/20 $\mu L$ reaction) can be reverse transcribed with the same efficiency as obtained with 1 $\mu g$ of MuLV. It is also very efficient in reverse transcribing long transcripts, which was particularly useful with our studies of the estrogen receptor *(14)*.
2. *Taq* DNA polymerase (Amplitaq, Applied Biosystems) is mixed with an equal volume of Taq-Start antibody (Clontech) and incubated at room temperature for 5–15 min. The mixture is stored at –20°C. The antibody inactivates the polymerase until the Taq-antibody complex is heat denatured in the first thermal step. Incorporating the antibody enhances the specificity and amplification efficiency of many templates and is now routinely used in our amplification reactions.
3. When less than 1 $\mu g$ of reverse transcribed RNA is used in the PCR reaction, the change in volume is compensated for by adding the appropriate reverse-transcription mixture that contains the buffers and dNTPs; water is substituted for the other components. When one or two competitors are used in the reaction, they are diluted with and added in water.
4. An external standard is added to most of our PCR reactions before analysis by CE. During the electrophoresis of many samples in the same electrode solutions, the retention times of the DNA components progressively shorten, as much as half a minute for large components, during overnight analysis of 20 samples. Shortening of peak retention times also occurs if the same buffers are used on subsequent days. This elution time-shortening effect does not detectably alter peak separation or their quantitation, because all the peaks are equally affected.

However, the effect does pose a problem when comparing peak elution patterns between electropherograms run at different times. The external standard provides a point of retention time reference that aids in the identification of closely eluting peaks using the System Gold offset feature. It also serves as an indicator of the extent of sample loading onto the capillary, and can be used to measure the quantity of PCR product or products produced in the reaction. Occasionally, 5-carboxyfluorescein, or any other suitable organic fluorescent molecule, is added in addition to or, in place of, the DNA external standard. It can be used only as an indicator of sample loading efficiency, however, and not as a standard for the quantity of DNA product formation.

5. A separation of the PhiX RF174 *Hae*III components under the CE conditions described is presented in **Fig. 2**. Notable features of the separation are the excellent separation between the 271 and 281 bp fragments, but only a near baseline separation between the 1078 and 1353 bp peaks. In the separation of fragments larger that 300 bp, the separation efficiency progressively decreases as a function of increasing fragment size. Comparison between electropherograms of the PhiX RF174 *Hae*III components run with YOYO-1 or ethidium bromide (detection with UV) as the intercalator show that separation of larger fragments in the presence of YOYO-1 is not as complete as obtained with ethidium bromide (data not shown). The greater number of YOYO-1 molecules bound per bp, and its larger molecular size, combine to produce DNA complexes that are comparatively larger and do not separate as efficiently.

The relationship between fragment size and concentration versus integrated area of the PhiX RF174 *Hae*III components is presented in the upper panel of **Fig. 3**. Note that the fragment size and concentration variables have the same area values. Linear regression analysis of these points yields a line that does not pass through zero, but whose slope is independent of the amount of sample loaded onto the capillary. This proportional detection response over the entire fluorescent range of the instrument can also be demonstrated by comparing the area ratios of two different-sized DNA fragments at several different concentrations. The reason that the regression line does not pass through zero is seen in the lower panel of **Fig. 3**, which shows that the amount per area value of the different size fragments is not a constant. That is, the 72 and 118 bp fragments bind many fewer YOYO-1 molecules per bp than do the larger ones, resulting in much lower fluorescence, and less peak area. This effect is less pronounced for the 194 and 234 bp fragments and finally reaches an essentially constant value for DNA fragments larger than 600 bp. It is important that this variability in the amount per area values be recognized and compensated for because significant error can be introduced when DNA fragments of different sizes are measured. Fortunately, calculating the area correction values is very simple. A conversion table is constructed from the integrated areas of an electropherogram of the PhiX RF174 *Hae*III components. The amount per area value for DNA fragments of 600 bp or greater is arbitrarily set to one. Corrected values for the smaller fragments are obtained by dividing their amount per area value by that of the larger

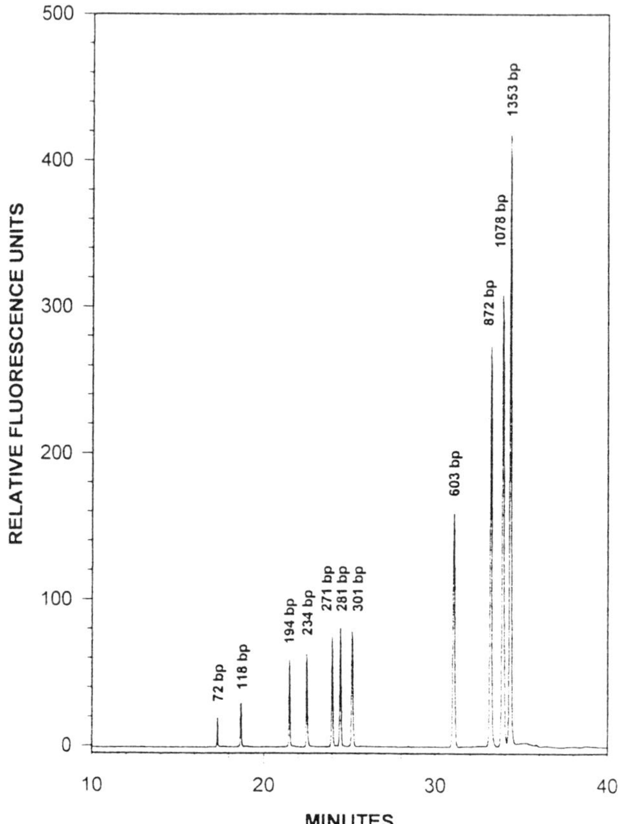



Fig. 2. Electropherogram of the PhiX RF174 *Hae*III components. The separation and LIF detection conditions were as described under methods. The PhiX RF174 *Hae*III component concentration was 20 ng/µL. Loading was by pressure for 15 s. The total length of the capillary was 47 cm and the separation voltage 200 v/cm.

fragments. The integrated areas of any peaks can then be normalized by multiplying their integrated area by the conversion factor that corresponds to their size.

6. Stock solution concentrations of the competitors are determined from integrated area comparison with the calibrated actin standard or, when size allows, with the phiX 174 RF DNA components. Dilution of the competitors from their stock solutions for use in competitive RT- PCR or PCR reactions is always done just before use. Very dilute solutions of the competitors diminish quickly even standing at 4°C. An electropherogram illustrating the separation and peak shapes of two GAPDH products is presented in **Fig. 4**. Also presented in the inset of the figure is a graph showing linearly regression of measured template GAPDH concentrations divided by competitor GAPDH concentrations and plotted as a function of the

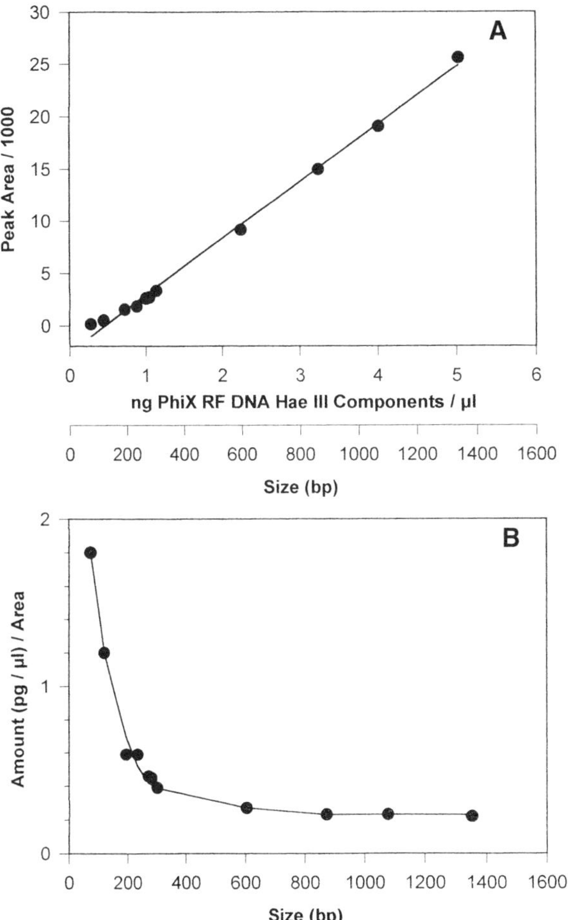



Fig. 3. **(A)** Relationship between the concentration (normal X axis) of the individual PhiX RF174 *Hae*III components and their integrated areas. The lower X axis demonstrates that the same relationship holds when the integrated area of each peak is plotted as a function of its size in basepairs. **(B)** The relationship between the amount/integrated area value for the each PhiX RF174 *Hae*III component as a function of basepair size.

reverse-transcribed total RNA concentration (cDNA). The GAPDH competitor concentration was fixed at 0.1 amol/0.1 mL of reaction mixture. The point on the X-axis where the template GAPDH concentration divided by the competitor GAPDH concentration is 1, the equal to the amount of total RNA that contains 0.1 amol of GAAPDH mRNA, assuming that the reverse transcription to xcDNA is 100%.

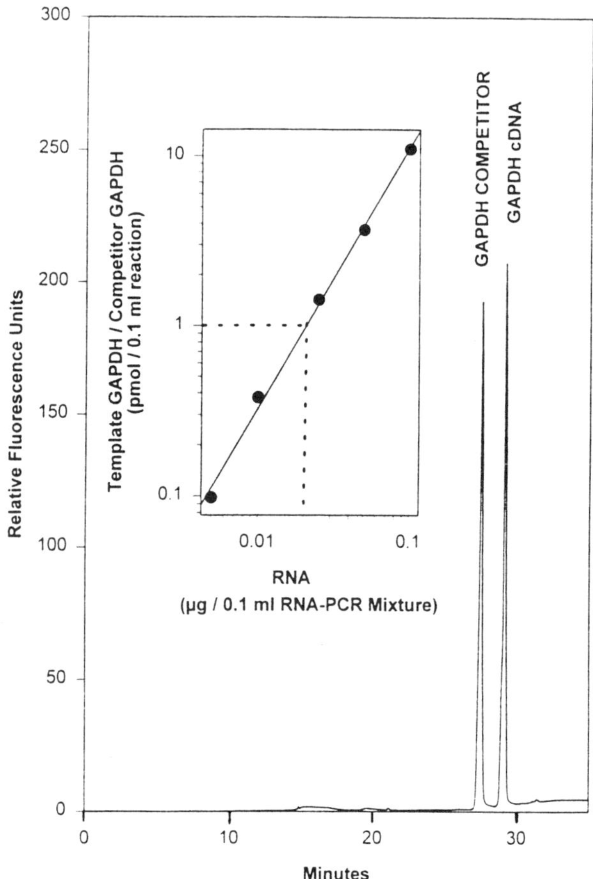


Fig. 4. Electropherogram illustrating the separation of the GAPDH competitor (elution time, 27.5 min) and the amplified GAPDH cDNA template (elution time, 29.2 min). The amplified reaction products were pressure loaded onto the capillary column (15 s) without any additional treatment as described. Inset, linear regression of data obtained by dividing the concentration of the amplified cDNA template GAPDH by the concentration of the cDNA competitor GADPH and plotting the value as a function of various reverse transcribed total RNA concentrations. The GAPDH competitor concentration was fixed at 0.1 amol/0.1 mL amplification reaction volume. The total was RNA isolated from HepG2 cells as described.

7. YOYO-1 containing solutions are used for 2 d or 30–40 samples.
8. Loading of a concentrated "plug" of YOYO-1 in TBE that does not contain HPMC allows complete mixing between the sample and the intercalator. When

current is applied, excess intercalator is separated from the sample. Because the plug solution is much less viscous than the separation buffer, the DNA components run rapidly though it and stack on the separation buffer, which produces the sharp peaks obtained.

9. A single water-dilution/concentration of amplified samples in Microcon 30 or Centricon 30 units can result in a 10 to 20% loss. If the initially concentrated sample is diluted again with water and concentrated, the loss can be as much as 80%.

## References


1. Murphy, L. D., Herzog, C. E., Rudick, J. B., Fojo, A. T., and Bates, S. E. (1990) Use of the polymerase chain reaction in the quantitation of *mdr*-1 gene expression. *Biochemistry* **29,** 10,351–10,356.
2. Wang, A. M., Doyle, M. V., and Mark, D. F. (1989) Quantitation of mRNA by the polymerase chain reaction. *Proc. Natl. Acad. Sci. USA* **86,** 9717–9721.
3. Gilliland, G., Perrin, S., Blanchard, K., and Bunn, H. F. (1990) Analysis of cytokine mRNA and DNA: detection and quantitation by competitive polymerase chain reaction. *Proc. Natl. Acad. Sci. USA* **87,** 2725–2729.
4. Apostolakos, M. J., Schuermann, W. H. T., Frampton, M. W., Utell, M. J., and Willye, J. C. (1993) Measurement of gene expression by multiplex competitive polymerase chain reaction. *Anal. Biochem.* **213,** 277–284.
5. Schwartz, H. E. and Ulfelder, K. J. (1992) Capillary electrophoresis with laser-induced fluorescence detection of PCR fragments using thiazole orange. *Anal. Chem.* **64,** 1737–1740.
6. Guttman, A., Wanders, B., and Cooke, N. (1992) Enhanced separation of DNA restriction fragments by capillary gel electrophoresis using field strength gradients. *Anal. Chem.* **64,** 2348–2351.
7. Landers, J. P., Oda, R. P., Spelsberg, J. A., Nolan, J. A., and Ulfelder, K. J. (1993) Capillary electrophoresis: a powerful microanalytical technique for biologically active molecules. *Biotechniques* **14,** 98–111.
8. Dveksler, G. S., Basile, A. A., and Dieffenbach, C. W. (1992) Analysis of gene expression: use of oligonucleotide primers for glyceraldehyde-3-phosphate dehydrogenase. *PCR Methods Appl.* **1,** 283–285.
9. Rye, H. S., Yue, S., Wemmer, D. E., Quesada, M. A., Haugland, R. P., Mathies, R. A., and Glazer, A. N. (1992) Stable fluorescent complexes of double-stranded DNA with bis-intercalating asymmetric cyanine dyes: properties and applications. *Nucleic Acids Res.* **20,** 2803–2812.
10. Srinivasan, K., Morris, S. C., Girard, J. E., Kline, M. C., and Reeder, D. J. (1993) Enhanced detection of PCR products through use of TOTO and YOYO intercalating dyes with laser induced fluorescence-capillary electrophoresis. *Appl. Theoretical Electrophoresis* **3,** 235–239.
11. Fasco, M. J., Treanor, C. P., Spivack, S., Figge, H. L., and Kaminsky, L. S. (1995) Quantitative RNA-polymerase chain reaction-DNA analysis by capillary electrophoresis and laser-induced fluorescence. Anal. Biochem. **224,** 140–147.

12. Fasco, M. J., Treanor, C., and Kaminsky, L. S. (1996) Cytochrome P450 mRNA induction: quantitation by RNA-polymerase chain reaction using capillary electrophoresis. *Methods Enzymol.* **272,** 401–412.
13. Huang, Z. Q., Fasco, M. J., and Kaminsky, L. S. (1996) Optimization of DNase I removal of contaminating DNA from RNA for use in quantitative RNA-PCR. *Biotechniques* **20,** 1012–1020.
14. Fasco, M. J. (1997) Quantitation of estrogen receptor mRNA and its alternatively-spliced mRNAs in breast tumor cells and tissues. *Anal. Biochem.* **245,** 167–178.
15. Forster, E. (1994) An improved general method to generate internal standards for competitive PCR. *Biotechniques* **16,** 18–20.

# 9

# Competitor Calibration and Analysis of Competitive Amplified PCR Products by High-Performance Liquid Chromatography (HPLC)

## Thomas Köhler

## 1. Introduction

To assay gene expression or virus genomes in tissues or body fluids, competitive polymerase chain reaction (cPCR) is now performed in many laboratories. cPCR is a quantitative adaption of the PCR method in which a known number of copies of a synthetic RNA *(1)* or DNA *(2–4)* is coamplified with the target sample and therefore compete for the common primers and reagents in the same reaction tube. After coamplification, both products are distinguished by characteristic features, e.g., size, mostly by electrophoretic methods, or by probe-specific hybridization, e.g., PCR-ELISA *(4,5)*.

As an alternative, automated high-perfomance liquid chromatography (HPLC) separation may be used for analysis of PCR products. A number of reports described the application of this technique for both rapid analysis of single DNA fragments, e.g., plasmids *(6, 7)*, or differential products obtained from enzymatic cleavage of DNA or competitive amplification *(5,8–10)*. A very recent application of HPLC is the extremely precise calibration of competitor DNA fragments (about 100–500 bp in length) using a Low DNA Mass Ladder *(5,10)*.

Recently, small columns filled with nonporous anion exchange resins (TSK-Gel, Tosoh Corp., Japan), which have their best resolution below 1000 bp have been widely succeeded for separation of DNA fragments *(5–10)*. This column material may be operated with high eluent pressure of about 100 bar, resulting in fast and efficient separation of DNA fragments according to their net charge. Alternatively, Gen-Pak FAX (Millipore Corp., Milford, MA) or Resource Q columns (Pharmacia LKB Biotechnology, Uppsala, Sweden) may be used for PCR product separation.

From: Methods in Molecular Medicine, Vol. 26: Quantitative PCR Protocols
Edited by: B. Kochanowski and U. Reischl © Humana Press Inc., Totowa, NJ

Exemplary for the measurement of cDNAs coding for the multidrug resistance-associated protein (MRP) and endogenous reference gene product glyceraldehyde-3-phosphate dehydrogenase (GAPDH), the method permits accurate quantitation of gene expression. Prognostically, MRP gene expression is an important parameter for patients suffering from acute myelogenous leukemia (AML). Increased levels of MRP mRNA were found in patients with relapsed *de novo* AML. Thus, overexpression of the gene might contribute to leukemic relapse *(11,12)*.

## 2. Materials

1. Cell lines and patient samples: The high-level multidrug resistant human T-lymphoblastoid cell line CCRF ADR5000, drug-selected by adriamycin as described earlier *(13)* was used. Mononuclear cells (MNC) from bone marrow (BM) aspirates of patients suffering from acute myelogenic leukemia (AML) were isolated by conventional density gradient centrifugation using Ficoll/Paque (Amersham Pharmacia Biotech, Uppsala, Sweden).
2. RNA and cDNA: Isolate whole RNA by conventional techniques or using RNAzolB (Biotecx, Houston, TX). Synthesize cDNA from 1 µg aliquots of whole RNA sample in a 20 µL standard reaction mixture containing AMV reverse transcriptase buffer (250 m$M$ Tris/HCl; pH 8.3, 250 m$M$ KCl, 50 m$M$ MgCl$_2$, 50 m$M$ dithiothreitol, 2.5 m$M$ spermidine), 5 U AMV reverse transcriptase, 0.5 m$M$ of each dNTP (Promega, Madison, WI), 10 U recombinant RNase inhibitor (AGS, Heidelberg, Germany), and 200 ng oligo(dT) (Amersham Pharmacia Biotech).
3. Primers and DNA competitors: Primers were designed by using the automated OLIGO 5.0 Primer Analysis Software (National Biosciences, Plymouth, MN), and checked for specificity by the Hitachi HIBIO DNASIS 2.1 DNA Sequence Analysis System (Hitachi Software Engineering Co., Yokohama, Japan) using available sequence information. A homologous competitor fragment developed for MRP quantitation *(5)* was generated by a modified site-directed mutagenesis protocol described by Förster *(3)*. Briefly, the competitor fragment for the chosen PCR product was generated by PCR amplification of the target DNA sequence with the appropriate 5' primer and a 3' linker primer (LP) carrying the original 3' primer sequence on its 5' end. Primers used in this procedure were MRP3 (5'-GCTCGTCTTGTCCTGTTTCT-3', 5' primer), MRP4 (5'-CTCCACCTCCTCATTCGCAT-3') and MRP-LP (5'-MRP4-CCTTCTTCCA GTTCTTTACC-3'). A purified 355 bp subsequence of the pMS1 plasmid carrying a heterologous (multifunctional) competitor DNA fragment, which was designed and cloned in our laboratory *(5)* was used for GAPDH quantitation. Primers hGAPDH1 (5'-CGTCTTCACC ACCATGGAGA-3') and hGAPDHrc (5'-CGGCCATCAC GCCACAGTTT-3') (sequences kindly given by Dr. H. Garn, Institute of Immunology, Philipps-University of Marburg, Germany) were used.
4. HPLC columns: TSK DEAE-NPR column (4.6 mm ID, length: 35 mm) protected by a DEAE-NPR guard column (4.6 mm ID, length: 5 mm) (TosoHaas GmbH,

Stuttgart, Germany); filter element (4.0 mm diameter, 0.5 micron), distributor disk (Alltech Associates, Deerfield, IL), 100 μL sample loop.

5. Stationary phase of HPLC columns: Hydrophilic DEAE linked anion exchanger, capacity: greater than 0.15 meq/mL, particle size: 2.5 μm diameter, pH-range: 2 to 12, $pK_a$ of anionic groups: 11.2.

6. HPLC-system: consisting of Gastorr GF103 degasser, PU-980 intelligent HPLC pump, low pressure gradient former, UV-975 UV/VIS detector, AS-950 intelligent sampler, 84 well plate capacity, suitable for 0.2 mL MicroAmp reaction tubes (Jasco Labor und Datentechnik GmbH, Gross-Umstadt, Germany.

7. Mobile phase for HPLC: Buffer A: 25 m$M$/L Tris-HCl, 1 $M$ NaCl; pH 9.0. Buffer B: 25 m$M$ Tris-HCl; pH 9.0.

## 3. Methods

### 3.1. Standard Competitive PCR Assay

1. Amplify aliquots of reverse transcribed RNA (2 μL for MRP and 1 μL for GAPDH) in 50 μL standard PCR reaction mixtures containing 2 μL of 3' and 5' primer (each 10 pmol/μL), 5 μL 10x Taq polymerase buffer (100 m$M$ Tris-HCl, 500 m$M$ KCl, 15 m$M$ MgCl$_2$, 0.01% (w/v) gelatin; pH 8.3), 1.5 U AmpliTaq polymerase (Perkin-Elmer, Norwalk, CT), and 8 μL dNTPs (0.2 m$M$ each, [Promega, Madison, WI]) using 2'-deoxyuridine 5'-triphosphate (dUTP) (Boehringer, Mannheim, Germany) instead of dTTP.

2. Add 0.2 U uracil-DNA glycosylase (UDG) (Boehringer, Mannheim, Germany) to each reaction tube in order to prevent carryover contamination.

3. Start amplification with an initial 15 min incubation step at 37°C to ensure destruction of contaminating DNA originating from previous PCRs by UDG, then perform a subsequent 10 min denaturation at 94°C, followed by addition of AmpliTaq polymerase at 72°C (manual "hot start" technique).

4. After an initial PCR cycle to allow second strand cDNA synthesis, add aliquots of diluted competitor fragment (appropriate volume 2–5 μL). Perform MRP amplification with 35 cycles (30 s 94°C, 30 s 53°C, 1 min 72°C) and GAPDH amplification with 22 cycles (30 s 94°C, 30 s 58°C, 45 s 72°C) using a GeneAmp 9600 thermal cycler (Perkin-Elmer, Norwalk, CT).

### 3.2. Purification and Storage of Competitor DNA Fragments

1. Gel-purify PCR amplified standard fragments by electrophoresis through a 2% (w/v) agarose gel, and cut out the ethidium bromide-stained bands from the gel.

2. Isolate DNA by using a Sephaglas Band Prep Kit (Amersham Pharmacia Biotech). Dilute the purified DNA competitor fragment to a final concentration of about 10 μg/μL with sterile H$_2$O.

3. Prepare the competitor working dilutions with an aqueous solution of 10 ng/μL HindIII digested λ-DNA (AGS, Heidelberg, Germany) or an equivalent carrier for competitor stabilization, store at –20°C in 1.7 mL Multi Twist Top Vials (Sorenson Bio Science, Salt Lake City, UT).

### *3.3. HPLC Separation, Modified Protocol (7)*

1. Use a mobile phase of binary composition consisting of buffers A and B (*see* **Subheading 2., item 7**). Only HPLC grade water and sterile filtered buffers should be used.
2. After equilibration with 10 mL 25% A in B run a blank gradient program.
3. Before processing the PCR samples analyze a 200 ng aliquote of pBR322-HaeIII-digested standard DNA (Sigma, St. Louis, MO) in order to prove the column separation capacity (at least 16 peaks should be distinguishable).
4. Load the column with 5 to 10 µL of the original PCR reaction mixture (volume depends on amplification yield). Avoid column overload that results in characteristic ascending peak tailing.
5. Employ a discontinuous gradient program as follows:
   a. Equilibration of the column with 25% A in B.
   b. 25% A in B: sample application up to 0.5 min.
   c. 25–43% A in B: linear gradient up to 4.5 min.
   d. 43–60% A in B: linear gradient up to 20 min.
   e. 60–100% A in B: linear gradient up to 22 min.
   f. 100–25% A in B: linear gradient up to 24 min.
   g. 25% A in B: equilibration.
6. HPLC operating conditions: operative pressure: 84–110 bar (maximum back pressure: 200 bar), flow rate: 1 mL/min, temperature: room temperature, UV-detection: 260 nm, separation time of each run: 25 min
7. Perform data aquisition and auto-peak search with the respective chromatographic software, e.g., NINA Chromato-Graphic-System software (Nuclear Interface GmbH, Münster, Germany). Integrate the areas under the individual peaks (mV · S), which serve as a relative measurement of DNA quantity.

### *3.4. Calibration of Competitor DNA Fragments by HPLC*

1. Separate aliquots of purified competitor stock solution and Low DNA Mass Ladder (Life Technologies, Gaithersburg, MD) composed of a mixture of six blunt-ended DNA fragments ranging from 100–2000 bp, containing 5–100 ng DNA, respectively, by the desired discontinuous gradient program. Perform separation of both mass ladder and purified standard fragment in duplicate on the same day using the same charge of eluent.
2. Plot the estimated peak integrals of the individual mass ladder components as a function of DNA amounts to generate a calibration graph (**Fig. 1**).
3. Compare the peak integral of the gel-purified fragment with the calibration curve to calculate the accurate competitor amount (**Fig. 1**).

### *3.5. Competitive Amplification of MRP and GAPDH cDNA, Quantitation of PCR Products by HPLC*

1. Prepare three vials containing the standard PCR reaction mixture supplemented with identical quantities of MRP cDNA (usually achieved by premixing and

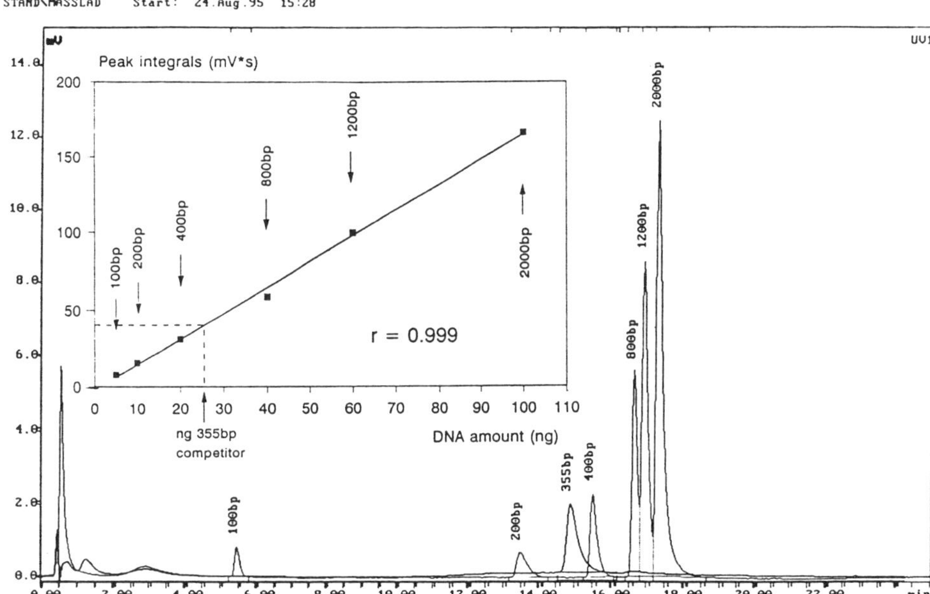



Fig. 1. Competitor calibration by HPLC. HPLC profiles of purified 355 bp multifunctional competitor DNA fragment and Low DNA Mass Ladder containing 5, 10, 20, 40, 60, and 100 ng DNA in the bands, respectively. Data obtained by using a TSK DEAE-NPR anion-exchange column. Insert: Calibration graph generated from the mass ladder.

    aliquoting cDNA with the PCR reaction mix), amplify in the presence of 9.75, 2.93, and 0.98 zepttomoles (zmol) of the homologous competitor DNA fragment.

2. Amplify GAPDH cDNA from the same source in duplicate using 2.13 and 1.07 attomoles (amol) of the 355 bp heterologous competitor fragment by the indicated cycle program. Be sure to record both the physiological and pathological range of gene expression by the introduced competitor amounts.

3. To check for day to day reproducibility of values of each run, perform **steps 1** and **2** likewise with cDNA from the MRP and GAPDH expressing cell line CCRF ADR5000. For this, prepare at least a 200 µL cDNA reference batch.

4. Dilute MRP and GAPDH PCR samples 1:4 and 1:2, respectively, with buffer B, and load the anion-exchange column with 20 µL of the mixture. Multiply the calculated ratios of competitor product to amplified target DNA with the initial amount of competitor added to the PCR reaction. Two representative chromatograms obtained from separation of competitive amplified MRP and GAPDH cDNA are shown in **Fig. 2**.

5. Before calculation of the product ratios, correct for UV absorbance differences resulting from different length of standard and target product by an experimentally determined multiplier. The correction factors that match the ratios of high and low mol weight products were 0.665 for MRP and 0.4 for GAPDH *(10)*.

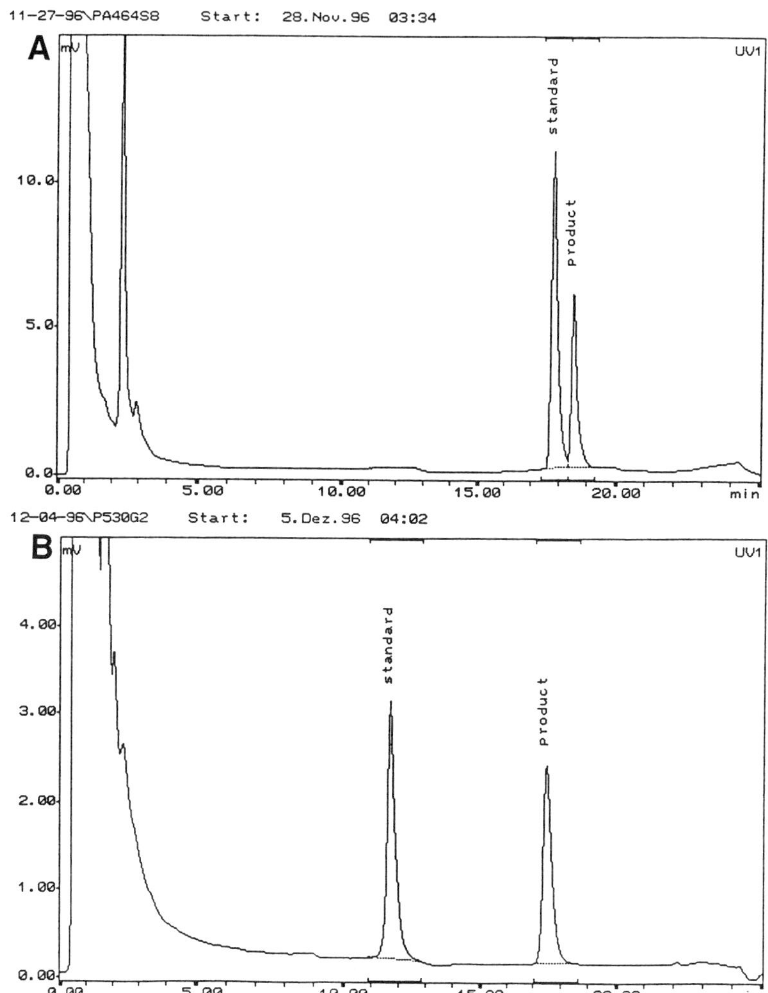



Fig. 2. Separation of competitive amplified PCR products. **(A)** 2 μL of MRP cDNA prepared from bone marrow MNC of an AML patient coamplified with 2.93 zmol competitor fragment. Competitor (standard) derived PCR product: 256 bp, target-derived PCR product (product): 340 bp; **(B)** GAPDH cDNA coamplified with 2.13 amol GAPDH competitor. Standard: 139 bp, product: 300 bp.

6. Calculate the initial cDNA amounts from the ratios of target to standard derived product corrected for UV absorbance differences and express the mean values in terms of zmol MRP cDNA per amol GAPDH cDNA (**Fig. 3**).

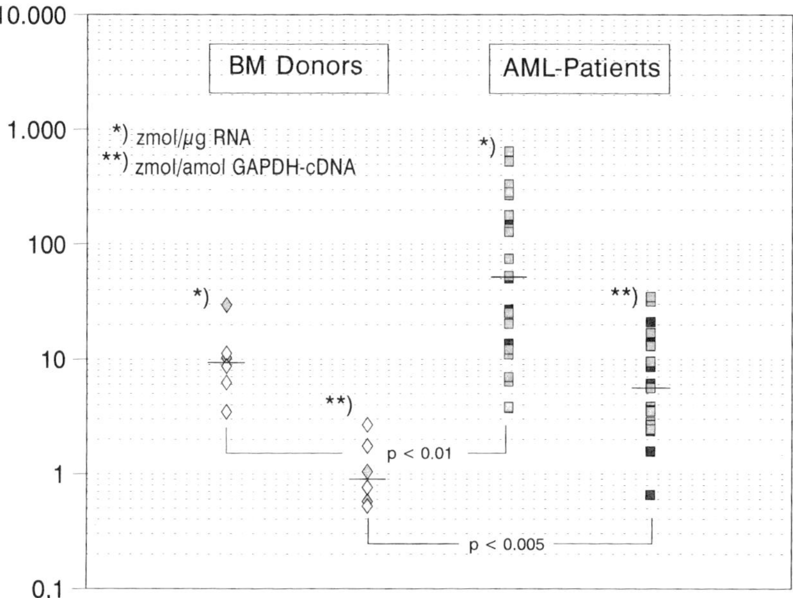



Fig. 3. MRP gene expression in normal and leukemic BM cells. cDNA samples from BM donors ($n$ = 6, rhombs) and AML patients with poor treatment response ($n$ = 21, squares) were assayed for MRP gene expression. Data normalized and expressed as zmol MRP cDNA per µg RNA and zmol MRP cDNA per amol GAPDH cDNA. There was a significant overexpression of the MRP gene in MNC of AML-minor/nonresponders compared to BM donors. Please note that the significance is more distinct when values normalized by the GAPDH reference gene transcripts were compared (P less than 0.005 vs P less than 0.01) applying the Mann-Whitney U-Test (horizontal lines: medians).

## 4. Notes

1. The conditions recommended above allow the separation of DNA-fragments between 20–2000 bp in length. Particularly short DNA fragments (< 80 bp) differing by approx 3 bp in length can be distinguished. Nevertheless, differential PCR products ranging from 100–500 bp should differ by at least 50 bp to achieve desired baseline resolution.

2. When PCR samples prepared without any mineral oil overlay are analyzed and a guard column is used, DNA extraction, e.g., with water-saturated chloroform as recommended in ref. *(8)*, is not required.

3. Column cleaning with 4x 50 µL of 0.2 mol/L NaOH should be frequently performed (i.e., every 30–50 runs). Problems caused by clogging of column will result in increased back pressure (> 110 bar). Partial clogging of the filter ele-

ment or distributor disk can result in tailing peaks caused by uneven sample distribution. Simple backflushing with half of the normal flow rate is often successful in cleaning the top filter. Nevertheless, precolumn filter and distributor disk should be replaced every 300 runs. For long-time storage of column at room temperature, we recommend equilibration with 20% acetonitril in water in order to protect from the growth of microorganisms. Following this guideline, up to 2000 separations may be performed with a single column.

4. Linear correlation between loaded sample volumes and corresponding peak integrals ( r > 0.998, not shown), as well as high precision signal recovery (< 2% as demonstrated by repeated analysis of an amplified MRP sample), both make HPLC optimally suited for DNA analysis.

5. The excellent and reproducible competitor calibration success by the proposed HPLC protocol was recently demonstrated in our laboratory. For this, GAPDH cDNA amounts measured in a unique sample by applying two separately and independently calibrated preparations of the competitor fragment were compared. The measured cDNA values differed by only 2% from the mean value *(10)*. Therefore, competitor calibration performed by this method was extremely precise taking into account the multitude of steps required for standard fragment synthesis, purification, calibration, dilution, application, and detection of the synthesized products. Therefore, the procedure is highly recommended and should be favored over traditional UV absorbance measurement *(1)* or calibration by densitometric means *(3)*.

6. In contrast, HPLC calibration of in vitro synthesized cRNA using nonporous TSK-Gel columns failed to be successful (unpublished observation). This may be caused by problems in separation of RNA mixtures (e.g., RNA mol wt markers). Individual RNAs are nonuniform because of varying conformations that may strongly influence the interaction with the anion-exchange resin. On the other hand, RNA mass standards, not to be confused with mol wt markers, are currently not commercially available.

7. If double-stranded DNA competitors are used to assay reverse transcribed mRNA in a sample, the template will be theoretically underestimated by factor 2. This discrepancy can be overcome by applying single-stranded DNA *(2)* or simple multiplication of the experimental values by two *(14)*. Alternatively, cDNA can be converted to double-stranded DNA by extension of thermocycling by one preceding PCR cycle before adding the standards, thus causing proper target start amounts.

8. Although the approach using double-stranded DNA competitors was shown to be valuable in quantifying cDNA, the method neglects RT variability. Therefore, using DNA competitors for gene expression studies absolutely requires data normalization in comparison to reference gene transcripts expressed in the same sample, e.g., GAPDH mRNA *(10)*. Data not normalized and expressed in terms of molecules per μg RNA may lead to misinterpretion of findings as mentioned in **Fig. 3.**

9. HPLC detection of PCR products is a time-consuming method. Even if automation using an autosampler allows processing of less than 60 PCR samples a day, this technique provides one of the best ways for clear separation of competitive amplified from nonspecific PCR products in order to quantify gene expression reliably.

## References


1. Wang, A. M., Doyle, M. V., and Mark D. F. (1989) Quantitation of mRNA by the polymerase chain reaction. *Proc. Natl. Acad. Sci. USA* **86**, 9717–9721.
2. De Kant, E., Rochlitz, C. F., and Herrmann, R. (1994) Gene expression analysis by a competitive and differential PCR with antisense competitors. *Biotechniques* **17**, 934–942.
3. Förster, E. (1994) An improved general method to generate internal standards for competitive PCR. *Biotechniques* **16**, 18–20.
4. Köhler, T., Laßner, D., Rost, A. K., Thamm, B., Pustowoit, B., and Remke, H., eds. (1995) *Quantitation of mRNA by Polymerase Chain Reaction: Nonradioactive PCR Methods.* Springer-Verlag, Heidelberg, Germany.
5. Alard, P., Lantz, O., Sabagh, M., Calvo, C. F., Weill, D. E., Chavanel, G., Senik, A., and Charpentier, B. (1993) A versatile ELISA-PCR assay for mRNA quantitation from a few cells. *Biotechniques* **15**, 730–737.
6. Henninger, H. P., Hofmann, R., Grewe, M., Schulze-Specking, A., and Decker, K. (1993) Purification and quantitative analysis of nucleic acids by anion-exchange high-performance liquid chromatography. *Biol. Chem. Hoppe-Seyler* **374**, 625–634.
7. Katz, E. D., Bloch, W., and Wages, J. (1992) HPLC Quantitation and identification of DNA amplified by the polymerase chain reaction. *Amplifications* **8**, 10–13.
8. Gaus, H., Lipford, G. B., Wagner, H., and Heeg K. (1993) Quantitative analysis of lymphokine mRNA expression by a nonradioactive method using PCR and anion exchange chromatography. *J. Immunol. Methods* **158**, 229–236.
9. Zeillinger, R., Schneeberger, C., and Speiser, P. (1993) Rapid quantitative analysis of differential PCR products by high-performance liquid chromatography. *Biotechniques* **15**, 89–95.
10. Köhler, T., Rost, A. K., and Remke, H. (1997) Calibration and storage of DNA competitors used for contamination-protected competitive PCR. *Biotechniques* **23**, 722–726.
11. Schneider, E., Cowan, K. H., Bader, H., Toomey, S., Schwartz, G. N., Karp, J. E., Burke, P. J., and Kaufmann, S. H. (1995) Increased expression of the multidrug resistance-associated protein gene in relapsed acute leukemia. *Blood* **85**, 186–193.
12. Hart, S. M., Ganeshaguru, K., Hoffbrand, A. V., Prentice, H. G., and Mehta, A. B. (1994) Expression of the multidrug resistance-associated protein (MRP) in acute leukemia. *Leukemia* **8**, 2163–2168.
13. Gekeler, V., Weger, S., and Probst, H. (1990) MDR1/P-glycoprotein gene segments analyzed from various human leukemic cell lines exhibiting different multidrug resistance profiles. *Biochem. Biophys. Res. Commun.* **169**, 796–802.
14. Pannetier, C., Delassus, S., Darche, S., Saucier, C., and Kourilsky, P. (1993) Quantitative titration of nucleic acids by enzymatic amplification reactions run to saturation. *Nucl. Acids Res.* **21**, 577–583.

# 10

# Quantifying Amplicons with ELISA

Olivier Lantz, Elizabeth Bonney, and Yassine Taoufik

## 1. Introduction

Among the numerous assays proposed for quantifying specific nucleic-acid sequences in biological samples, PCR offers the greatest sensitivity and versatility. The assay for quantifying the amount of polymerase chain reaction (PCR) products is a crucial step in any quantitative PCR method. It should be sensitive and specific, able to display a wide dynamic range, nonradioactive, easy to do, and inexpensive. The results of the assay should also be easily digitalized. Quantification of amplicons with enzyme-linked immunosorbent assay (ELISA) fulfills these criteria. It can be automatized and readers are already available in most research and clinical laboratories. This assay can be accomplished by using colorimetry, fluorometry, or luminometry, depending on the substrate used. Luminometry displays the best sensitivity and has the widest dynamic range of these three methods (*1* and *see* **Subheading 1.2.3.**). In this chapter, we will describe some of the available formats, the one we have been using this past few years, and its use in kinetic quantitative PCR or with internal standard.

## 1.1. The Different Formats

According to the way the amplicons are captured and how their amount is measured, several ELISA formats have been described in the literature: **Fig. 1** outlines some of them. Amplicons can be captured onto microtiter plates in two ways (*see* Chapter 3.3.2). The first is to use biotinylated primers that label the resulting amplicons with biotin and allow them to be captured on avidin-coated microtiter plates *(2)*. The second one is to hybridize the amplicons to a capture oligonucleotide *(3)*. Another format which has been recently described *(4)* is a kind of "ELISA run-off" with internal standard.

From: *Methods in Molecular Medicine, Vol 26: Quantitative PCR Protocols*
Edited by: B. Kochanowski and U. Reischl © Humana Press Inc., Totowa, NJ

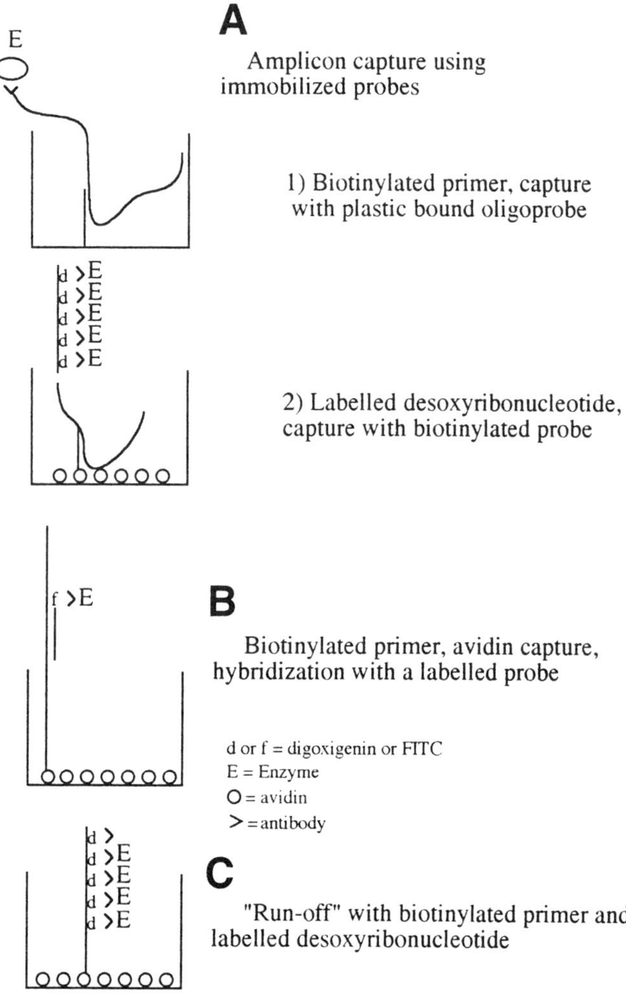



Fig. 1. The different ELISA formats. *See* text for explanation.

### 1.1.1. Format "Biotinylated Primer, Capture to Bound Avidin and Hybridization with a Probe" (*Fig. 1A*)

When using biotinylated primers to capture the amplicons, a hybridization step with a tagged oligo-probe is required because the PCR products cannot be labeled during the PCR. Otherwise nonspecific products would also be labeled, captured, and detected. After the amplicons are captured onto avidin coated plates, the next step is alkaline denaturation, which will leave a single strand of

DNA bound to the plate. Heat denaturation of the amplicon is possible but much less convenient (O. L., unpublished results).

The single-strand DNA is hybridized to a digoxigenin or FITC-labeled oligo-probe. Usually, the probe is tagged with only one label because tailing with Terminal transferase greatly increases the background (O. L., unpublished results).

The main problem in this format is the biotin-binding capacity of the wells; if not sufficient, there will be a competition between the nonincorporated biotinylated primers and the amplicons. Because the amplicons are much bigger, they may not be captured as efficiently as the primers. Some authors have suggested using avidin-coated microbeads *(5)* or avidin-bound to specially treated plastic in order to increase the binding capacity. In our experience, however, the usual high-quality plastic from Nunc or Dynatech allows enough avidin binding to capture 1 pmol of biotin/well *(2)*. In this format, a potential problem can be secondary structures or breaks of single-strand DNA, which may prevent probe hybridization.

### 1.1.2. Format "In Cycle Labeling and Capture by a Bound Probe" *(Fig. 1B)*

The capture oligonucleotide can be either directly coupled to plastic or bound to avidin-coated plates through a biotin moiety. The amplicons can be tagged with several molecules of the chosen label during the amplification (using digoxigenin- or FITC-dUTP for instance). Biotinylated primers can also be used to label the amplicon. In this case, one has to use capture oligo-probes, which are directly coupled to plastic. These are available in some proprietary kits. The amount of labeled desoxynucleotide used during the PCR is limited by the somewhat lower PCR efficiency that most of them induce. Thus, only a few tagged nucleotides can be incorporated per amplicon. The main problem of the oligo-probe capture format is the competition between the probe and the second nonhybridized strand of the amplicons, which may decrease the sensitivity.

### 1.1.3. Format "ELISA-PCR Run Off" *(Fig. 1C)*

An internal standard is coamplified during the PCR and, at the end of the amplification, the PCR tube is divided in two aliquots in which a specific biotinylated primer for either the target or the standard is added together with digoxigenin-dUTP and amplified again for one cycle. After revelation in parallel with an external scale of known amount of purified biotinylated products either for the target or the standard, one can compute a ratio of unknown over standard in molecules/well despite potentially different efficiency in the detection assay of the standard or of the target. This assay is able to detect around

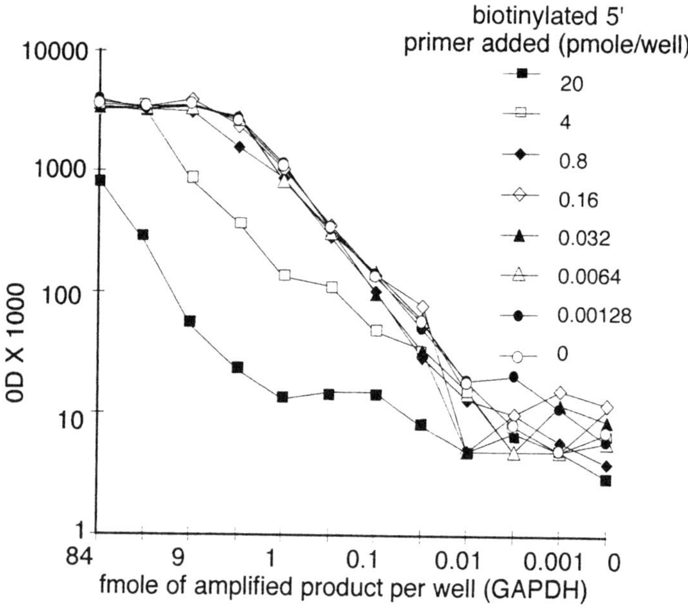


Fig. 2. Biotinylated primers do not significantly compete with biotinylated amplicons if their concentration is below 1 pmol/well. Purified amplified products (GAPDH) were avidin-captured with or without several concentrations of biotinylated 5' primers and then hybridized with homologous probes. Reprinted with permission from **ref. 2**.

0.1 fmol of amplicon/well. There is a quite cumbersome step after the PCR (the run-off reaction on duplicates of the PCR reaction), but no hybridization step during the ELISA.

## 1.2. Critical Parameters When Quantifying Amplicon Using ELISA

As shown in Chapter 5, sensitivity and dynamic range of the ELISA assay are the two key characteristics that will determine reliability in kinetic quantitative PCR or the number of reactions to be done in end-point PCR with internal standard. In the main ELISA formats previously described, the parameters that will determine the sensitivity and the dynamic range of the ELISA are the binding capacity of the avidin-coated microplates, the efficiency of hybridization, the sensitivity and the dynamic range of the revelation system used. The sensitivity and the dynamic range of the hybridizing step can be checked by hybridizing serial dilutions of known amounts of amplicons as shown in **Figs. 2** and **4**.

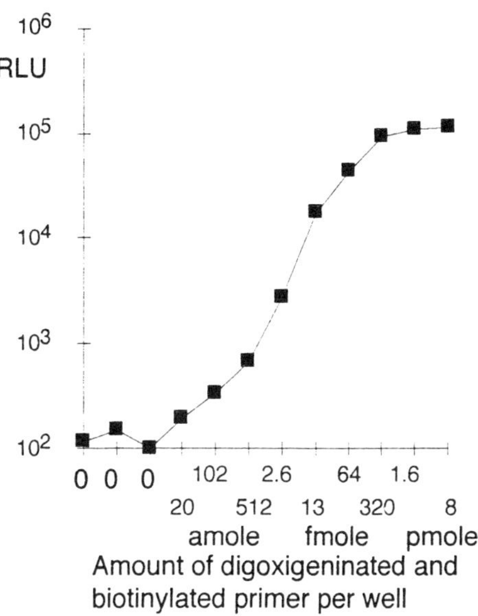


Fig. 3. Test of the biotin binding capacity of avidin-coated microtiter plates: serial dilutions of a digoxigeninated biotinylated primer were captured and revealed with anti-digoxigenin alkaline phosphatase conjugated antibody and CSPD substrate. Luminometry reading was done after 15 min.

### 1.2.1. Binding Capacity of Avidin-Coated Microplates

As shown in **Fig. 2**, unincorporated biotinylated primers can compete with the amplicons for binding to the microtiter plates. Therefore, one has to make sure that the avidin-coated microplates have a high enough biotin-capture capacity. The easiest way is to label a biotinylated primer with digoxigenin and to capture serial dilutions of this digoxigeninated biotinylated primer on the microtiter plate to be tested. The plateau should be reached at or above 1 pmol of digoxigeninated biotinylated oligonucleotide/well. An example of such experiment is displayed in **Fig. 3**, where it can be seen that above 320 fmol of oligo/well a plateau begins. In that case, one can reduce the amount of biotinylated primer in the PCR in order to not saturate the plates during the capture step, because primers are much smaller than the amplicons and could be captured preferentially.

### 1.2.2. Efficiency of Hybridization

To achieve specificity when detecting PCR products with ELISA, a hybridization step is usually performed. Compared to membrane hybridization, liq-

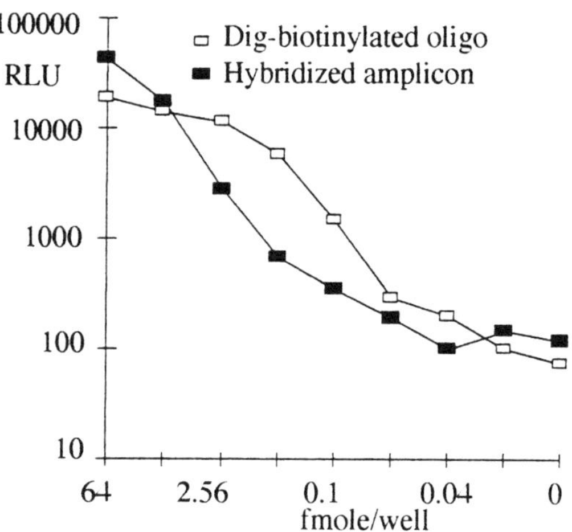



Fig. 4. Liquid hybridization is very efficient: serial dilutions of digoxigeninated biotinylated primer or of PCR products were captured onto avidin-coated microplates. The PCR products were then hybridized with a digoxigenin-labeled oligo probe (*see* **Fig. 7**). Both series were then revealed with antidigoxigenin alkaline phosphatase conjugated antibody and PNP substrate.

uid hybridization is much faster and goes to completion. A high sensitivity can then be achieved. As shown in **Fig. 4**, the hybridization step is very efficient because the curve obtained with a digoxigenin-labeled biotinylated oligonucleotide is almost identical to that obtained by hybridizing biotinylated PCR products with a digoxigenin-labeled oligo-probe. The plateau observed at high PCR product concentrations is related to the quite low amount (0.2 pmol) of probe in the wells.

In order to increase sensitivity, the background should be minimal. Plates should be saturated with BSA and herring sperm DNA can be used during the hybridization step. In these conditions, nonspecific binding can be reduced to almost nothing. **Figure 5** shows a comparison of the results obtained with plates coated with avidin using two different protocols. Protocol A gives the best sensitivity because the background is low.

## 1.2.3. Sensitivity and Dynamic Range of the Revelation Step

This part of the assay can be accomplished by using colorimetry, fluorometry or luminometry, depending on the substrate used. Fluorometry using alkaline phosphatase and methylbelliferon as substrate is not better than colorimetry (O. L., unpublished results). As shown in **Fig. 6**, luminometry displays the best sensitivity and the widest dynamic range, when compared with colorimetry.

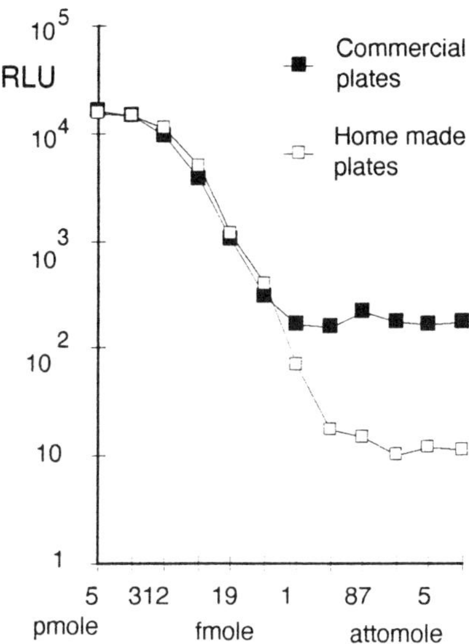

Fig. 5. Comparison of two procedures for coating microplates with avidin. Serial dilutions of a digoxigeninated biotinylated primer were captured in both kind of plates and revealed as above.

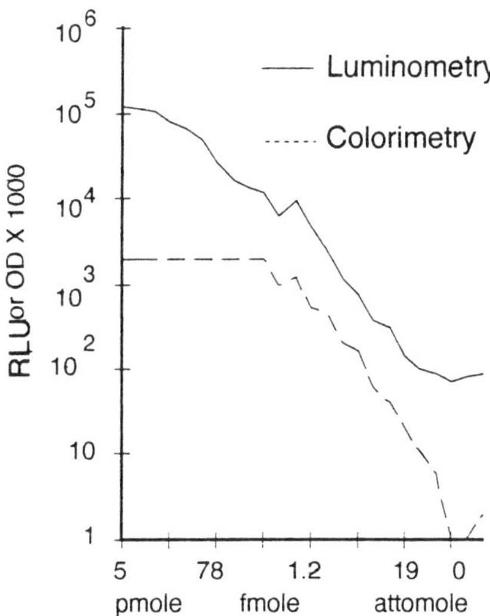

Fig. 6. Comparison of colorimetry vs luminometry for quantifying amplicons. Serial dilutions of a digoxigeninated and biotinylated primer were revealed with an anti-digoxigenin alkaline phosphatase conjugated antibody and either PNP or CSPD was added. Incubation time was respectively 15 min for CSPD and luminometry reading, and 24 h with PNP and OD reading.

One has to use alkaline phosphatase and CSPD as substrate because peroxidase with luminol displays a high background and a poor sensitivity (O. L., unpublished results).

## 1.3. Critical Assessment of the Different ELISA Formats

There is no direct comparison of the sensitivity of the different ELISA formats in the literature, especially between the two main formats "in cycle labeling and capture by a bound probe" (**Fig. 1A**) and "biotinylated primer, capture to bound avidin and hybridization with a probe" (**Fig. 1B**). In format 1B, it is below 0.1 fmol (19 pg of a 300 bp amplicons) of amplicons/well, whereas it is about 0.5 fmol in format 1A2 *(3)*. As previously stressed, luminometry gives the best sensitivity and the widest dynamic range. Using probes directly coupled to alkaline phosphatase increases the background (O. L., unpublished results).

ELISA for amplicon quantitation compares favorably with any other method of PCR product quantitation such as ethidium bromide staining or Southern blot of agarose or acrylamide gels, hot PCR followed by agarose gel and autoradiography, or dot blot. In this discussion, we do not give a detailed comparison to other methods such as electroluminescence or proximity fluorescence with acridium ester or rare earth metals. Although their sensitivity is of the same magnitude and they are homogenous phase methods, they require either special equipment or reagents.

## 2. Materials

1. High-binding capacity microtiter plate (Maxisorp, Nunc). For luminometry: luminite 2 (Dynatech; Chantilly, VA) or FluoroNunc Maxisorp (Nunc).
2. Avidin (Sigma, St. Louis, MO, cat. no. A9275).
3. Bovine serum albumin (BSA) containing neither alkalin phosphatase nor biotin (Sigma, cat. no. A6793).
4. 1 $M$ Carbonate buffer (per 1 L): 320 mL sodium carbonate (1 $M$) and 680 mL sodium bicarbonate (1 $M$) pH 9.6.
5. Phosphate-buffered saline (PBS) (per 1 L): 8.23 g $Na_2HPO_4$, 2.35 g $NaH_2PO_4$, 4 g NaCl.
6. TE (per 1 L): 10 m$M$ Tris HCl, pH 7.5, and 1 m$M$ EDTA.
7. 20X SSPE (per 1 L): 175.3 g NaCl, 35.88 g $NaH_2PO_4$, 7.444 g EDTA adjust pH to 7.4.
8. Buffer 1: PBS with 0.1% Tween-20.
9. Hybridization mix: 6X SSPE with 100 µg/mL herring sperm DNA (Euromedex, Souffelweyersheim, France), 0.5% sodium didecylsulfate (SDS), and labeled probe, 10 pmol/mL.
10. Buffer 2: TE, 0.1% Tween-20.

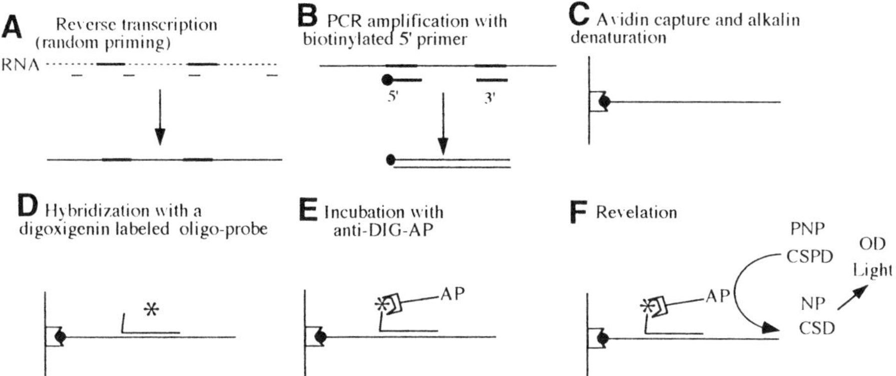

Fig. 7. Outline of the PCR-ELISA assay described in this chapter.

11. Buffer 3: PBS, NaCl 0.15 $M$, 0.1% Tween-20.
12. Buffer 4: PBS, 0.15 $M$ NaCl, 1% BSA.
13. Substrate buffer: 0.1 $M$ diethanolamine buffer, pH 9.8. Stock solution (1 $M$): 97 mL diethanolamine (Sigma, cat. no. D2286), 400 mg $MgCl_2$ (2 m$M$), 0.2 g sodium azide, $H_2O$ to 1 L.
14. 3' end oligonucleotide labeling kit (Boehringer Manheim, Indianapolis, IN, cat. no. 1362372) or FITC coupled oligonucleotide (ON CPG column, Clontech, Palo Alto, CA, cat. no. 5227).
15. Alkaline phosphatase conjugated anti-digoxigenin antibody (Boehringer Manheim, cat. no. 1093274) or anti-FITC (Boehringer Manheim, cat. no. 1426338).
16. PNP (Sigma, substrate 104–105 ) or CSPD (Tropix, Bedford, MA, cat. no. MC005).
17. Luminescence enhancer Saphire II (2 %) (Tropix, cat. no. LAX250).

# 3. Methods

## 3.1. ELISA

The PCR is carried out with a pair of primers, of which one is biotinylated. The PCR products are captured onto avidin-coated microplates and alkaline denaturated. A digoxigenin or FITC-labeled oligo-probe is hybridized to the captured DNA strand. The amount of probe is then estimated by addition of an alkalinephosphatase coupled anti-digoxigenin or -FITC antibody and of PNP or CSPD and measuring OD or luminescence. In this assay, there is no competition between the probe and the second strand of the DNA product and there is only one digoxigenin or FITC molecule for each PCR product. The assay we are using is outlined in **Fig. 7**.

### 3.1.1. Making Avidin-Coated Microtiterplate

Dilute avidin in carbonate buffer, pH 9.6, 0.1 *M* (0.1 mg/mL) and pipet 100 μL of the solution into each well of the microtiter plate. Incubate 2 h at 37°C. Recover the avidin solution, which can be used four times. Wash the well twice with the wash buffer 1. Block the plates with 1% BSA in carbonate buffer (300 μL) for 2 h at 37°C. Store the plate –20°C.

### 3.1.2. PCR

PCR is performed in 50 μL of amplification buffer (Perkin-Elmer) containing 1 μL of cDNA, 0.25 μ*M* of each primer (one primer/pair is biotinylated), 2 m*M* MgC1$_2$, 0.2 m*M* dNTP (Promega, Madison, WI) and 1.25 U of thermoactivable *Taq* DNA polymerase (Amplitaq Gold, Perkin-Elmer). Samples are overlayed with mineral oil and amplified in 96-well plates (Costar, Cambridge, MA). Amplification is performed in a PTC-100 thermal cycler from MJ Research (Watertown, MA). The thermal profile is initially 95°C for 8 min, followed by 42 cycles at 45 s, 60°C and 72°C both for 1 min (*see* **Note 1**).

### 3.1.3. Amplicon Capture and Alkaline Denaturation

Thaw the plates, and wash three times with PBS. Pipet 100 μL TE into the wells and add 5 μL of the PCR reaction. Incubate 1 h at 4°C. Add 100 μL/well 0.1 *N* NaOH, incubate for 10 min room temperature. Wash three times with wash buffer 2.

### 3.1.4. Liquid Hybridization

Add 100 μL per well of the hybridization mix. Incubate 2 h at 42°C. Wash twice with wash buffer 3 and once with buffer 4 (*see* **Notes 4** and **5**).

### 3.1.5. Revelation

Incubate for 1 h with 100 μL/well of anti-digoxigenin antibody coupled to alkaline phosphatase (diluted 1/7500 in buffer 4). Wash four times with buffer 2 and twice with substrate buffer. Incubate for 1–24 h at room temperature with substrate PNP (1 mg/mL in diethanolamin buffer) and read optical density (OD) at 405 nm (*see* **Notes 2, 3,** and **6**)

## 3.2. PCR Quantitation

### 3.2.1. ELISA Used in Kinetic Quantitative PCR

Sample an aliquot of every PCR reaction every three to four cycles beginning around cycle 20 and measure the OD of each sample.

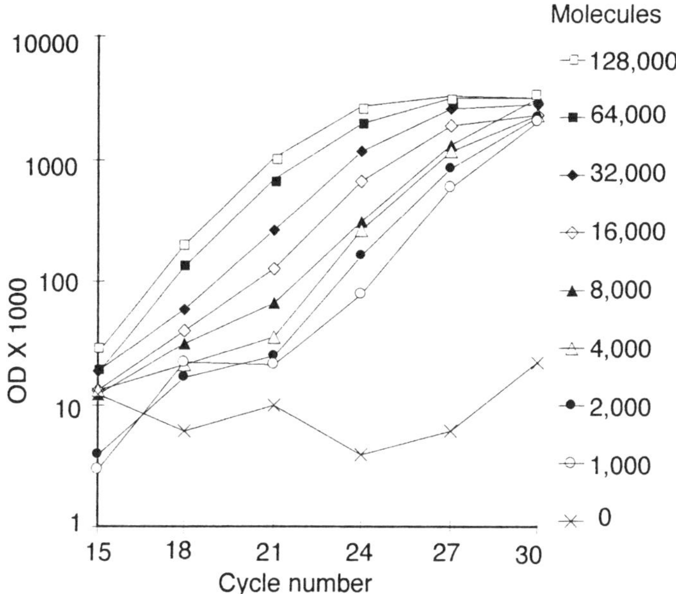


Fig. 8. Example of kinetic quantitative PCR-ELISA: Serial dilutions of purified and quantified PCR products were amplified and every reaction was sampled at the indicated number of cycles and the amount of amplicons measured by ELISA.

A calibration curve is obtained by the amplification of serial dilutions of known amount of PCR products as shown in **Fig. 8**. Use this curve as an external scale and compare curves obtained with unknown samples amplified in the same experiments (*see* Chapter 5 for the calculations).

### 3.2.2. ELISA in End-Point Quantitative PCR

In end-point quantitative PCR, it is useful to use an internal standard. To be suitable for ELISA, it should have an homologous sequence to the segment to be amplified and it should be hybridized with a different probe. The construction of such a standard is displayed in **Fig. 9**. The standard is identical to the target except for the 18 nt sequence used for hybridizing either the cDNA or the standard probe (for a detailed protocol of a similar standard, *see* **ref. 6** and **Notes 7** and **8**).

## 4. Notes

1. Always keep the PCR to check products on an agarose gel if needed.
2. To test whether there are problems with the avidinated plate binding or the revelation step use serial dilution of a 5'-biotinylated and 3' digoxigeninated oligo-

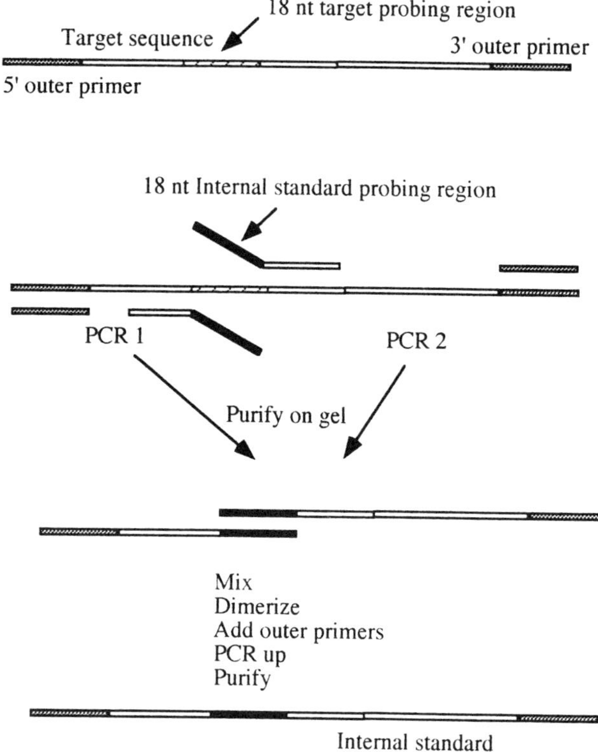



Fig. 9. Construction of internal standard suitable for ELISA-PCR. Together with the two primers used for amplifying the cDNA, two overlapping primers corresponding to the probe sequence are used in two parallel PCR reactions. The two products are purified, mixed, allowed to dimerize by doing a few PCR cycles and then the outer primers are added and a few PCR cycles are again carried out. The products are gel purified, quantified and aliquoted. The standard is identical to the target except for the 18 nt sequence used for hybridizing either the cDNA or the standard probe.

nucleotide (from 10 pmol/well to 1 amol/well). Do an early reading of OD to estimated the binding capacity of the plate (it should not saturate at a concentration below 1 pmol). Do a late reading to verify the sensitivity of the revelation step; it should be below 0.1 fmol/well.

3. If there are problems with the probe or the hybridizing step: pool PCR products, quantify on ethidium bromide-stained agarose gel, make serial dilutions from 1 pmol/well to 10 amol/well and transfer to avidin-coated microplates. Use replicates and hybridize with several batches of the relevant probe.

4. If there have been a problem with a given batch of probe, one can wash the plates, redenaturate the PCR products with sodium hydroxide and rehybridize with a new batch of probe.

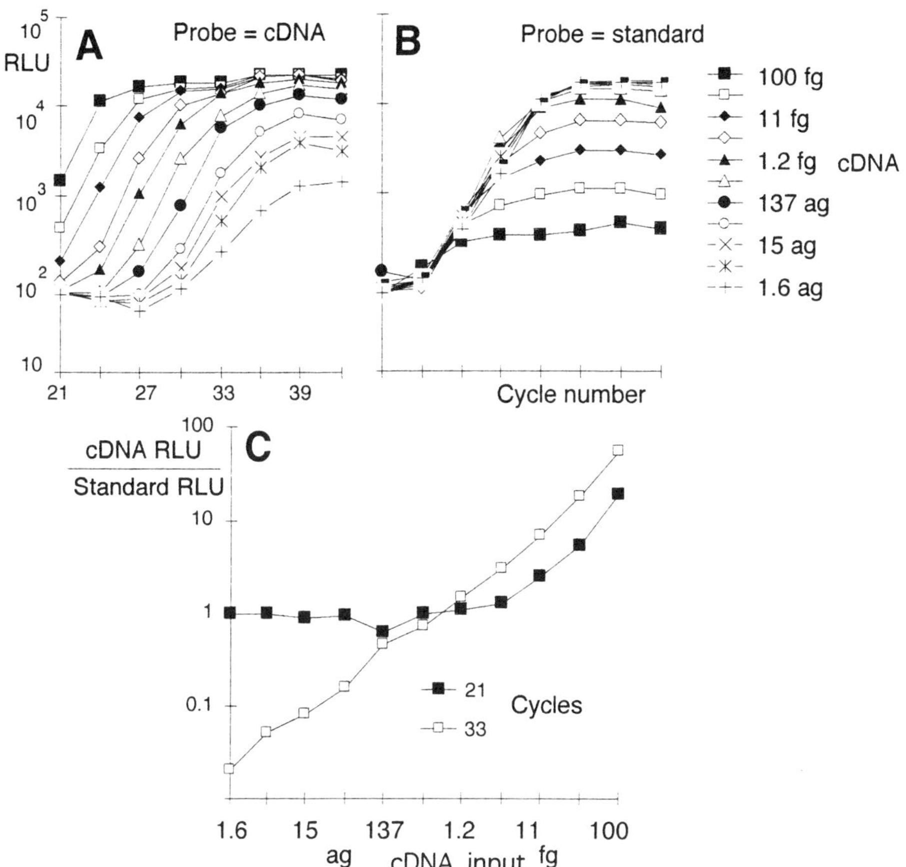



Fig. 10. ELISA-PCR with internal standard. Serial dilutions of a GAPDH PCR products were amplified together with 1 fg of an homologous internal standard. Every reaction was sampled at the indicated number of cycles and hybridized either with a cDNA **(A)** or an internal standard probe **(B)**. The ration of both signals is plotted in C for cycle 33 showing the linear relationship between the cDNA input and this ratio over four orders of magnitude.

5. In most cases, the problem comes from the probe. For some as yet unknown reason, certain probes have problems to be stored once labeled with digoxigenin.
6. When using luminometry, white or black plates should be used and the substrate is CSPD (250–30 μ*M*) with the 2% enhancer. A microtiter plate luminometer is also required.
7. The results obtained using an internal standard are shown in **Fig. 10**. Serial dilutions of known amount of a purified GAPDH PCR products ranging from 100 fg to 1.6 ag/reaction have been amplified together with 1 fg of the internal standard.

At serial number of cycles, 5 μL of every reactions have been sampled and divided in two parts to be captured onto two avidin-coated plates and hybridized either with the cDNA or the internal standard probe. Panel A displays the results obtained with the cDNA probe and panel B those with the internal standard. At high cDNA input, the amplification of the internal standard is inhibited, and the opposite is true at low cDNA concentrations where the internal standard inhibits the amplification of the cDNA. The ratio of the signal obtained with the cDNA over the internal standard at cycle 33 is shown in panel C, displaying linear relationship between the input and the ratio of the signals cDNA/ internal standard: these results demonstrate that, using an external scale and an internal standard, true quantitation is possible in end-point quantitative PCR over four orders of magnitude using ELISA and luminometry. One cannot use straight colorimetry because of the too low dynamic range of this method (data not shown). Indeed, when using colorimetry as done in the Monitor kits, several dilutions of the PCR reactions must be assayed: four for the cDNA and two for the standard.

8. There are several formats for assaying the amount of PCR products by ELISA. Luminometry can be used in all of them and improves the assay reliability especially when using internal standards. ELISA is a nonhomogenous phase assay and is still somewhat labor intensive compared with the homogenous phase assays. The Perkin ELMER 7700 allows homogenous phase detection of the amount of amplicons during the PCR itself. It remains to be seen, if procedures with two colors (one for the cDNA and one for the standard) will be easily implemented.

## References


1. Martin, C. S., Butler, L., and Bronstein, I. (1995) Quantitation of PCR products with chemiluminescence. *Biotechniques* **18,** 908–913.
2. Alard, P., Lantz, O., Sebagh, M., Calvo, C. F., Weill, D., Chavanel, G., Senik, A., and Charpentier, B. (1993) A versatile ELISA-PCR assay for mRNA quantitation from a few cells. *Biotechniques* **15,** 730–737.
3. Janezik, A., Semper, A., Holloway, J., and Holgate, S. (1995) Detection of cytokine mRNA expression by a sensitive RT-PCR ELISA detection system. *Biochemica* 30–32.
4. Zou, W., Durand-Gasselin, I., Dulioust, A., Maillot, M. C., Galanaud, P., and Emilie, D. (1995) Quantification of cytokine gene expression by competitive PCR using colorimetric assay. *Eur. Cytokine Network* **6,** 257–264.
5. Holodny, M., Katzenstein, D. A., Sengupta, S., Wang, A. M., Casipit, C., Schwartz, D. H., Konrad, M., Groves, E., and T. Merigan. (1991) Detection and quantification of human immunodeficiency RNA in patient serum by use of the polymerase chain reaction. *J. Infect. Diseases* **163,** 862–866.
6. Taoufik, Y., Froger, D., Benoliel, S., Dussaix, E., Delfraissy, J. F., and Lantz, O. (1998) Quantitative ELISA-PCR at saturation using homologous internal DNA standards and chemiluminescence revelation. *Eur. Cytokine Network* **9,** 197–204.

# 11

## Competitive PCR Quantitation Utilizing a Microtiter Plate Based Format for the Detection of PCR Products

Bernd Kochanowski and Wolfgang Jilg

## 1. Introduction

In microbiology, the polymerase chain reaction (PCR) has become an important tool for the analysis of clinical samples. For example, it led to the detection of Hepatitis-B Virus DNA (HBV-DNA) in patients with serological patterns not previously associated with active infection *(1)*. PCR-based assays are also more sensitive than conventional tests for the detection of many other infective agents, like Hepatitis-C Virus (HCV), herpes viruses, or Legionella pneumophila *(2)*. In some cases, however, the detection of the nucleic-acid sequences of a certain infectious agent is not very informative. Here is a quantitative information of the amount of nucleic acid needed. DNA quantitation is useful for monitoring the viral load during antiviral therapy *(3)*, in estimating the degree of infectiousity of individuals *(4,5)*, and (e.g., for herpes viruses) to differentiate between latent and active infection. Owing to its sensitivity, quantitative PCR is the most sensitive technique for the quantitation of nucleic acids and therefore has been investigated in detail.

Quantitation by PCR is often impaired by the amplification process itself, the presence of inhibitors in the DNA samples, the geometric position of the samples in the thermal cycler, and pipetting variations *(6)*. An increasing number of reports have been published that apply different techniques for PCR quantitation. To control the amplification process, a number of strategies like external, internal noncompetitive, and internal competitive quantitative PCR have been devised. The first is especially easy to perform and gives sufficient reliable results, but does not control for inhibitors in the individual DNA preparations. The latter two strategies monitor the different steps of PCR (although

From: *Methods in Molecular Medicine, Vol 26: Quantitative PCR Protocols*
Edited by: B. Kochanowski and U. Reischl © Humana Press Inc., Totowa, NJ

usually not the nucleic-acid preparation) but are technically more demanding. For a comparison of both strategies see Chapter 6.

The second step of a quantitative PCR test is the product quantitation. Many different techniques have been applied for the quantitation of PCR products. Products can be detected with a variety of techniques, i.e., on the gel or in microtiter plates, with or without hybridization, with the introduction of radioactive or nonradioactive labels, or without any labeling.

A more recent technique for the quantitation of PCR products is the so called enzyme-linked oligonucleotide sorbent assay (ELOSA). It allows the processing of large series of samples and is much faster than Southern-blot hybridization. A number of protocols of nonradioactive ELOSA techniques have been published so far. Variations concern the binding of the product or the probe to the microtiter plate *(7–9)*, the kind of detection label *(6,9,10)*, detection marker incorporation into the amplificon or the probe *(9–11)*, the length of the probe *(6)*, and detection with chemiluminescent, colorimetric, or fluorimetric procedures *(11–13)*.

There have been only a limited number of studies comparing different ELOSA protocols with respect to sensitivity and quantitative ability. Recently, we compared a long detection probe with a short one and colorimetric detection with chemiluminescence for PCR quantitation of HBV DNA. In summary we were able to show that the sensitivity of the long probe (nested PCR product, 250 bp) is better than that of an 30-mer-oligonucleotide. Moreover chemiluminescent detection showed no significant advantage compared to colorimetry. The latter results compare to those of Yank et al. *(13)*, demonstrating that detection with fluorencence is as sensitive as colorimetry. In Chapter 10, Lantz et al. on the other hand proposed that chemilumenescence has a greater sensitivity. Different results might be owing to different protocols and parameters such as amount of primers used.

Because the long probe comprises the greater part of the amplicon, it cannot be used for the development of a competitive assay. On the other hand, the short probe used for the detection of amplification products of the surface region of the HBV genome shows a sufficient sensitivity. We therefore present a protocol that combines competitive PCR with ELOSA technology. A competitor is used that differs from the wild-type target only by the probe-recognition site, whose 5'-3' orientation has been inverted. Competitor and wild-type product are specifically hybridized with individual probes. Owing to a similar distribution of nucleotides along the probe sequence, a similar hybridization efficiency of the probes to their respective targets can be assumed. By co-amplifying clinical samples with two or three different dilutions of the competitor, a precise quantitation of samples is possible. The ratio of the optical density (OD) measured in the colorimetric detection process with both products of the two reactions is plotted vs the initial amount of competitor and the point sought where the ratio is 1:1. The virtual amount of competitor at that point corresponds to the actual amount of wild-type target.

Owing to the ease of detection and the linear range of ELOSA of at least three magnitudes of order, we show a variation of the aforementioned protocol, which can be used for series of samples where the amount of specific nucleic acids does not differ by more than four orders of magnitude and which allows the use of one single competitor per sample.

## 2. Materials

1. All reagents used are of standard molecular biology grade.
2. Diethyl pyrocarbonate (DEPC)-treated water. Add 0.1% DEPC to water, incubate overnight at 37 C and autoclave.
3. Yeast-carrier tRNA (Boehringer Mannheim GmbH, Mannheim, Germany). Used as a carrier for the dilutions of the reference samples. Store as a 10 µg/µL solution at $-20°C$.
4. Aerosol resistant tips (ART; Molecular BioProducts, San Diego, CA).
5. Hinged 1.5-mL reaction tubes (Eppendorf, Hamburg, Germany) sterilized by autoclaving.
6. Thin-walled 0.5 mL GeneAmp reaction tubes (Perkin Elmer, Weiterstadt, Germany).
7. Primer Sag 2 (5'-CATCATCCATATAGCTGAAAGCCAAACA), the 5'-Biotinlabeled primer Sag1 (5'-bio-CTCGTGTTACAGGCGGGGTTTTTC) and the 5'-digoxigenin labeled probes (wild type, 5'-digACTGAGCC-AGGAGAAACGGACTCAGGCCCA; competitor, 5'-Dig-ACCCGGAC-TCAGGCAAAGAGGACCGAGTAC) have been purchased high-pressure liquid chromotography (HPLC)-purified from MWG Biotech (Ebersberg, Germany).
8. Mutagenesis primers Sag1, CS1 (5'-GAGTGACTCGGTTCTCTTTGCCT-GATCCGGGTTTACTAGTGCCATTTGTTCAG), CS2 (5'-TAAACCCGGA-GTCAGGCAAAGAGAACCGAGTCACTCCCATAGGAATTTCCG AA), and CS3 (5'-CCAATTATGTAGCCCATGAAG) were synthesized on a Millipore 8909 Expedite system (Millipore, Hamburg, Germany) and purified with ready-to-use gel-filtration chromatography columns (NAP 10; Pharmacia Biotech, Freiburg, Germany).
9. PCR reagents and *Taq* DNA polymerase are purchased from Perkin Elmer.
10. Low electroendo-osmosis (LE) agarose (Biozym GmbH, Hess, Oldenburg, Germany).
11. DNA molecular-weight marker VIII, (Boehringer Mannheim, Penzberg, Germany).
12. 10 mg/mL Ethidium bromide from Sigma (Deisenhofen, Germany).
13. 10X TBE solution: Dissolve 108 g Tris(hydroxymethyl)aminomethane and 55 g borate in 1 L of ddH$_2$O. Add ethylenediaminetetraacetic acid (EDTA), pH 8.0 to 0.5 *M*.

14. 20X SSC: Dissolve 175.3 g NaCl and 88.2 g Na-citrate in 1 L of ddH$_2$O. Adjust to pH 7.0 with 10 $N$ NaOH.
15. TE buffer: 10 m$M$ Tris-HCl, 1 m$M$ EDTA, pH 7.4.
16. Blocking solution: Add 1% bovine serum albumin (BSA) into TE buffer.
17. Alkaline solution: 0.5 $M$ NaOH, 1.5 $M$ NaCl.
18. Wash buffer: 0.1X SSC, 0.05% Tween-20.
19. TSE Buffer: 10 m$M$ Tris-HCl pH 7.6, 150 m$M$ NaCl, 1 m$M$ EDTA *(12)*.
20. Hybridization buffer: 1X TE, 1X SSC, 2X Denhardt's solution *(10)*.
21. Anti-digoxigenin/peroxidase (Boehringer Mannheim).
22. 2,2'-Azino-di-[3-ethylbenzthiazoline sulfonate (6)] (ABTS, Boehringer Mannheim) in ABTS buffer (Boehringer Mannheim). Prepare fresh prior to use (1 mg/mL).
23. 3 $M$ Na-acetate, pH 5.2.
24. Absolute ethanol: Store at 4°C.
25. 75% Ethanol: Store at 4°C.
26. TA cloning kit (Invitrogen BV, Leek, The Netherlands) (*see* **Note 7**).
27. Luria's broth (LB) agarose plates: Dissolve 10 g tryptone (Unipath Ltd, Basingstoke, UK), 5 g yeast extract (Unipath Ltd), 10 g NaCl, and 15 g agarose (Merck, Darmstadt, Germany) in 1 L of ddH$_2$O. Add X-Gal (40 µg/mL) and ampicillin (50 µg/mL).
28. LB medium: Dissolve 10 g tryptone, 5 g yeast extract, and 10 g NaCl in 1 L of ddH$_2$O, and add Ampicillin (50 µg/mL).
29. Lysis buffer: 20 m$M$ Tris-HCl, pH 8.0, 2 m$M$ EDTA, 1% Triton X-100.
30. Scalpel.
31. Ready-to-use gel-filtration chromatography columns (NAP 10; Pharmacia Biotech).
32. Qiaquick (Qiagen plasmid, Qiagen GmbH, Hilden, Germany), spin columns with silica gel matrix.
33. Qiagen plasmid purification kit (Qiagen GmbH) with an anion-exchange chromatography matrix.
34. Streptavidin-coated microtiter plates (MicroCoat, Penzberg, Germany) incubated with ELOSA blocking solution, overnight at 37°C.
35. Ultrafree MC filter units (Millipore, Bedford, MA) containing low binding membranes.
36. Spectral photometer (Kontron Instruments, Milan, Italy) to determine DNA concentrations.
37. Multichannel pipet and microtiter plate washer (optional).
38. Shaker at 37°C.
39. Standard agarose-gel electrophoresis equipment, including ultraviolet (UV)-transilluminator.
40. Photometer (SLT Instruments, Crailsheim, Germany).
41. Centrifuge (Heraeus Medifuge 15000, Heraeus Sepatech, Germany) for 1.5-mL reaction tubes.
42. A PCR thermocycler (for example, PTC 200, MJ Research, Watertown, MA).

# 3. Methods

## 3.1. Isolation of DNA

1. Isolate the DNA from 400 µL of serum with columns from the QIAamp blood kit used according to the instructions of the manufacturer (*see* **Note 1**).
2. Elute the DNA from the columns with 100 µL TE buffer (*see* **Note 2**).

## 3.2. PCR (16)

Perform the amplification using the GeneAmp kit (Perkin Elmer). The PCR mixture (50 µL) contains: 5 µL 10X PCR buffer (provided with the kit); 1 µL of each dNTP; 0.25 µL *Taq* DNA polymerase (1.25 U); 0.1 µ*M* of each primer (one of which is biotinylated); 5 µL of the DNA-preparation; and DEPC treated water up to 50 µL.

Thirty cycles were performed as follows: Denaturation at 94°C for 60 s, annealing at 55°C for 30 s, and extension at 72°C for 90 s. The extension after the last cycle was prolonged to 7 min. For procedures to avoid false positive PCR results see **ref.** *14*.

## 3.3. ELOSA Protocol

For ELOSA, PCR is performed with first-round PCR primers. One of the primers is 5'-labeled with biotin. The 5'-digoxigenin-labeled probe is complementary to a specific sequence of the 5'-biotin-labeled PCR strand (*see* **Fig. 1**).

1. Prior to usage block the streptavidin-coated microtiter plates overnight at 37°C with blocking solution.
2. Remove the blocking solution and let the wells drain in an upside-down position on paper towels. Blocked plates can be stored for 14 d at 4°C.
3. Add 10 µL of each PCR sample to 100 µL blocking solution into the wells of the microtiter plate to bind the PCR products via the biotinylated primer to the coated microtiter plate.
4. After incubation for 1 h at 37°C, empty the wells and add the alkaline solution for 15 min at 37°C to remove the nonbiotinylated strand.
5. Remove the alkaline solution and wash the wells twice with wash buffer and once with blocking solution by adding 200 µL of each solution into the wells.
6. Dilute either the wild-type or the competitor probe with the hybridization buffer to a concentration of 1 pmol/µL (*see* **Note 3**). Pipet 100 µL of the diluted probe into the wells and incubate for 2.5 h at 50°C to hybridize the probe to the target.
7. Wash the wells containing the hybrids twice with washing buffer and once with solution 1.
8. Dilute the Anti-digoxigenin/peroxidase 1:1000 in TSE, add 100 µL of the solution and let the antibody bind to the digoxigenin for 30 min at 37°C.
9. Then, for the last time wash thrice with the washing buffer. Do not use the blocking solution, because its use would lead to a high background.

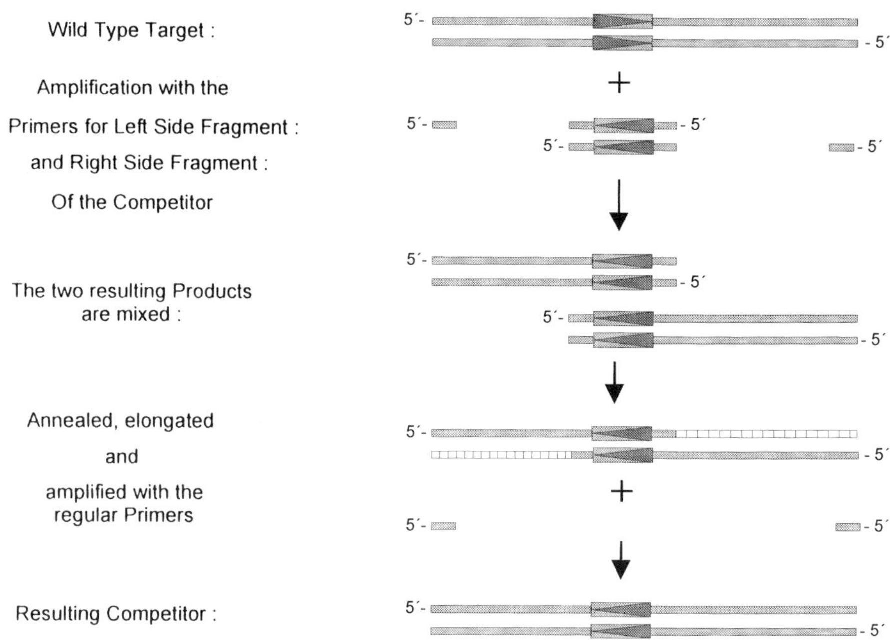


Fig. 1. Schematic flowchart of the construction of the competitor used for ELISA detection of PCR products.

10. Prepare a dilution of ABTS in ABTS buffer (1 mg/mL) and add 100 μL into the wells.
11. After 1 h of incubation at 37°C measure OD at 405 nm with a suitable microtiter plate photometer.

## 3.4. Preparation of the Competitor

A schematic flowchart of the construction of the competitor is shown in Fig. 1. The competitor differs from the wild-type target only by the probe recognition site, whose 5'–3' orientation is inverted. The wild-type target is amplified with two primer sets so that a left-side fragment incorporates the new probe sequence at the 3'-end and a right-side fragment incorporates it at its 5'-end, respectively. The two products are isolated, mixed, annealed, elongated, and amplified with the regular primer pair Sag1 and Sag2 (*see* **Note 4**).

1. To construct the left-side and the right-side fragments of the competitor, amplify your target of interest twice; once with the regular 5'-primer (Sag1 ) and the first mutagenesis primer (CS1, new probe recognition site incorporated), and once with the second mutagenesis primer (CS2, new probe recognition site incorporated) and the regular 3'-primer (CS3, *see* **Note 5**).

2. Combine the two different PCR samples and remove the primers with the aid of QIAquick spin columns as described by the manufacturer. Elute the column with 50 µL of TE buffer.
3. Amplify 5 µL of the eluate with the regular primers Sag1 and CS3. The left and right competitor fragments anneal and will be elongated to give the full-length competitor, which is subsequently amplified by the two primers.
4. Afterwards, separate the reaction mixture on a 1% agarose gel containing ethidium bromide. Cut the desired product out of the gel with a scalpel (*see* **Note 6**).
5. Cut the gel fragments with the same scalpel into very small pieces and put these into Ultrafree MC Filter Units and store at –20°C, over night. Centrifugate (3000*g*, 15 min, 4°C) and measure the volume by weighting the tube before the addition of gel fragments and after centrifugation and removal of the filter unit. Add adequate amounts of acetate buffer (1/10 of the vol) and ethanol (three times the vol) and let the nucleic acids precipitate for 8 h (or over night) at –80°C *(15)*. Centrifugate (3000*g*, 30 min, 4°C) and wash the pellet with cold 70% ethanol. After centrifugation, take the pellet to almost complete dryness in a safety cabinet.
6. Dissolve the pellet in an appropriate volume of TE buffer and check the amount of extracted DNA on 1% agarose gel.

## 3.5. Cloning into a TA-Vector

For subsequent cloning of the PCR product into a plasmid, we use the TA cloning kit (*see* **Note 7**).

Perform the cloning procedure according to the instructions of the manufacturer. Use about 50–200 ng competitor DNA for the ligation reaction. First white clones on LB agar plates were viable after 24 h. Positive clones were selected at best after 48 h of incubation.

## 3.6. Detection and Preparation of Recombinant Clones

1. Pick a small part of each white colony and suspend it into 50 µL of lysis buffer in 500 µL PCR reaction tubes. Subculture the remainder of the selected white colonies on LB agar plates.
2. Heat the tubes containing the bacterial suspension in lysis buffer for 5 min at 95°C.
3. Centrifugate at 3000*g* for 1 min.
4. Amplify 5 µL of the supernatants by PCR using primers Sag1 and Sag2.
5. Rise PCR positive colonies in 100 mL LB medium (containing 50 µL/mL of ampicillin) and isolate the plasmid with the Qiagen plasmid medi Kit, according to the instructions of the manufacturer.
6. Quantitate the DNA concentration of the plasmid preparation by measuring absorbance at 260 nm. Amount of plasmid (molecules/µL) = OD × Dil × 1.5 × $10^{14}$ (OD is the absorbance, Dil the dilution of the plasmid for measurement).
7. Prepare a 10-fold dilution series of the plasmid DNA in TE buffer.

### *3.6. Competitive PCR (see Note 8)*

Coamplify the DNA from clinical samples with three different dilutions of the competitor and quantitate the products as previously outlined (*see* **Note 9** and **10**). In the assay described herein we used $10^3$, $10^{4,}$ and $10^5$ competitor molecules/sample.

1. Add to each PCR reaction mixture 5 µL of the competitor, reduce the amount of added water in the master mix by 5 µL, and amplify both the wild-type target and the competitor simultaneously with the regular set of primers.
2. Quantitate the products of wild-type target and competitor separately.
3. Plot the OD of both products separately against each competitor dilution (**Fig. 2**).
4. The point of intersection of both graphs represents the number of wild-type molecules present in the investigated sample.

## 4. Notes

1. For a higher sensitivity, isolate DNA from 400 µL serum. In this case, also use twice as much volume from buffer A1 and A2 of the Qiagen kit.
2. Nucleic acids can be eluted from columns with a variety of buffers. Elution with water should be avoided because nucleic acids possess an autocatalytic activity that might lead to degradation. This particular activity is not observed at the mild alkaline condition of the TE buffer.
3. The actual amount of probes used in an assay has to be established individually. Depending of the affinity of the probes to the target molecules the concentration needed might differ. The amount of target nucleic acid usually present in the samples could be another reason why a different concentration of probes might be useful. Therefore different concentrations of the probe should be evaluated.
4. The 5'-end of primers CS1 and CS2 are complementary to each other so that the 3'-ends of the opposite strands of the resulting two amplicons can anneal and prime an elongation by the Taq DNA polymerase used in the consecutive PCR (in **Subheading 3.4., step 3**). This leads to the desired full-length competitor sequence.
5. For the construction of the right-side fragment, we did not use the regular primer Sag2, but a primer further downstream. The resulting amplicon with Sag2 would have been too short for isolation by electrophoresis on a 1% agarose gel.
6. Visualize the product under UV only for a short time. UV radiation can cause damage to nucleic acids and can therefore can lead to problems in the cloning procedure or induce sequence alterations.
7. TA cloning kits are provided by a variety of suppliers. A prerequisite for TA cloning is the usage of a polymerase producing single A extensions at the 3'-end of the respective amplicon. Polymerases possessing a 3'–5' exonuclease activity can not be used, for TA cloning. If such a polymerase has been used an alternative cloning strategy should be applied.

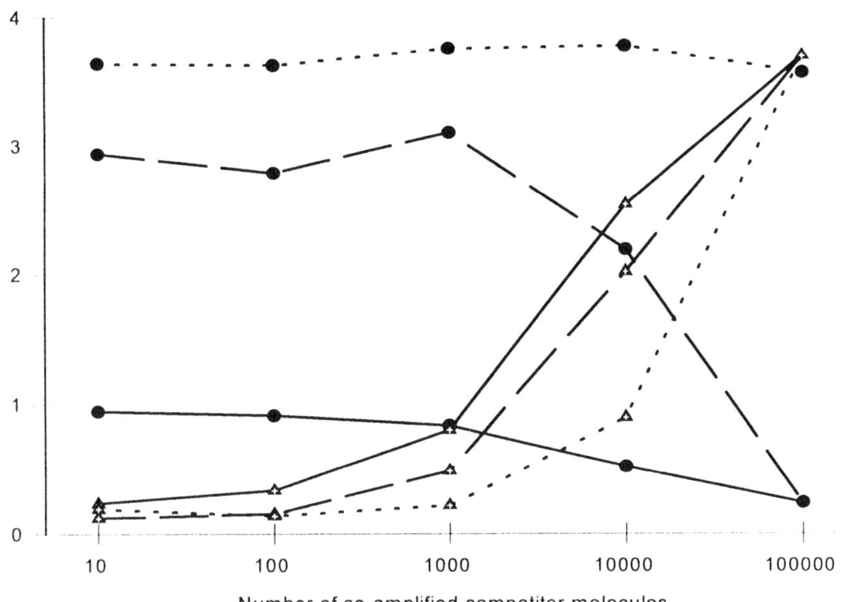


Fig. 2. Constant amounts of wild-type plasmid were coamplified with dilution series of competitor. Full line shows experiments with $10^3$; broken line, $10^4$; dotted line, $10^5$ molecules of wild-type target. Wild-type products are represented by circles and competitor products by triangles.

8. In certain cases, it is possible to work with one single competitor dilution for all samples. Because a calibration curve has to be set up in each PCR run, a single competitor dilution is only useful if a lot of samples are to be quantitated in the same PCR run. This curve is defined by competitive amplification of a dilution series of the wild-type plasmid with the dilution of the competitor, which is also used for the samples. The ratio of the OD of wild-type to competitor are blotted vs the initial amount of wild-type plasmid. The amount of initial number of wild-type at the ratio of both products for the individual samples corresponds to the amount of nucleic acid present in the samples.

9. Prior to the usage of the competitor, an equal amplification efficiency of wild-type target and competitor have to be demonstrated. For this system (as shown in **Fig. 3**) $10^3$–$10^5$ molecules of plasmids containing the two target species were amplified for 25–35 cycles, the products separated on agarose gel and their amount measured by densitometry. The depicted data support a similar amplification efficiency for both targets.

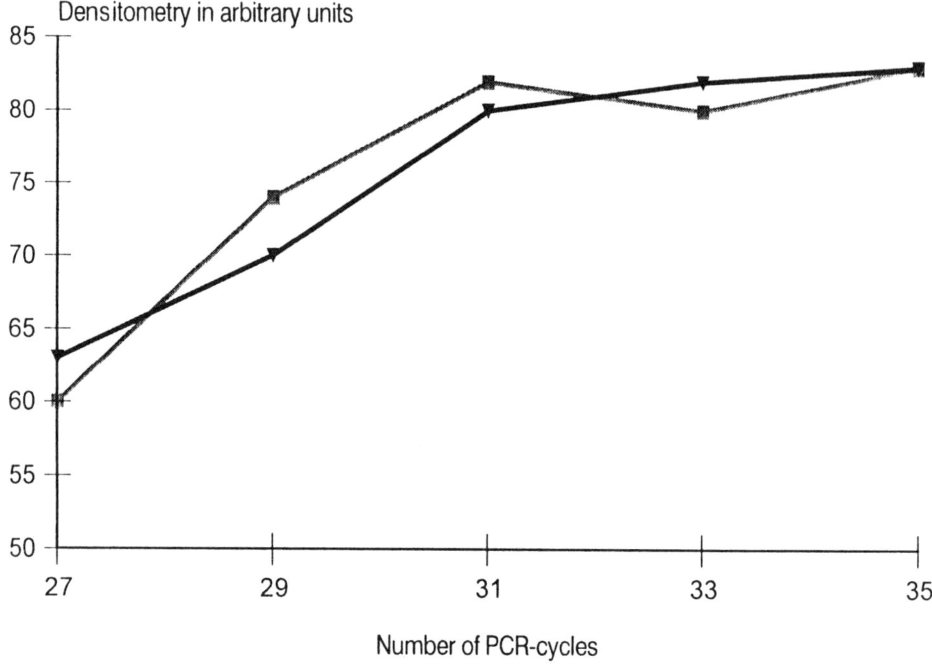


Fig. 3. Exponential phase of the PCR of competitor and wild-type target. $10^4$ molecules of both targets were coamplified for a varying number of cycles and the products quantitated densitometrically on an ethidium bromide stained agarose gel.

10. Of major importance is the magnitude of order of the estimated amount of wild-type target present in the samples. **Figure 4** shows the analysis of two different sera. Owing to primary results by limited dilution, we knew that the samples of interest contained between $10^2$ and $10^6$ HBV molecules/sample. Constant amounts of each serum were coamplified with three different dilutions of the competitor ($10^3$, $10^4$, and $10^5$ molecules/samples). The OD of both products is blotted vs the initial amount of competitor. Using suitable software (e.g., TableCurve, Jandel Scientific) the point can be sought were the ratio between both products would have been equal.

## References

1. Sumazaki, R., Motz, M., Wolf, H., Heinig. J., Jilg, W., and Deinhardt, F. (1989). Detection of hepatitis B virus in serum using amplification of viral DNA by means of the polymerase chain reaction. *J. Med. Virol.* **27,** 304–308.
2. Ng, D. L., Koh, B. B., Tay, L., and Heng, B. H. (1997) Comparison of polymerase chain reaction and conventional culture for the detection of legionellae in cooling tower waters in Singapore. *Lett. Appl. Microbiol.* **24(3),** 214–216.

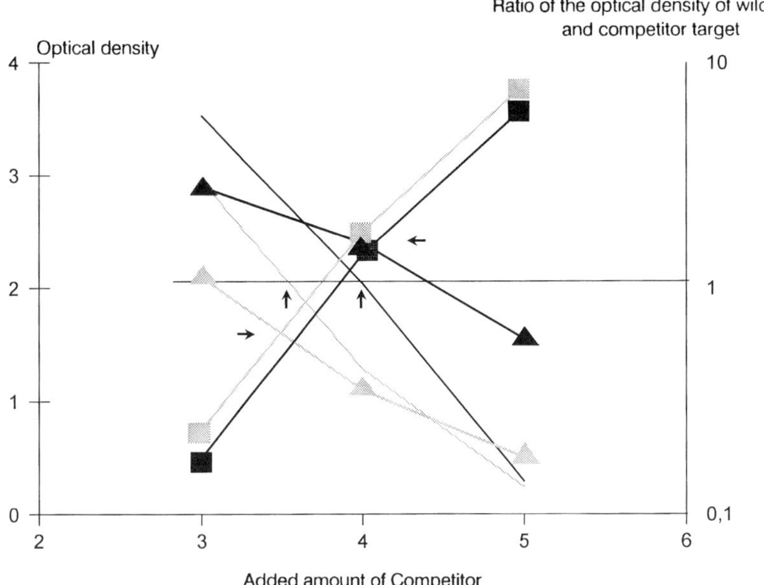


Fig. 4. Results of the competitive HBV-PCR of two patients sera. Constant amounts of DNA from the sera were amplified with $10^3$, $10^4$, and $10^5$ molecules of competitor. The OD of wild-type and competitor products are indicated by triangles and squares, respectively (*see* left y-axis). The thin lines represent the ratio between both products for each serum (*see* right y-axis). Both sera are represented by different shades of gray. The arrows indicate the point of equal amounts of both targets.

3. Toyoda, M., Carlos, J. B., Galera, O. A., et. al. (1997) Correlation of cytomegalovirus DNA levels with responds to antiviral therapy in cardiac and renal allograft recipients. *Transplantation* **63(7),** 957–963.
4. Uchida, T., Shimojima, M., Gotoh, K., Shikata, T., Tanaka, and Kiyosawa, B. (1994) "Silent" hepatitis B virus mutants are responsible for non-A, non-B, non-C, non-D, non-E hepatitis. *Microbiol. Immunol.* **38(4),** 281–284.
5. Jilg, W., Sieger, E., Zachoval, R., and Schätzl, H. (1995) Individuals with antibodies against hepatitis B core antigen as the only serological marker for hepatitis B infection: high percentage of carriers of hepatitis B and C virus. *J. Hepatol.* **23,** 14–20.
6. Reischl, U. and Kochanowski, B. (1995) Quantitative PCR: a survey of the present technology. *Mol. Biotechnol.* **3,** 55–71.
7. Kohsaka, Tanagushi, A., Richman, D. D., and Carson, D. A. (1993) Microtiter format gene quantification by covalent capture of competitive PCR products: application to HIV-1 detection. *Nucleic Res.* **21,** 3469–3472.
8. Hataya, T., Inoue, A. K., and Shikata, E. (1994) A PCR-microplate hybridisation method for plant virus detection. *J. Virol Methods* **46,** 223–236.

9. Maza, C., Mantero, G., and Primi, D. (1991). DNA enzyme immunoassay: a rapid and convenient colorimetric method for diagnosis of cystic fibrosis. *Mol. Cell. Probes* **5,** 459–466.

10. Iitä, A., Dahlen, P., Nunn, M., Mukkala, V.-M, and Siitari, H. (1992). Detection of amplified HTLV-I/-II viral sequences using time-resolved fluorometry. *Anal. Biochem.* **202,** 76–81.

11. Suzuki, K., Okanomoto, N., Watanabe, S., and Kano, T. (1992) Chemiluminescent microtiter method for detecting PCR amplified HIV-1 DNA. *J. Virol. Methods* **38,** 113–122.

12. Lehtovaara, P., Uusi-Oukari, M., Buchert, P., Lakksonen, M., Bengtstrom, M., Ranki, M. (1993) Quantitative PCR for hepatitis B virus with colorimetric detection. *PCR Methods Appl.* **3(3),** 169–175.

13. Yang, B., Viscidi, R., and Yolken, R. (1993) Quantitative measurement of nonisotopically labeled polymerase chain reaction products. *Anal. Biochem.* **213,** 422–425.

14. Kwok, S. and Higushi, R. (1989) Avoiding false positives with PCR. *Nature* **339,** 237–238.

15. Tautz, D. and Renz, M. (1983) An optimized freeze squeeze method for the recovery of DNA fragments from agarose gels. *Anal. Biochem.* **132,** 14–19.

# 12

# Competitive and Differential RT-PCR (CD-RT-PCR) for Measurement of Normalized Gene Expression Using Antisense Competitors


**Eric de Kant**


## 1. Introduction

### 1.1. Gene Expression Analysis by PCR

Quantitative mRNA characterization by reverse transcription (RT) of RNA and subsequent polymerase chain reaction (PCR) (RT-PCR) is, compared to qualitative RT-PCR detection of RNA, more complicated because of two features inherent in in vitro amplification. First, during the exponential phase, minute differences in a number of variables can greatly influence reaction rates, with substantial effect on the yield of PCR products. Second, as a consequence of reaction components consumption and generation of inhibitors, the amplification enters a plateau phase. At this point, the reaction rate declines to an unknown level. Another source of errors in quantitative RT-PCR analysis lies in the determination of the amount of RNA to be analyzed for each sample. In small samples, the total amount of RNA may even be beyond the limit of detection. The sample loading problem can be solved by presenting the level of expression of the gene of interest in reference to a constitutively expressed gene. In a PCR, this can be done by the simultaneous amplification of two different genes in one reaction vessel, which was called differential PCR *(1)*. However, in many cases, a quantitative PCR assay is desired that is internally controlled both for errors in comparison between samples and for the efficiency of the amplification reaction. To that end, a technique was devised that combines competitive PCR and differential RT-PCR by coamplification of two genes and their corresponding competitive templates *(2)*. This chapter describes the working procedures for the complete assay called competitive and differential RT-PCR (CD-RT-PCR) and concomitant techniques.


From: *Methods in Molecular Medicine, Vol 26: Quantitative PCR Protocols*
Edited by B. Kochanowski and U. Reischl © Humana Press Inc., Totowa, NJ

## *1.2. Synthesis of Mutant Antisense Competitive DNA Fragments*

CD-RT-PCR requires site-specific mutagenized single-stranded DNA fragments as competitors *(2)*. These competitive DNA fragments can easily be generated without subcloning (**Fig. 1**). For that purpose, two successive rounds of PCR are performed for both genes. In the first, mutagenic primers are used in two separate PCRs to create or remove unique restriction sites. The resulting two mutant fragments, which are homologous around the mutated sequence, are recombined by overlap extension in the second round of PCR *(3)*. The biotinylated sense strands can be physically separated from the required antisense strands by alkali denaturation following immobilization of the biotinylated PCR fragments that are directly bound to streptavidin-coupled polystyrene magnetic beads.

## *1.3. Outline of the Complete CD-RT-PCR Assay*

A schematic outline of the concept of CD-RT-PCR is shown in **Fig. 2**. CD-RT-PCR comprises random-primed RT followed by coamplification of first-strand cDNAs from both the gene of interest and the reference gene, and their matching artificially mutated antisense competitive DNAs. The competitive templates are mutagenized in a way that either the competitor DNA or the cDNA sequence contains a unique restriction site exactly in the center of the segment that is amplified during CD-RT-PCR. The high homology between wild-type and mutant sequences promotes perfect competition not only during PCR but also during an essential final reannealing step. Note that only mutant antisense (not double-stranded) DNA acts as perfect competitor for first-strand cDNAs *(2)*. After CD-RT-PCR, restriction enzyme digestion enables visualization of digestible homoduplexes and nondigestible homo- and heteroduplexes. Separation of the digested DNA by gel electrophoresis or high performance liquid chromotography (HPLC) results in a pattern of four different fragment sizes. HPLC analysis provides the quantitative information concerning the ratios of digested and undigested amplification products that is required for calculation of the amounts of the two specific cDNAs prior to PCR and finally the normalized expression level of the gene of interest.

## *1.4. Conclusion*

CD-RT-PCR combined with discrimination of amplified cDNAs and competitor DNAs via restriction enzyme digestion and HPLC analysis provides a sensitive assay for reproducible and accurate measurement of normalized gene expression without the need for extensive competitor titrations for every single sample.

## 2. Materials

1. Cells or tissue from which RNA will be extracted.
2. RNA purification kit: RNAzol (Biotecx Laboratories, Houston, TX).

**competitor synthesis**

| | | see |
|---|---|---|
| | site-directed mutagenesis | 3.3.1.(1-4) |
| | recombination by overlap extension | 3.3.1.(5-6) |
| | amplification and purification | 3.3.1.(7-9) |
| NaOH | immobilization and alkali denaturation | 3.3.2.(1-7) |
| | magnetic separation | 3.3.2.(8,9) |
| NaOH | solid-phase synthesis and alkali denaturation | 3.3.2.(10-16) |

= mutagenic primers; = gene specific primers
sense anti-sense  sense anti-sense

Fig. 1. A schematic outline of antisense competitor DNA synthesis.

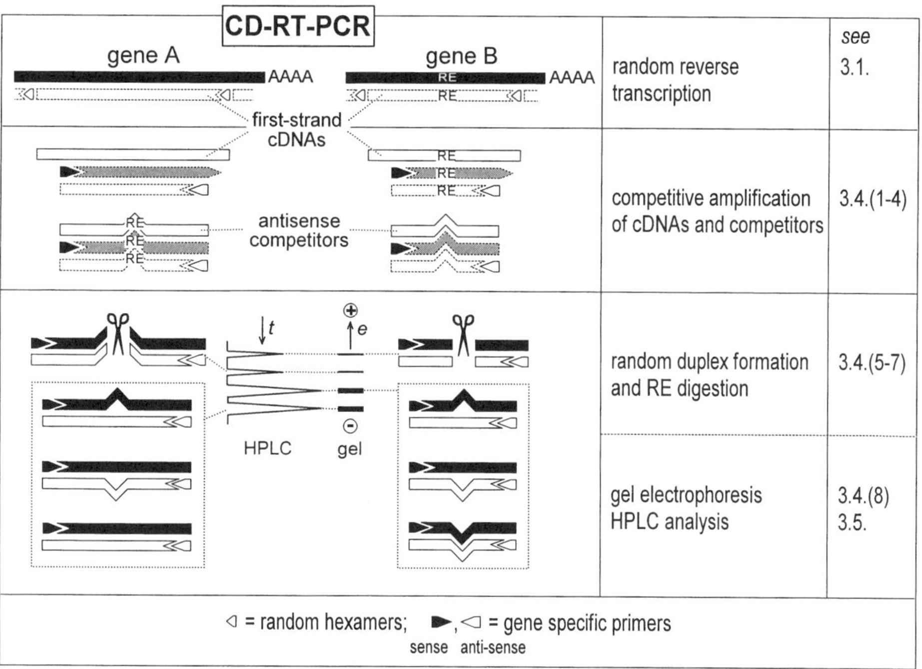


Fig. 2. A schematic outline of CD-RT-PCR. An example is shown with restriction enzyme (RE) sites in the competitor of gene A and in the wild-type cDNA sequence of gene B.

3. SuperScript RNase H- Reverse Transcriptase (200 U/μL; Life Technologies, Gaithersburg, MD) with manufacturer-recommended buffer and DTT (0.1 $M$).
4. rRNasin® ribonuclease inhibitor (40 U/μL; Promega, Madison, WI).
5. Random Primer oligonucleotides (hexamers; Life Technologies, Gaithersburg, MD).
6. A mixture of 2.5 m$M$ of each deoxynucleotide (4dNTP mix: dATP, dGTP, dCTP, and dTTP).
7. 10X PCR buffer: 100 m$M$ Tris-HCl, pH 8.3, 500 m$M$ KCl, 15 m$M$ MgCl$_2$, 0.01% (w/v) gelatin.
8. AmpliTaq® DNA Polymerase (5 U/μL; Perkin-Elmer).
9. Two sets of gene-specific primers (10 μ$M$): A nonmodified and a biotinylated primer complementary to the 5' end (sense primer) and a nonmodified primer complementary to the 3' end (antisense primer) of the sequence to be amplified during CD-RT-PCR; Mutagenic (sense and antisense) primers for preparation of a mutant fragment with a newly created or removed restriction enzyme recognition site. Mutagenic primers are complementary to the central portion of the CD-RT-PCR fragments. (*see* **Note 1** and **Fig. 1**).
10. Light mineral oil.
11. An automated thermal cycler (we used a Perkin-Elmer machine).
12. GeneClean (Bio101, La Jolla, CA) or any other method for purifying DNA from agarose gels.
13. Streptavidin-coated magnetic beads: Dynabeads(R) M-280 Streptavidin, and a Magnetic Particle Concentrator: Dynal MPC(R) for eppendorf tubes (Dynal, Oslo, Norway).
14. 2X Binding and Washing buffer (B&W): 10 m$M$ Tris-HCl, pH 7.5, 1 m$M$ EDTA, 2 $M$ NaCl.
15. TE buffer: 10 m$M$ Tris-HCl, pH 8.0, 1 m$M$ EDTA.
16. 0.1 $N$ NaOH for alkali denaturation, 1 $N$ HCl for neutralization, and 1 $M$ Tris-HCl, pH 8.0 for buffering.
17. Siliconized tubes.
18. Carrier DNA (10 ng/μL): Sonicated Salmon Sperm DNA (Stratagene, La Jolla, CA)
19. Antisense competitor DNA (*see* **Subheadings 3.3.** and **3.4.**).
20. Appropriate restriction enzyme for digestion of CD-RT-PCR products (*see* **Note 1**).
21. Potassium glutamate buffer (6X KGB): 600 m$M$ potassium glutamate, 150 m$M$ Tris-acetate, pH 7.5, 60 m$M$ magnesium acetate, 300 μg/mL bovine serum albumin (Fraction V; Sigma), 3 m$M$ mercaptoethanol.
22. Materials for gel electrophoresis and staining of DNA *(4,5)*: Electrophoresis buffer; 6x nondenaturing gel-loading buffer and 95% Formamide-gel-loading buffer; EDTA/SDS solution: 10 mM EDTA, 0.1% SDS; DNA mol wt marker; DNA grade agarose (normal and low-melting-point agarose); Acrylamide, *bis*-Acrylamide, Ammonium persulfate and TEMED (N,N,N',N'-Tetramethylethylenediamine) for 10% nondenaturing polyacrylamide gels; Ethidium bromide; Silver Stain Kit (Bio-Rad, Hercules, CA).
23. An automated HPLC system. In this work, the chromatographic system was from Knauer and consisted of a solvent delivery system, a UV detector (260 nm) and

an autosampler equipped with a 20 µL loop. Data were collected using a Shimadzu integrator.

24. Analytical anion-exchange DEAE-NPR column (Perkin-Elmer), 35 mm × 4.6 mm I.D., packed with 2.5 µm particles.
25. HPLC buffer A: 1 *M* NaCl, 25 m*M* Tris-HCl, pH 9.0, and buffer B: 25 m*M* Tris-HCl, pH 9.0.

## 3. Methods

### 3.1. RNA Purification and First-Strand cDNA Synthesis

All manipulation must be carried out in an RNase-free environment.

1. Isolate RNA using the RNAzol method or any other established procedure *(6)* depending on the type of tissue. RNAs must be stored at –80°C.
2. Combine the following to perform a 20 µL reverse transcriptase (RT) reaction:
   a. RNA (0.2 to 2 µg; *see* **Note 2**).
   b. 2 µL 10X PCR buffer (results in equal or slightly higher yields than 4 µL 5X RT buffer).
   c. 2 µL 0.1 *M* DTT.
   d. 1 µL 0.1 m*M* random primers (*see* **Note 3**).
   e. 5 µL 2.5 m*M* 4dNTP mix.
   f. Ultrapure H$_2$O to 18 µL.
3. Denature RNA at 94°C for 3 min in this mixture, and cool on ice.
4. Add 1 µL RNasin and 1µL reverse transcriptase (2 µL of a 1:1 mixture).
5. Incubate at room temperature for 10 min and at 42°C for 40 min.
6. Heat reaction to 94°C for 3 min, and chill on ice for 1 min to inactivate the enzyme and to denature the nucleic acids.
7. Proceed directly to PCR (**Subheading 3.2.**) or store the cDNA at –20°C.

### 3.2. Basic Protocol for PCR Amplification

The conditions for PCR amplification of DNA depend on the combination of template and primers that is used. A few variables may have to be optimized to achieve a fair result with high enough sensitivity and specificity: MgCl$_2$, template and primer concentrations, and cycle profile parameters. To allow detection of contaminations, always prepare a blank PCR that contains all reaction components but DNA.

1. Combine the following in a 500 µL tube to perform a 25 µL PCR:
   Template DNA: cDNA representing 10–100 ng RNA or amplification products (e.g., competitors) derived from a previous PCR (1 fg–1000 pg):
   a. 2.5 µL 10X PCR buffer.
   b. 2 µL 2.5 m*M* 4 dNTP mix.
   c. 0.1–5 µL 10 µ*M* oligonucleotide primers (0.01–0.5 µ*M* final; *see* **Subheading 3.4.**).
   d. H$_2$O to 24.9 µL.
   e. 0.1 µL *Taq* DNA polymerase. (*see* **Note 4**).
2. Overlay reaction with mineral oil.

3. Carry out 25–35 PCR cycles in an automated thermal cycler under conditions that are best suited for the type of PCR. A typical cycle profile using PCR primers of about 20 bases in length with a GC content close to 50% yielding amplification products of < 500 bp (*see* **Note 1**) may look like this:

     5 min, 94°C (initial denaturation).

   Followed by 25–35 cycles of:

     30 s, 94°C (denaturation).

     30 s, 50–60°C (annealing).

     30 s, 72°C (extension).

   Followed by:

     10 min, 72°C (final extension).

4. Remove PCR product by inserting a micropipet through the mineral oil layer and drawing up the sample.
5. Analyze 5–10 μL of the reaction mix by agarose gel electrophoresis to verify the size, purity, and yield of PCR products and/or analyze by HPLC (*see* **Subheading 3.6.**).

## *3.3. Synthesis of Antisense Competitor DNA for CD-RT-PCR*

### *3.3.1. Site-Directed Mutagenesis and Recombination of cDNA by PCR*

The purpose of this PCR technique *(3)* is to mutagenize the center of a cDNA fragment, that will afterwards serve as competitive template in CD-RT-PCR, so that a unique restriction enzyme recognition site is either created or removed from that sequence segment. To avoid time-consuming DNA purification, the mutant products of a first round of PCR are excised from low-melting-point agarose and directly used for recombination in a second round of PCR. The protocol is schematically outlined in the first part of **Fig. 1** and applies to both the gene of interest and the reference gene. PCR is carried out basically as described in **Subheading 3.2.**

1. Perform a first round of two separate PCRs, each with a mutagenic and a normal primer, to create two partly overlapping mutated amplification products.
2. Separate mutant amplification products by gel electrophoresis on a preparative low-melting-point agarose gel.
3. Locate bands of interest using UV-illumination of the ethidium bromide-stained gel (*see* **Note 5**).
4. Excise bands of interest with a razor blade, and transfer to a microfuge tube.
5. Melt gel slices at 68°C (5–10 min), and dilute in 1 mL water.
6. Mix 1 μL of each of the two excised and diluted intermediate products from the first round of PCR in a second round of PCR of which the first cycle is run without primers. Half of the overlapping mutant strands act as primers on one another and generate mutant recombined fragments.
7. During the denaturation step of the second cycle, insert a micropipet tip through the mineral oil layer to add the biotinylated sense and the nonmodified antisense flanking primers in less than 1/10 the volume of the reaction. The recombined mutant fragments are amplified during subsequent cycles.

8. Separate biotinylated PCR products on a preparative agarose gel (*see* **steps 2–4**).
9. Purify fragment of interest using GeneClean.

### 3.3.2. Solid-Phase Synthesis and Magnetic Separation of Antisense Strands

Only antisense mutant fragments act as perfect competitors for first-strand cDNA *(2)*. A simple protocol for repeated synthesis and separation of the antisense strands from double-stranded biotinylated DNA fragments is listed below and schematically shown as part of **Fig. 1**.

1. Wash Streptavidin-coated magnetic beads (Dynabeads) with 1X Binding and Washing (B & W) buffer using a Magnetic Particle Concentrator (MPC) for Eppendorf tubes.
2. Resuspend Dynabeads in 2X B & W buffer.
3. Mix equal volumes (50 μL) of the resuspended beads and the purified PCR. Use approximately 0.3 mg Dynabeads per μg PCR product.
4. Incubate at room temperature for 30 min, keeping the beads suspended by gentle rotation of the tube.
5. Using the MPC, collect the immobilized DNA, and remove the supernatant.
6. Wash beads extensively using the MPC with 3X 100 μL 1X B & W and 3X 100 μL TE to remove uncoupled PCR primers and products.
7. Resuspend beads in 20 μL of freshly prepared 0.1 $N$ NaOH solution, and incubate at room temperature for 10 min to melt the DNA duplex.
8. Transfer alkali supernatant containing the nonbiotinylated antisense strands to a clean tube.
9. Neutralize supernatant with 4 μL of a freshly prepared 1:1 mixture of 1 $N$ HCl and 1 $M$ Tris-HCl, pH 8.0 (*see* **Note 6**).
10. Using the MPC, wash beads once with 0.1 $N$ NaOH (50 μL), once with 1X B&W (50 μL), and once with TE (50 μL) (*see* **Note 7**).
11. Resuspend beads in a prewarmed PCR mixture including all reaction components (*see* **Subheading 3.2**) except for the sense primers.
12. Incubate for 10 min at a temperature between 55°C and 60°C to allow both specific annealing and extension of the antisense primers (even below 55°C the polymerization rates of *Taq* DNA polymerase are significant). Shake gently every 2 min or use a shaking water bath in order to keep the beads in suspension.
13. Using the MPC, collect beads, and keep the supernatant containing the reaction mix at the desired annealing/extension temperature for successive rounds of solid-phase DNA synthesis. The immobilized sense strands can be reused several times for repeated solid-phase synthesis of antisense DNA.
14. Wash beads extensively using the MPC with 3 × 100 μL 1 × B&W and 3 × 100 μL TE to remove free primers.
15. Again, resuspend and incubate beads in 20 μL 0.1 $N$ NaOH to separate the immobilized sense strands and the antisense strands (**steps 7–10**).

16. The procedure may be repeated several times to increase the yield of antisense strands: Wash beads, incubate beads in the PCR mixture (**steps 10–12**) that was kept at the desired annealing/extension temperature (**step 13**) and proceed as described in **steps 14** and **15**.
17. Dilute 1 μL neutralized antisense DNA in 24 μL EDTA/SDS.
18. Mix 10 μL of Formamide-gel-loading buffer and 10 μL of 25x diluted antisense DNA, denature at 95°C for 10 min and cool on ice.
19. Apply denatured antisense DNA on a 10% polyacrylamide gel and verify and quantify antisense competitors by electrophoresis under nondenaturing conditions and subsequent silver-staining and densitometric scanning using a concentration standard (*see* **Note 8**).
20. Make a serial dilution of antisense competitors down to a concentration of 0.1 fg/μL, and add carrier DNA to a final concentration of 1 ng/μL to each competitor concentration (*see* **Note 9**).
21. Heat-denature these DNA samples at 94°C for 5 min.
22. Cool down slowly to hybridize the single-stranded competitor DNA to denatured carrier DNA thus stabilizing the DNA in double-stranded structures.
23. Aliquotize and store at –20°C in siliconized tubes.

## *3.4. CD-RT-PCR and Restriction Enzyme Digestion*

This section describes how to perform CD-RT-PCR up to the quantification of the digested reaction products. Before doing so, it is helpful to gain some information about the primer concentrations that have to be applied for successful coamplification of the gene of interest and the reference gene and about the approximate amount of competitor DNAs that have to be used for detectable competition with both cDNAs. Hence, it is recommended to perform a few simple pilot experiments (*see* **Note 10**) in order to minimize the number of reactions required for gene expression analysis of a single sample. In many cases, a single reaction will do since quantification by CD-RT-PCR is accurate within a wide range of different competitor concentrations. After tuning the CD-RT-PCR in a few pilot experiments, measurement in samples with a 35-fold difference in expression levels can be performed with a single primer and competitor concentration *(2)*. PCR is carried out basically as described in **Subheading 3.2.** The protocol below describes a CD-RT-PCR for a series of samples with a wider range of expression levels (greater than 35-fold) where a single competitor concentration is expected to yield inconclusive results (*see* **Note 11**).

1. Set up a PCR master mix with a fixed amount of cDNA and other components for separate reactions with different competitor concentrations. Primers for both the gene of interest and the reference gene are added to this mixture. Leave out the volume necessary to add competitors to the reactions (*see* **Note 12**).
2. Divide the master mix into the different reactions.

3. Add exactly defined amounts of antisense mutant competitor DNA of both the gene of interest and the reference gene to the different reactions. It is recommended to choose competitor dilutions with a 100-fold difference in concentration, and add equal volumes of these dilutions to the different reactions (*see* **Notes 12** and **13**).

4. Run PCRs.

5. Heat-denature amplification products at 94°C for 3 min.

6. Cool down slowly to room temperature within 1–2 h to promote stabilization of heteroduplexes based on random reannealing of nearly homologous mutated and non-mutated sequences (*see* **Note 14**).

7. Digest DNA (*see* **Note 15**) without precipitating it by adding the appropriate restriction enzyme in a mix with KGB to a portion (5–10 µL) of the PCR (*see* **Note 16**).

8. To assess the success of the reactions, load a few samples on an ethidium bromide-stained gel system. Use high percentage (6%) agarose gels (NuSieve 5:1 agarose) if the samples were digested in KGB, as the buffer negatively influences electrophoretic properties. A pattern of four bands should appear representing the undigested fragments and the overlapping halves of digested fragments of the two different genes and their corresponding competitors.

## 3.5. HPLC Analysis of CD-RT-PCR Products after Restriction Enzyme Digestion

A trustworthy estimation of the ratios of the DNA products after CD-RT-PCR and restriction enzyme digestion relies on a technique that offers efficient and quantitative discrimination of the resulting four different fragment sizes. HPLC analysis of DNA comprises direct UV absorbance measurement resulting in a high linearity over a large range of DNA concentrations *(7)*. For this reason and because of the sensitivity of the technique, HPLC is highly suitable for measurement of CD-RT-PCR products.

1. Prepare HPLC system including the analytical column as was described by Katz in the Methods in Molecular Biology series *(8)*.

2. Program a gradient profile. For a fast and efficient separation of 50–300 bp fragments that allows the injection of consecutive samples every 15 min, a linear mobile phase change of buffer B in A is recommended as follows:
   a. from 70–60% B in 10 s.
   b. from 60–48% B in 3 min.
   c. from 48–42% B in 5 min.
   d. from 42–0% B in 10 s, and hold for 1 min for clean-up.
   e. from 0%–70% in 10 s, and hold for 5 min for reequilibration.

3. Inject the digested CD-RT-PCR samples, and run the gradient profile.

4. Verify whether the retention times of the appearing peaks are as expected by comparison with the retention times of fragments of an appropriate standard that is injected periodically.

5. Collect integrated data of the area under the curve of relevant peaks.

### 3.6. Calculation of Normalized Expression Levels

1. Calculate the proportion of digested fragments (*d*) for both genes separately from the area under the curve (AUC) of the peaks that correspond to the digested (dig) and undigested (undig) amplification products of a gene:

$$d = \text{AUCdig}/(\text{AUCdig} + \text{AUCundig})$$

2. Calculate the amount of specific cDNA (Y) prior to PCR for both genes separately. The initial amount of competitor (C) for each cDNA species that was added to the reaction is known. Fill in C and *d* into the following equation (*see* **Note 17**), where n = +1 or -1 depending on whether the competitor possesses (+) or lacks (−) the unique restriction enzyme recognition sequence:

$$Y = C[(1-\sqrt{d})\sqrt{d}]^n$$

3. Present the expression level (X) of the gene of interest (i) in a molar relation to the internal standard or reference gene (r) by inclusion of the size (S) of the amplified fragments (in bp) into the following equation:

$$Xi = (Yi/Yr) \cdot (Sr/Si)$$

## 4. Notes

1. Typical primers have a GC content as close to 50% as possible. Primers used for amplification of cDNA and competitor DNA during CD-RT-PCR were 20–22 bases in length. Pairs of mutagenic primers are recommended to be 30–35 bases in length, with base substitutions (one to four) centrally located in a 3' region of 17 to 20 bases in length that confers opposite primers complementary to one another (*see also* **Fig. 1**).

   To avoid problems caused by amplification of contaminating genomic DNA, the sense and antisense primers for CD-RT-PCR should reside in different exons.

   The position of CD-RT-PCR primers should also be carefully considered with respect to the size of the fragments to be amplified. CD-RT-PCR analysis requires adequate measurement of both the full-size amplification products of two genes (A and B) and the fragments that result from cutting these products in two equal halves. To achieve equal intervening distances between the resulting four bands on gel or HPLC peaks (*see* **Fig. 2**), the size (S) of gene A and B should relate as follows: SA = $\sqrt{2}$ × SB. It is recommended to choose 200 bp < SA < 300 bp. SA will then be small enough for amplification with randomly primed cDNA, whereas the template and SB/2 will not be too small for gel electrophoresis and HPLC.

   Primer design and the choice of an appropriate restriction enzyme for CD-RT-PCR are mutually dependent. Mutagenic primers may be used to create or remove a unique restriction enzyme site within the sequence region that is to be amplified during CD-RT-PCR. Choose these primers in a way that the unique restriction enzyme sites reside exactly in the middle of the amplification products of CD-RT-PCR (*see also* **Fig. 2**). The two halves of digested fragments will then elute at (about) the same time during HPLC analysis, which lowers the detection limit as a result of increased AUCs and reduces the complexity of the chromatograms.

2. It is not required to know the RNA concentration of the samples as long as quantifiable amounts of specific DNA can be amplified from it. CD-RT-PCR may therefore be applied to measure RNA expression on even minute amounts of mRNA from biopsies, needle aspirations and so forth.

3. In CD-RT-PCR, expression levels are presented as cDNA ratios of the target and the reference gene. Linearity between these ratios and normalized mRNA expression levels is achieved by using randomly annealing hexamer primers in RT. In contrast to oligo(dT) or gene specific oligo priming, randomly primed reverse transcription results in equally efficient cDNA synthesis of different RNA species and overcomes problems caused by sequence complexity and mRNA secondary structure.

4. Specificity and sensitivity may be further improved by performing a Hot-Start PCR *(9)* and/or a nested primer strategy. It is recommended for both RT-reactions and PCRs to prepare a master mix of all invariable components of a series of reactions and add aliquots of this mix to the variable reaction components in each separate reaction tube.

5. Beware of UV-radiation induced DNA mutagenesis. Work as quickly as possible and/or use a low-energy UV trans-illuminator.

6. Perform a titration of HCl(1:1)Tris-HCl, against NaOH before using the solutions for neutralization of the DNA samples. Always use the same pipet for both solutions. Calibration differences between different pipets may cause neutralization problems.

7. The immobilized sense strands can be stored in TE at 4°C for several weeks.

8. The antisense DNA can easily be distinguished from the sense strands based on the principle that these complementary single-stranded DNA fragments have different sequence dependent conformations that influence electrophoretic mobility *(5)*. As a consequence, denatured PCR products appear as 2 separate bands on a nondenaturing gel. After the magnetic separation procedure, only the band that represents the antisense strands remains *(2)*. It is recommended to set aside antisense DNA from the first time of alkali denaturation and pool DNA from subsequent cycles of solid-phase synthesis and denaturation, as the purity of antisense DNA from the first harvesting is sometimes not as required.

9. Use a single well-calibrated pipet for the preparation of a competitor DNA dilution series to avoid concentrations errors. We used a Gilson P200 pipette and combined 50 μL competitor DNA (from a 10x concentration), 50 μL carrier DNA (10 ng/μL) and 2 × 200 μL water.

10. First, to find the conditions that yield quantifiable amounts of PCR products both for the gene of interest and the reference gene, perform a series of differential RT-PCRs (without competitors) by adding the primers for these genes in different concentration ratios to the reaction mixtures. Second, titrate the competitors against a fixed amount of cDNA. The four bands/peaks comprising the digestible

and nondigestible DNA fragments of both the gene of interest and the reference gene should all be present in quantifiable amounts after CD-RT-PCR and restriction enzyme digestion (*see* **Fig. 2.**).

11. To cover extremely wide ranges of expression levels, it may sometimes be necessary to also use two different primer concentration ratios.

12. Use a single well-calibrated pipet for the preparation of a CD-RT-PCR to reduce experimental variation. To ensure adequate pipeting, the volumes should not be too low. A 25 µL CD-RT-PCR could be made up by 15 µL of a master mix that was divided into separate reactions and 10 µL of competitors.

13. The reference gene can be fine-tuned more precisely, provided the reactions are roughly standardized with respect to the input of cDNA amounts by taking dilutions with a 10-fold difference in concentration.

14. Most PCR machines permit the programming of a temperature ramping rate to slowly cool the samples. Alternatively, a water bath may be used.

15. As a control for restriction enzyme digestion, also digest a PCR that solely contains fully digestible DNA. One could also add a DNA fragment carrying the restriction enzyme site as an internal control to every sample after denaturation and reannealing.

16. KGB should be diluted to a final reaction concentration varying between 0.5 and 2.0 KGB depending on the optimal reaction conditions for the restriction enzyme used. Digestion with EcoRI was performed in 0.5x KGB: 7 µL PCR + 5 µL enzyme mix containing 3 µL $H_2O$ + 1 µL 6x KGB + 1 µL EcoRI.

17. The equation is based on the quadratic distribution principle of duplex formation. **Fig. 2** shows an example where a one to one ratio of first-strand cDNA to competitor DNA before PCR renders one out of four duplexes distinguishable from the others by restriction enzyme digestion.

## References


1. Rochlitz, C. F., de Kant, E., Neubauer, A., Heide, I., Bohmer, R., Oertel, J., Huhn, D., and Herrmann, R. (1992) PCR-determined expression of the MDR1 gene in chronic lymphocytic leukemia. *Ann. Hematol.* **65**, 241–246.

2. de Kant, E., Rochlitz, C. F., and Herrmann, R. (1994) Gene expression analysis by a competitive and differential PCR with antisense competitors. *BioTechniques* **17**, 934–942.

3. Higuchi, R., Krummel, B., and Saiki, R. K. (1988) A general method of in vitro preparation and specific mutagenesis of DNA fragments: study of protein and DNA interactions. *Nucleic Acids Res.* **16**, 7351–7367.

4. Maniatis, T., Fritsch, E. F., and Stambrook, J. (1989) *Molecular Cloning. A Laboratory Manual* (2nd ed.), Cold Spring Harbor Laboratories, Cold Spring Harbor, NY.

5. Orita, M., Suzuki, Y., Sekiya, T., and Hayashi, K. (1989) Rapid and sensitive detection of point mutations and DNA polymorphisms using the polymerase chain reaction. *Genomics* **5**, 874–879.

6. Chomzynski, P., and Sacchi, N. (1987) Single-step method of RNA isolation by acid guanidinium thiocyanate-phenol-chloroform extraction. *Anal. Biochem.* **162,** 156–159.

7. Katz, E. D., and Dong, M. W. (1990) Rapid analysis and purification of polymerase chain reaction products by high-performance liquid chromatography. *BioTechniques* **8,** 546–555.

8. Katz, E. D. (1993) Quantitation and purification of polymerase chain reaction products by high-performance liquid chromatography, in *Methods in Molecular Biology, Vol. 15: PCR Protocols: Current Methods and Applications* (White, B.A., ed., Humana, Totowa, NJ, pp. 63–74.

9. Chou, Q., Russell, M., Birch, D. E., Raymond, J., and Bloch, W. (1992) Prevention of pre-PCR mis-priming and primer dimerization improves low-copy-number amplifications. *Nucleic Acids Res.* **20,** 1713–1723.

# 13

# Amplified Assay for Specific Dual-Labeled DNA Using the Coagulation Cascade (EDNA-ELCA)

George J. Doellgast, G. Alan Beard, Margaret Sheehan, Nathan Iyer, Louis S. Kucera, and Stephen H. Richardson

## 1. Introduction

### 1.1. Colorimetric Detection of DNA

Detection of the products of the PCR reaction using nonisotopically labeled DNA molecules containing biotin, fluorescein, or digoxigenin has become a popular method for identification of specific products of polymerase chain reaction (PCR) *(1,3)*. These labeled molecules are prepared either by PCR synthesis in the presence of labeled deoxyuridine triphosphate *(1,3)* or by hybridization of labeled probes to the unlabeled PCR product *(1,2)*. Detection is then in a format very similar to enzyme-linked immunosorbent assays (ELISA) using similarly labeled antigens and antibodies, i.e., by capture on the receptor for one ligand (streptavidin or antibody) and using the other ligand for detection.

Sensitivity of detection of these labeled products is an issue when the objective is the use of a labeled probe or PCR product to detect low concentrations of specific DNA/RNA *(4–7)*, or to measure small copy numbers of important viruses and bacteria *(8–12)*. Examples would be proviral DNA in HIV-positive patients *(13)*, detection of viral RNA by reverse transcriptase (RT)-PCR *(10)*, or for small numbers of bacteria isolated by magnetic immunoseparation techniques from relatively crude samples of foods *(4–6)*.

We have been developing an assay amplification system based on the clotting cascade. We developed this technique initially as a way of measuring coagulation factors in plasma *(14–17)*, but have since expanded its use to any analyte that can be labeled with the coagulation-activator obtained from the venom of the Russell's viper, known as Russell's viper venom factor X activator (RVV-XA) *(18–21)*. In one approach, we used antibodies dual-labeled with

From: *Methods in Molecular Medicine, Vol 26: Quantitative PCR Protocols*
Edited by: B. Kochanowski and U. Reischl © Humana Press Inc., Totowa, NJ

biotin or fluorescein and RVV-XA for the detection of immune complexes containing botulinum toxin *(21,22)*. That led to the use of similar reagents and labels for the measurement of DNA molecules, which were dual-labeled with biotin and fluorescein *(23)*. In this work, we describe the system and how it is used to measure labeled DNA molecules.

## *1.2. Measurement of Coagulation Using the Enzyme-Linked Coagulation Assay (ELCA) System*

Coagulation is the classical cascade amplification system. It exists as a cascade to utilize rapidly the small signal of damage to a blood vessel wall in order to stop blood flow from that vessel by formation of a fibrin plug or "clot." It accomplishes this by a variety of proenzyme–enzyme conversions of clotting "factors" or inactive proteins into active proteases or protease cofactors (reviewed in **ref.** *24*).

Coagulation is most commonly measured by taking advantage of the formation of a clot in solution, owing to the polymerization of fibrin, and this is measured using light scattering. We developed a solid-phase alternative to the measurement of coagulation, in which an enzyme-labeled fibrin polymer is attached to a fibrinogen-coated microtiter plate. This technique, because of its similarity to ELISA assays, is known as enzyme-linked coagulation assay, or ELCA *(14–18)*.

In **Fig. 1**, we see the basis for the ELCA system for the measurement of a specific snake venom enzyme. Russell's viper venom factor X activator (RVV-XA) is able to convert the proenzyme factor X, present in plasma at a concentration of about 5–10 µg/mL, into its active enzyme form Xa. This then binds to factor Va (a cofactor protein present in plasma at concentrations roughly equal to factor X) in the presence of lipid and calcium to form the "prothrombinase complex," an enzyme complex that can convert the proenzyme factor II (also known as prothrombin) into the active enzyme thrombin. Prothrombin is present in plasma at a concentration of about 100–200 µg/mL. Thrombin is then able to hydrolyze the molecule fibrinogen, present in plasma at a concentration of 2–4 mg/mL, into the readily polymerizable form fibrin. Thus, a small initiating signal results in the production of milligram quantities of fibrin when reactions occur in blood. We have adapted this powerful cascade to measurements carried out in microtiter plates.

The ELCA system is built around the hydrolysis of two distinct forms of fibrinogen by thrombin. One of the forms is bound to a solid phase such as a microtiter plate, and the other is a labeled (e.g., enzyme-labeled) molecule present in solution. When the two forms are both hydrolyzed by thrombin, they associate with one another and the label is bound to the solid phase. The visible result of the assay is very similar to an ELISA assay in which enzyme-labeled analyte is bound to a solid-phase proportional to the amount of labeled complex formed.

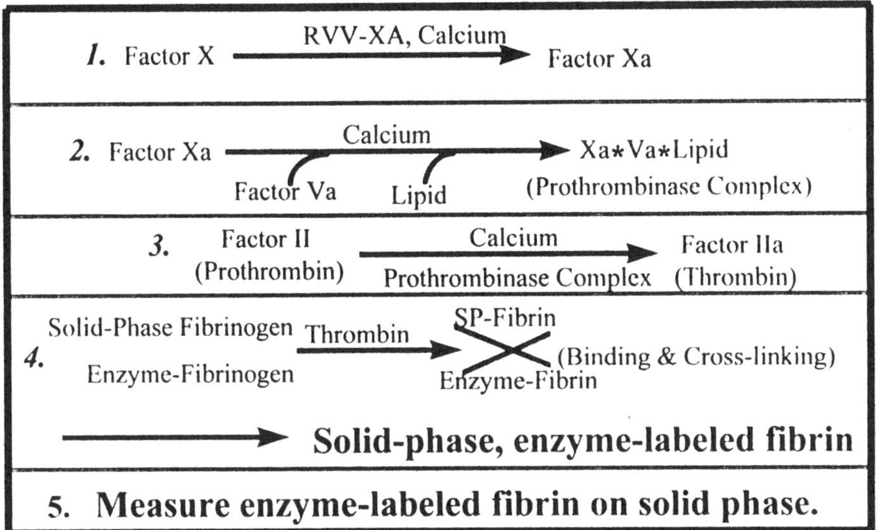



Fig. 1. Stages in ELCA assay for RVV-XA. The clotting cascade proceeding from the activation of factor X by RVV-XA and ending with the generation of solid-phase, enzyme-labeled fibrin (ELCA).

## 1.3. ELCA Reactions: Differences from Standard ELISA Formats

The process of formation of solid-phase fibrin is very different from a standard ELISA format. ELCA uses the clotting cascade, the reactions of which are specific proteolytic reactions. Assembled as a sequence, the reaction cascade results in a nonlinear increase in the amount of labeled fibrin deposited (signal) with time. However, the individual reactions of the sequence can be separated functionally. One can incubate RVV-XA with factor X, and the amount of Xa formed over time can be measured by adding the rest of the components of the cascade to an aliquot of the Xa generated. Also, if one adds nonionic detergent (e.g., Triton, Tween) to the mixture at some time period after initiating the reaction, the only enzyme that continues to be active is thrombin, and the amount of labeled fibrin developed can be assayed as a function of thrombin concentration. Finally, by removing the fibrinogen-coated solid phase at various intervals and measuring the bound enzyme-labeled fibrin, one can assess the amount of labeled fibrin generated at these distinct intervals and progressively increase the signal by these steps.

In our application of the system to measurement of dual-labeled analytes including DNA, we typically separate the ELCA reaction for detection of RVV-XA into three phases. First, the RVV-XA, bound to a solid phase as part of a specific immunological or DNA binding reaction, is mixed with a substrate consisting of factors II, V, X, alkaline phosphatase-labeled fibrinogen, and lipid in a calcium buffer. The reaction proceeds through **step 3** (**Fig. 1**) for a set period of time. Next, nonionic detergent and a solid phase with fibrinogen bound (polystyrene "pegs" in the microtiter plate format) is added to the microtiter well, and reaction 4 is allowed to proceed for a set period of time. The pegs are removed from the well and washed with deionized water, then are put into a substrate consisting of phenolphthalein monophosphate (PMP) at pH 9.8 in a second, flat-bottom microtiter plate. The color that develops is bright red, readily apparent to the eye and measurable in a microtiter plate reader set at 550 nm.

The sensitivity of the ELCA reaction can be manipulated by varying the incubation time at each stage of the reaction. In **Fig. 2** we see three times for the initial reactions (**steps 1–3, Fig. 1**), i.e., incubation for 20, 30, and 40 min. Doubling the reaction time for this stage, one gets an increase of sensitivity of 20-fold, down to a limit of detection of less than 10 fg/mL ($10^{-14}$ g/mL, or 7.5 × $10^{-16}$ g/75 µL sample). RVV-XA has a molecular weight of approx 100,000, so this is 7.5 × $10^{-21}$ mol/sample. Multiplying this by Avogadro's number (6.023 × $10^{23}$ mol/mol) = 4.5 × $10^3$ mol/sample, or 60 mol/mL. This approach can be put to use in estimating how much bound analyte is present at very low limits of detection. It approaches a useful range for detection of specific DNA/RNA in the absence of PCR amplification, if labeled conjugate can be detected at a sensitivity equivalent to the detection of this labeling enzyme.

It should be noted that the dynamic range of the assay as performed here is narrow. That is a characteristic of clotting reactions, resulting in a critical point for accumulation of fibrin to a level allowing its rapid polymerization. In order to extend the dynamic range of the assay to accommodate a wider range of bound analyte, it is useful to test the assay at several sensitivities during a single ELCA determination. To accomplish this objective, one can stop the generation of thrombin after a set period of time by addition of detergent buffer, then place the fibrinogen-coated pegs in for a limited time (e.g., 30 min), wash them with distilled water and place them in PMP substrate for 30 min. Color is obtained proportional to the sensitivity of detection at that assay period. After this, the pegs are washed off with water and placed back into the microtiter plate containing the thrombin generated in **steps 1–3** and alkaline phosphatase-fibrinogen for an additional 30 min, then back into the PMP for 30 min, then an additional 60 min in each. As the sequence proceeds, more labeled fibrin is generated by thrombin and deposited onto the pegs, and more color is generated. The result of using this approach is seen in **Fig. 3**, stopping the generation of thrombin at 25 min and continuing the measurement for several hours. The sensitivity of the assay increases at each step in the reaction. The assay can therefore proceed for up to 4 h as in this case, achieving the same sensitivity (down to 10 fg/mL) as was obtained by dou-

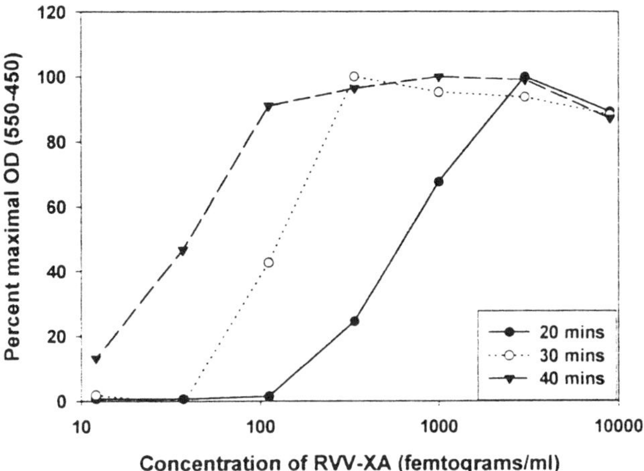


Fig. 2. ELCA reaction for RVV-XA, varying the incubation time with the factor II-V-X mixture from 20–40 min.

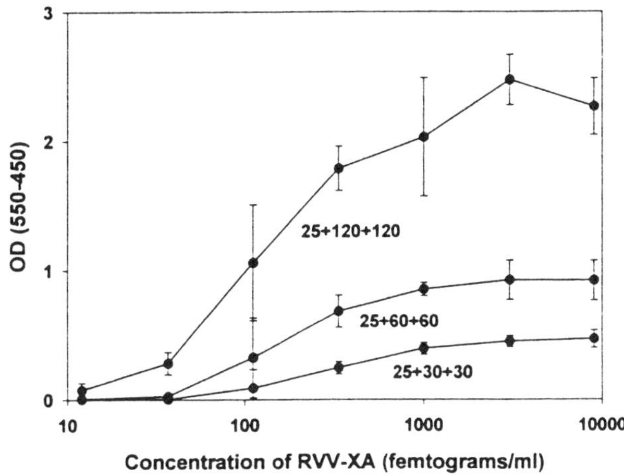


Fig. 3. Assay of RVV-XA using the ELCA system, with increased incubation times in **steps 4** and **5**.

bling the reaction time in **steps 1–3** (**Fig. 2**), and assessing the amount of complex formed at lower levels of sensitivity.

At the end of a reaction in which the sensitivity is assessed, the complex is still bound to the capture plate, and can be reassayed for the longer time in **steps 1–3**. Using a single protocol for binding, one can thus determine the optimal conditions for measuring the amount of product by ELCA in a single

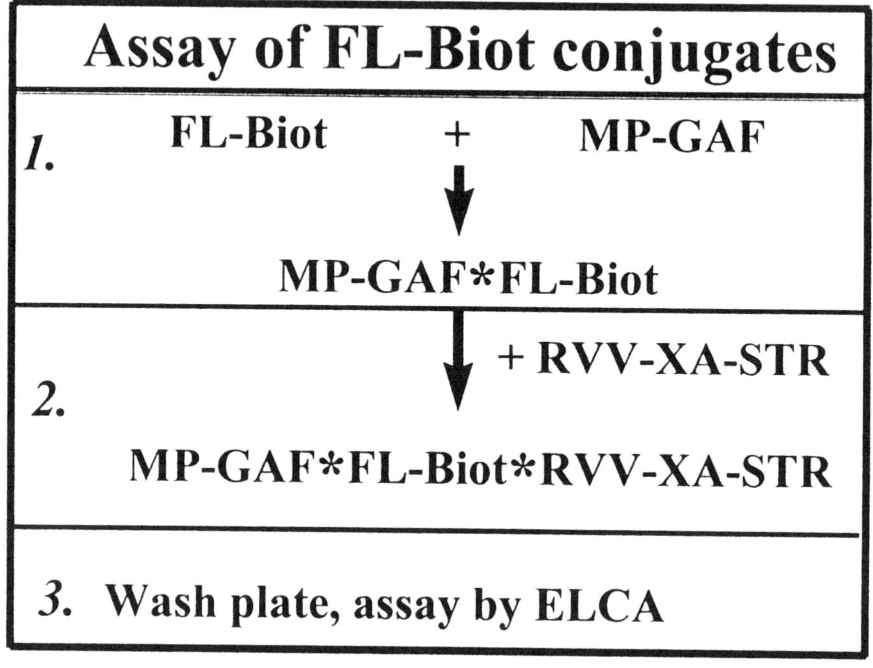

Fig. 4. Assay protocol for dual-labeled molecules by EDNA-ELCA. Fluoresceinated, biotinylated conjugates (FL-Biot) are bound to a microtiter plate coated with goat anti-fluorescein (MP-GAF, **step 1**). After washing off the unbound conjugate, RVV-XA-labeled streptavidin (RVV-XA-STR) is incubated with the plate and binds to the complex on the plate (**step 2**). The plate is again washed, and the ELCA detection system is used to measure the bound complex (**step 3**).

*experiment. Subsequent experiments in which this protocol is used for the detection of a given range of concentrations of analyte can be performed at a single, defined set of ELCA incubation conditions.*

### 1.4. Application of ELCA for Detection of Fluorescein-Biotin (FL-Biot) and FL-Biot-DNA

This assay has been applied for measurement of specific dual-labeled DNA sequences. **Figure 4** shows the approach we have used for this purpose. A molecule having the two functional groups fluorescein and biotin is bound to an anti-fluorescein matrix and then the RVV-XA-conjugate of streptavidin is bound to this complex. (The alternative approach of binding the dual-labeled molecule to streptavidin and using RVV-XA-goat anti-fluorescein as a detection label is much less sensitive assay approach in our hands.) **Figure 5** shows results obtained using dual-labeled DNA and a simple, low molecular weight compound (Fluorescein-biotin [FL-Biot], *see* **Materials**).

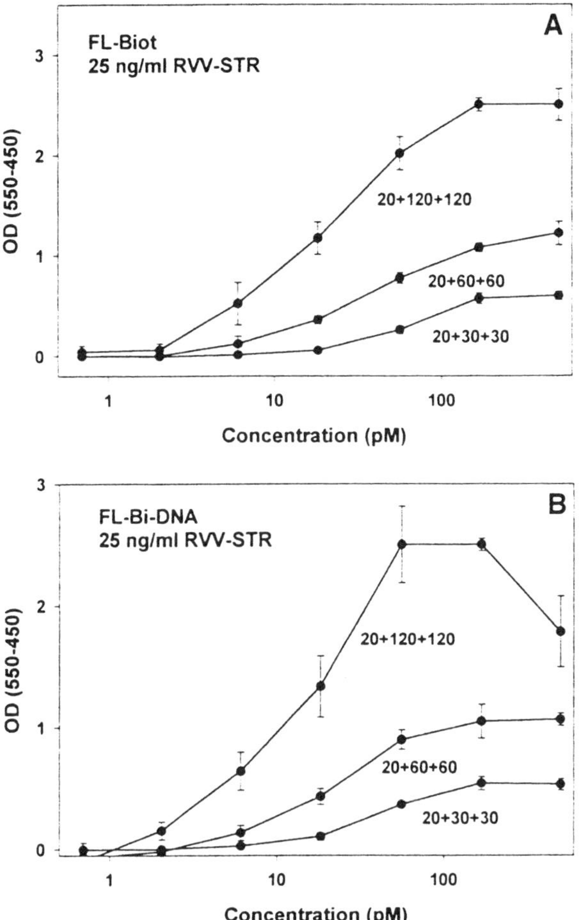



Fig. 5. Assay of FL-Biotin and FL-Biotin-DNA from Salmonella. FL-Biotin is a simple derivative not containing any nucleotide sequences (*see* **Materials**, mol wt 831) The Salmonella PCR product is described in **Materials**. It is a double-stranded 457 bp fragment encompassing the junction of the invA and invE regions, and labeled by PCR using mono-substituted biotinylated and fluoresceinated primers (mol wt approx 310,000).

In **Fig. 5**, we see that the length of derivative that is assayed does not appear to be as important as the number of moles of the derivative which are present. FL-Biot, a simple molecule, and FL-Biot-DNA labeled at the termini are detected with equivalent sensitivity. The detection limit in both cases is about 1–2 p$M$, despite a nearly 400-fold difference in their molecular weights.

## 2. Materials

1. Salmonella PCR primers and reagents: We have been using the sequences presented by Stone et al. **(8)** for Salmonella *inv* genes. We prepared two 20 bp primers designed

to bridge the junction between *inv*-E and *inv*-A genes to yield a 457 bp PCR product that hybridizes with the 3' end of the *inv*-E gene and the 5' end of the downstream *inv*-A gene. These are: AAACTGGACCACGGTGACAA for the *INV* A-B region and TGCCTACAAGCATGAAATGG for the *INV* E-F region. These are prepared by Operon Technologies, either unlabeled or 5' labeled with either biotin or fluorescein. When labeled PCR primers are used, they have single residues added onto the primers. An internal probe, specific for a region within the 457 bp sequence, was the 20-mer oligonucleotide CTGGTTGATTTCCTGATCGC.

2. HIV probes and reagents: We have been using the SK38/SK39 primers which are 28-bp primers specific for a 115 bp region of the gag region of HIV-1 *(13)*; they are: SK38: ATAATCCACCTATCCCAGTAGGAGAAAT and SK39: TTTGGTCCTTGTCTTATGTCCAGAATGC. These are also obtained from Operon Technologies, unlabeled or labeled with either fluorescein or biotin. An internal 41-bp probe labeled with fluorescein, SK19F, had the sequence ATCCTGGGATTAAATAAAATAGTAAGAATGTATAGCCCTAC. A single fluorescein residue was added onto the 5' end.

3. Salmonella DNA was extracted and purified from overnight cultures of either Salmonella enteritidis strain CDC5str or Salmonella typhimurium strain SL1344. Extraction was done using the DNA-G-NOME DNA extraction kit supplied by BIO 101 (La Jolla, CA). Used as PCR templates, "crude" DNA preparations are prepared from similar cultures containing from $10^6$–$10^9$ bacteria/mL, boiled for 10 min.

4. HIV-1 (strain HTLV-IIIB) *(26)* was obtained from Dr. Robert Gallo, University of Maryland, through Dr. Dani Bolognesi, Duke University. It was propagated in persistently infected H9IIIB cells. Virus titers are determined as described previously *(26)*.

5. PCR reagents: 100 m*M* stock solutions of deoxynucleotide triphosphates including biotinylated and fluoresceinated (dUTP form), Taq DNA polymerase and 10X buffer are obtained from Boehringer-Mannheim (Mannheim, Germany).

6. Reverse transcriptase reagents: RNase inhibitor from *Escherichia coli* (cloned) was obtained from Gibco-BRL (Gaithersburg, MD); Reverse transcriptase from avian myeloblastosis virus (AMV) and deoxy nucleotide triphosphate solutions are obtained from Boehringer Mannheim.

7. "End-tailing" reagents including Terminal Transferase (DNA deoxynucleotidylexotransferase, EC 2.7.7.31) from calf thymus and fluoresceinated and biotinylated dUTP are obtained from Boehringer Mannheim.

8. Goat anti-fluorescein is prepared by injecting goats with 2 mg fluorescein isothiocyanate-conjugated keyhole limpet hemocyanin (FL-KLH), subcutaneously in Complete Freund's Adjuvant. After 2 mo, booster injections of 1 mg each FL-KLH and fluorescein isothiocyanate-conjugated bovine serum albumin (FL-BSA) are injected intramuscularly in saline and monthly collections of serum are made 7 d after booster injections. The serum is passed through a 2.5 × 10 cm column of fluoresceinamine coupled to Pierce Aminolink gel (Pierce, Rockville, IL) according to the instructions of the manufacturer. After washing with several column volumes of 1.0 *M* NaCl in 0.05 *M* potassium phosphate buffer, pH 7.8, the antibody is eluted using a 0.8 *M* acetic acid solution (5% glacial acetic acid in water). The

collected eluate is brought to pH 5.5 by addition of Tris base, and precipitated by addition of 0.24 g/mL of ammonium sulfate (approx 40% saturated). The precipitated antibody is dissolved in 0.05 $M$ potassium phosphate buffer, pH 7.6 and dialyzed against this same solution, then is concentrated by placing the dialysis sac in a tube containing an equal volume of glycerol and rotated overnight at 4°C. The concentrated antibody solution is stored in this form at –20°C until used.

9. Anti-fluorescein-coated microtiter plates: NUNC (U96 Maxisorp, U-bottom plates, Nalge-NUNC International, Rochester, NY) microtiter plates are coated with 100 μL of a solution of 5 mL of goat anti-fluorescein in 0.2 $M$ sodium bicarbonate-carbonate, pH 9.5, for 1 h at 24°C, then overnight at 4°C. The plates are "blocked" by addition of 50 μL of a solution of 50 mg/mL casein in 0.1 $M$ Imidazole-HCl buffer, 0.5 $M$ NaCl, 0.5% Triton X-100, pH 7.5. (This solution is autoclaved and then brought to room temperature before use.) The antibody coating solution is not removed from the plate while it is being blocked. Total volume in the plate at this point is 150 μL. This is allowed to remain in the plate for at least 1 h at room temperature, but can be allowed to remain for 1–2 d at 4°C, or if convenient the plates can be stored at –20°C with the solution frozen in the plate. For large batches (100–300 plates) to be used for a period of several months, it is very convenient to use a commercial coating stabilization solution known as Stabilcoat (Sur Modics, Inc., Eden Prairie, MN). The coated, blocked plates are washed with distilled water. We use the Elcawash system (Elcatech, Winston-Salem, NC) designed to rapidly wash microtiter plates in the same format as a cuvet washer. They are then "tamped" onto a dry paper towel and 125 μL of Stabilcoat solution is put into the plate. The plate is treated with Stabilcoat for 30 min to 1 h, and the solution is then removed from the plate. It can be reused several times, in our experience. After the plates are treated, they are dried in a freeze-drier overnight and packaged either in sealed bags or with aluminum foil. If convenient, the plates can be stored overnight at –20°C and dried the next day. The manufacturer suggests that the stabilcoat-treated plates can also be dried at room temperature or in a 37°C incubator with equivalent results. Once prepared in this form, the plates are usable for months-years when stored at 4°C, consistent with the claims of the Stabilcoat manufacturer. We can provide such stabilized plates.

10. Gelatin-imidazole-saline-Tween (GIST) buffer used for binding of labeled conjugates and RVV-XA-streptavidin to the plate consists of: 10 mg/mL bovine gelatin (Sigma, St. Louis, MO) in 0.05 $M$ Imidazole-HCl, pH 7, 0.5 $M$ NaCl, 0.5% Tween 20. The solution is autoclaved after preparation. Note that gelatin is chosen for this application because the casein solutions we use for other immunoassays appear to contain enough biotin to block binding of the RVV-XA-streptavidin at the dilutions used in these experiments.

11. RVV-XA-streptavidin: This is prepared as described previously *(23)*. It is used as a 10 μg/mL stock by adding 0.5 mL of GIST buffer to a microtube containing the dried, stabilized RVV-XA-Streptavidin. The gelatin solution is autoclaved. The stock solution (10 μg/mL) is prepared fresh for each 1–3 d use, stored at 0–4°C, and diluted to 50 ng/mL (1:200) just before addition to each microtiter plate.

12. ELCA substrate: This is provided as a kit for the measurement of bound RVV-XA-Streptavidin (Elcatech). It includes a mixture of factors II, V, X, alkaline phosphatase-fibrinogen, and lipid lyophilized in a vacuum-sealed vial, a vial containing 9 mL of assay buffer, a vial containing 3 mL of detergent buffer, a set of fibrinogen-coated pegs, a flat-bottomed NUNC microtiter plate, and a vial of phenolphthalein monophosphate solution. Details on its use are provided in **Subheading 3.**

13. FL-Biot: This is a compound of fluorescein and biotin with a spacer, (5-((N(5-(N-(6-(biotinoyl)amino)hexanoyl)amino)pentyl)thioureidyl)fluorescein), molecular weight 831, available from Molecular Probes (Eugene, OR). This is useful as a control for assays using dual-labeled DNA/RNA molecules, especially because it is a simple molecule containing both ligands and it is not hydrolyzed by DNase or RNase.

# 3. Methods

## 3.1. PCR Protocol

The PCR protocol for Salmonella DNA is carried out using standard conditions, including 80 $\mu M$ of each nucleotide triphosphate, 25 U/mL of Taq DNA polymerase, 10 m$M$ Tris-HCl, 1.5 m$M$ MgCl$_2$, 50 m$M$ KCl, pH 8.3. Primers are added at a concentration of 1 $\mu M$. Each cycle is for 15 s of denaturation at 94°C, 15 s of primer annealing at 52°C, 15 s of polymerization at 72°C, and at the end of the cycling the mixture is kept at 72°C for 155 s and at 4°C for 15 min. Cycling is in a Delta Cycler II system from Ericomp, Inc. (San Diego, CA). For random labeling with biotinylated or fluoresceinated dUTP, the labeled nucleotide triphosphate is added as 1/3 of the total dTTP in the PCR mixture. When the (5') end-labeled primers are used, they are added in place of the unlabeled primers at the same concentrations.

## *3.2. RT-PCR Protocol for HIV-RNA*

The RT phase of this reaction is performed in 20 µL of total reaction volume, with 50 m$M$ Tris-HCl, 8 m$M$ MgCl$_2$, 30 m$M$ KCl, 1 m$M$ dithiothreitol (DTT), pH 8.5, 750 U/mL of RNase inhibitor (15 U/20 µL), 5 $\mu M$ Primer SK39 (*see* **Subheading 2.**), 1250 U/mL AMV reverse transcriptase (25 U/20 µL), 2 m$M$ nucleotide triphosphates. Incubation is at 42°C for 60 min, then at 96°C for 7 min to inactivate the RT. Twenty microliters of this reaction mixture is then added to 80 µL of PCR mixture containing 1 $\mu M$ primer SK38 and cycled. Each cycle is for 1 min of denaturation at 95°C, 1 min of primer annealing at 55°C, 1 min of polymerization at 72°C, and at the end of the cycling the mixture is kept at 72°C for 10 min and at 25°C for 10 min. When primers are 5' labeled with either biotin or fluorescein, they are added in place of the unlabeled primer; when the DNA produced is randomly labeled with biotin or fluorescein, the substituted nucleotides are added in a ratio of 1:2 with the unlabeled dTTP.

### 3.3. End-Tailing of Probes

"End-Tailing" of the hybridization probe for detection of Salmonella employed the 20-mer oligonucleotide internal invA probe. The reaction mixture contained 2 µg of probe in 100 µL of reaction mix containing 0.2 $M$ potassium cacodylate, 25 m$M$ Tris-HCl, 0.25 mg/mL bovine serum albumin (BSA), 1.5 m$M$ CoCl$_2$, pH 6.6, 30 µ$M$ dUTP-biotin, and 500 U/mL of Terminal Transferase. Incubation is for 2.5 h at 37°C. The reaction is stopped by addition of 10 µL of a solution containing 0.2 m$M$ ethylenediamine tetraacetic acid (EDTA) and 1 µg/mL glycogen, 12.5 µL of 4 $M$ LiCl, and 375 µL of 100% ethanol. The mixture is kept overnight at –20 °C and is centrifuged at 16,000g max in an Eppendorf table-top centrifuge for 10 min. The pellet is washed with 250 µL of cold 70% ethanol, and is then dried in a vacuum desiccator. The pellet is suspended in a buffer consisting of 10 m$M$ Tris-HCl, pH 7.4, 0.1 m$M$ EDTA (TE buffer).

### 3.4. Hybridization of Reaction Products with Internal Probes

Hybridization is in a volume of 50 µL. Probe concentrations are 200 n$M$ when the hybridization mixture is diluted 1:100 into the microtiter plate for EDNA-ELCA assay, and 8 or 40 n$M$ when the PCR mixture is less concentrated and so the hybridization mixture is diluted 1:1 in the microtiter plate. The amount of PCR product added varies, but is at most 4 µg/mL. The hybridization is performed in a buffer containing 0.1 mg/mL salmon sperm DNA (ssDNA) and 6X SSC (Final concentration 0.9 $M$ NaCl, 0.09 $M$ sodium citrate, pH 7.0). The mixture is heated at 95°C for 10 min, removed to an ice bath for 5 min, then incubated at 40°C for 1 h.

### 3.5. Estimation of DNA Concentrations in PCR Mixtures

Estimation of DNA concentrations in PCR mixtures is accomplished by visual comparison of intensity of ethidium bromide-stained, electrophoretically separated DNA with DNA quantitation standards obtained from Gibco-BRL. For the 457 bp Salmonella PCR product, electrophoresis is carried out in 2% agarose (Sigma) and for HIV-1 PCR product, electrophoresis is in 4% agarose (Continental Lab Products, San Diego, CA).

### 3.6. Extraction and Concentration of Viral RNA from HIV-1

Supernatants from H9IIIB cells containing $5 \times 10^4$ plaque-forming units (PFU) HIV-1/mL are extracted using the Gibco-BRL TRIZOL LS reagent, which is added to 250 µL of sample in a ratio of 3:1, reagent:sample (750 µL of TRIZOL). RNA is precipitated from the aqueous phase of the samples, dried, and dissolved in 50 µL of buffer. Complete recovery of viral RNA is assumed, or a concentration of viral RNA equivalent to $2.5 \times 10^2$ PFU/µL.

### 3.7. Binding of Dual-Labeled Product
### to Anti-Fluorescein-Coated Microtiter Plates

This is accomplished in the GIST buffer described in Materials. A dilution of the labeled DNA or FL-Biot standard is diluted to the appropriate concentrations and bound to the anti-fluorescein plate for 1 h at room temperature or overnight at 4°C. As a routine, we prefer overnight binding at 4°C for convenience, but have found that binding at room temperature for 1 h or up to 47°C for 30–60 min are all usable for this purpose. After binding, the plates are washed on the Elcawash microtiter plate washer using saline, and then are filled with 200 µL of GIST buffer and incubated for 10 min.

### 3.8. Binding of RVV-XA-Streptavidin

For this stage, the plate from the binding step is washed with saline, tamped on a paper towel, and 100 µL of diluted (50 ng/mL) RVV-XA-Streptavidin in GIST buffer is dispensed into the plate. Note that the concentration of the RVV-XA-Streptavidin (50 ng/mL, or 300 p$M$) used in the binding to the biotinylated analyte is over a million times greater than the minimal detectable level of RVV-XA by ELCA (10 fg/mL, or 0.1 f$M$). It is therefore essential to minimize the nonspecific binding of this conjugate by carefully controlling the incubation time with the plate and preventing drying of the anti-fluorescein plate. (This could result in the deposition of some RVV-XA-streptavidin onto the plate in a form that does not wash out in subsequent steps.) The plate is tightly wrapped in aluminum foil, and placed in a 37°C incubator for 20 min. After incubation, the plate is again washed with saline and filled with 200 µL of GIST buffer and incubated for 10 min. The plate is again washed with saline and tamped on a clean paper towel immediately before ELCA assay. The plate must be washed thoroughly with saline, because the Tween-20 detergent in the GIST buffer inhibits stages 1–3 of the clotting reaction.

### 3.9. Performance of the ELCA Assay

As seen in Fig. 2, the ELCA assay is remarkably sensitive to the incubation time for reactions 1–3. It is therefore necessary to dissolve the substrate mixture quickly in the reconstitution buffer and dispense it into the plate. The background activity from the substrate accumulates with time, so one should dissolve the substrate and dispense it into the plate within a period of 1–3 min. The factor mixture is supplied in a vacuum-sealed, 10-mL vial. Remove the aluminum tear-off caps on all the vials to be reconstituted, remove the caps from the 9 mL stock of ELCA buffer, and—when it is appropriate to start the reaction—remove the rubber seals from the substrate vials, carefully pour the buffer into the vial, recap, and mix until dissolved, then dispense 75 µL into each well of the microtiter plate. Place the plate into an incubator at 37°C for

15–30 min depending on the experiment, and then add 25 µL of detergent buffer and the fibrinogen-coated pegs. The plate is returned to the 37°C incubator for 30 min, during which time the phenolphthalein monophosphate substrate is added into the flat-bottom plate supplied with the pegs. After the incubation, wash the pegs with distilled water, tamp on a clean paper towel and place into the phenolphthalein monophosphate solution at room temperature. One should see a pink tint within several min from the positive samples. Depending on the length of the initial incubation and the sensitivity desired, one can either allow the color to continue to develop in the phenolphthalein monophosphate solution, or remove the pegs after 30 min, wash again with distilled water, and place the pegs back into the antibody plate. Meanwhile the plate would be read in a microplate reader set at 550 nm, or by difference at 550–450 nm. We have also photographed through a green filter, which gives excellent representation onto a color film (particularly when the photographic processor can adjust color composition in favor of red prints), or onto black and white film without special processing. More recently we have found digital photography to give excellent resolution. The pegs that are placed back into the antibody plate are incubated another 30 min, washed, and placed back into the phenolphthalein monophosphate substrate plate which is read and/or photographed. This is incubated another 30 min at room temperature, during which time the color continues to develop. If the color is adequate for the purposes of the experiment, then the reaction is stopped at this point. If not, then another cycle of incubation of pegs with antibody plate and plate containing phenolphthalein monophosphate will increase the sensitivity of the assay. This process can be repeated until the background color—owing to the nonspecific binding of the RVV-XA-streptavidin to the microtiter plate—obscures the specific signal owing to the specific binding to the bound biotinylated analyte. The total incubation time is represented by adding each incremental stage to the total. If the plate is allowed to incubate 20 min before the detergent buffer is added and the pegs are incubated for 30 min with the antibody plate and 30 min in the phenolphthalein monophosphate solution, then the assay is defined as 20+30+30. If there are two 30 min cycles in the antibody plate and two 30-min cycles in the phenolphthalein monophosphate, then the assay is 20+60+60. If there are two 30-min cycles and one 60-min cycle in each of the two last stages, then the assay would be 20+120+120. A comparison of the sensitivity of this approach for measurement of RVV-XA is shown in Fig. 3.

## 4. Notes

1. Dilution and comparison of PCR products with standard preparations: The FL-Biot preparation is a very useful standard for evaluating the binding and detection phase of the ELCA assay. As seen in **Fig. 5**, the length of the molecule

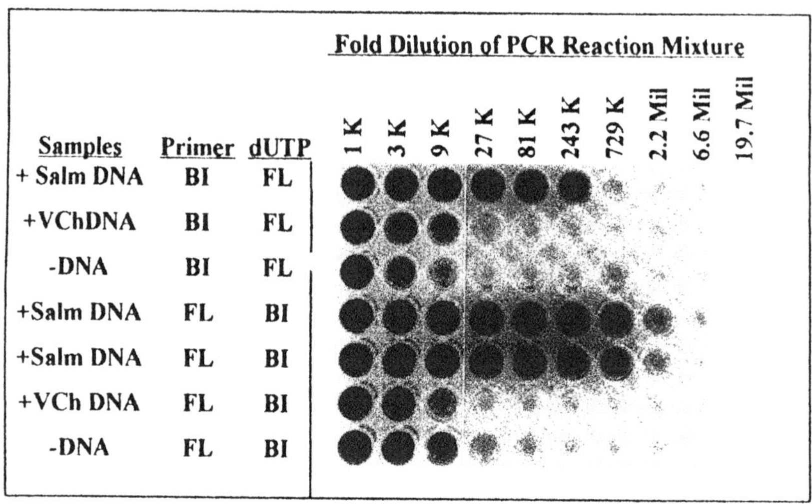



Fig. 6. Photograph showing assay of PCR products obtained from Salmonella DNA using random labeling with either fluoresceinated or biotinylated dUTP, and the complementary labeled primer.

is not critical for the detection of dual-labeled molecules, because dual-labeled 457-mer double stranded Salmonella PCR product DNA with single molecules of fluorescein and biotin at the extremes are detectable with the same sensitivity as the simple molecule FL-Biot. This suggests that the standardization of the EDNA-ELCA assay by use of FL-Biot is a useful index of the amount of dual-labeled product generated and bound to the microtiter plate. It can be used as a standard in any experimental protocol in which the amount of bound PCR-DNA labeled with both biotin and fluorescein is tested. It is to be expected that it will not serve as an accurate index of the concentration of dual-labeled DNA with multiple residues incorporated into the polymer, but it is a useful reagent for comparison of day-to-day sensitivity of the assay. Also, it is to be recognized that it may be useful in identifying the presence of nucleases that may hydrolyze double-labeled nucleic acid polymers, since it will not be hydrolyzed by such nucleases.

2. Random incorporation of dual labels during PCR. Rothschild et al. *(23)* suggested that the "geometry" of labeling, i.e., the number of residues of label incorporated into the product, is a critical determinant of the sensitivity of the assay for dual-labeled DNA. In **Fig. 6**, we show the results for two distinct PCR products from Salmonella, using the 457 bp PCR product. In one case, both primers are 5' labeled with fluorescein and biotin is randomly incorporated by PCR; in the other, biotin is incorporated into the 5' end of the primers and fluorescein is randomly incorporated by PCR. The labeling results in a detection limit of approx 42 pg/mL (500,000-fold dilution of 20.8 µg/mL, or 132 fM) for the product with random incorporation of fluorescein and 3.8–3.9 pg/mL (4 million-fold dilution of 15.7 µg/mL

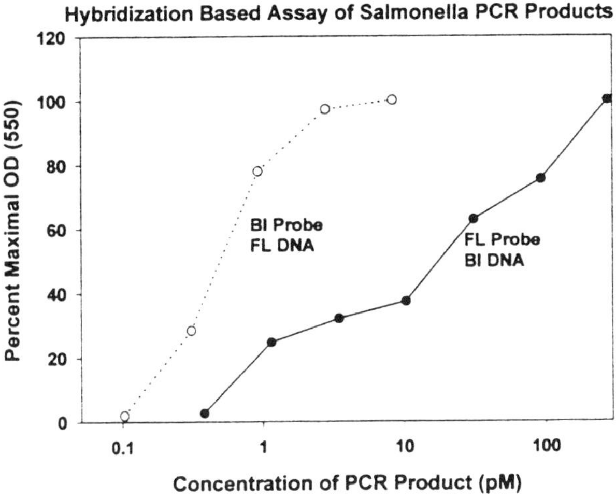


Fig. 7. Detection of Salmonella DNA end-labeled with either biotin or fluorescein by hybridization with the complementary end-tagged probe.

and 2.2 million-fold dilution of 3.8 µg/mL, or 12 fM) for two products with random incorporation of biotin. There are 437 bp (874 nucleotides) produced by the PCR reaction with approx 25% thymidine incorporation, or approx 218 thymidine residues. If 1/4–1/3 of these contain the biotinylated dUTP analog, then there are 55–73 biotin residues in this dual-labeled product, compared with 2 in the product in which fluorescein is incorporated. It is notable that the sensitivity of detection for these products of random labeling is much higher than that for single-labeled derivatives (**Fig. 5**). As would be expected, multiple binding sites for biotin and fluorescein therefore substantially increase the sensitivity of detection for labeled DNA prepared by PCR by binding more efficiently to the anti-fluorescein matrix and/or by binding more RVV-XA-streptavidin/mole of DNA.

There is an artifact which is generated when dual-labeled products are generated by PCR. That is seen in **Fig. 6**, in that the control PCR mixtures, i.e., mixtures in the absence of target DNA, yield detectable signal to a dilution of approximately 1% that achieved in the presence of specific DNA. For these controls, no DNA band is seen on electrophoresis, and the artifactual product is generated either with or without the addition of irrelevant DNA (*Vibrio cholerae* DNA in this case). In other control experiments, we found that generation of this artifactual product required the presence of *Taq* DNA polymerase in the PCR mixture, in that no product is detectable by EDNA-ELCA when the mixture is cycled 35 times in the presence of all reagents excluding the *Taq* DNA polymerase. In order to obtain specificity for detection of labeled product generated by PCR, it is therefore necessary to use an additional index of presence of specific DNA, for example by using hybridization probes.

3. Hybridized Salmonella PCR products: An alternative to incorporation of dual labels by PCR is the preparation of PCR product containing a single label and then hybridizing with an internal probe containing the complementary label. We tested this alternative for the Salmonella gene PCR product, using both primers 5' labeled with single residues of either biotin or fluorescein and an internal probe that is 3' end-tailed with the complementary labeled dUTP derivative. **Figure 7** shows the results for this experiment. Using a hybridization probe end-tailed with multiple biotin residues and DNA prepared using fluoresceinated primers, the limit of detection is between 100-300 fM PCR-derived DNA; using the complementary pair, the limit of detection is approximately 0.5–1 p*M*. Biotinylated product did not bind to the plate in the absence of the appropriate fluoresceinated hybridization probe, and the biotinylated probe did not bind in the absence of fluoresceinated DNA.

4. Hybridized RT-PCR products; HIV-1 viral RNA: For detection of HIV-RNA, we attempted to determine the limit of detection of PCR-generated DNA from small numbers of virus particles. The SK38 and SK39 primers used are 5' labeled with single residues of biotin, and the internal hybridization probe SK19 is 5' labeled with a single residue of fluorescein. In order to assess sensitivity of detection, from 0.88–880 PFU equivalents are replicated by RT and then by 15, 25, and 35 cycles of PCR. The products are diluted 1:5 in the hybridization mixture and this mixture is diluted on a microtiter plate 1:100 and serially threefold and assayed by EDNA-ELCA, or for the lower concentrations of PCR product 1:1 and serially twofold. The results are shown in **Fig. 8**. Using this protocol, after 15 cycles of PCR, the product from 880 and 88 PFU equivalents are detectable at dilutions of 1:400 and 1:40, respectively. Neither of these mixtures contained any DNA detectable by ethidium-bromide staining. After 25 cycles, the four dilutions tested yielded measurable product, with EDNA–ELCA curve midpoints of 1:30, 1:80, 1:1000, and 1:80,000 for the 0.88, 8.8, 88, and 880 PFU equivalents. Only the PCR products from the 88 and 880 PFU equivalent mixtures contained DNA detectable by staining, with approx 13 and 17 µg/mL of DNA, respectively. With 35 cycles of replication, all samples containing virus contained detectable DNA (2–25 µg/mL) and the product was detectable at dilutions from 20,000-fold (2 µg/mL, PCR product from 0.88 PFU equivalents) up to 200,000-fold (25 µg/mL, PCR product from 880 PFU equivalents). This is a detection limit of approx 1.5 p*M*, consistent with the detection limit of fluorescein biotin and the mono-derivatized Salmonella DNA (**Fig. 5**) and the equivalent hybridization reaction with the biotinylated Salmonella DNA (**Fig. 7**). A control consisting of an RT-PCR mixture that contained no viral RNA and was carried through 35 cycles of PCR showed no reactivity in this assay.

5. The dual-label approach outlined in this work is a very convenient way to identify specific DNA produced by PCR. A single-capture matrix is used, i.e., goat anti-fluorescein-coated microtiter plates. Using only biotinylated primers to generate specific DNA and mono-derivatized fluoresceinated hybridization probe, one can detect very low concentrations of PCR product. This sensitivity of detection should be improved by random incorporation of biotinylated dUTP in PCR

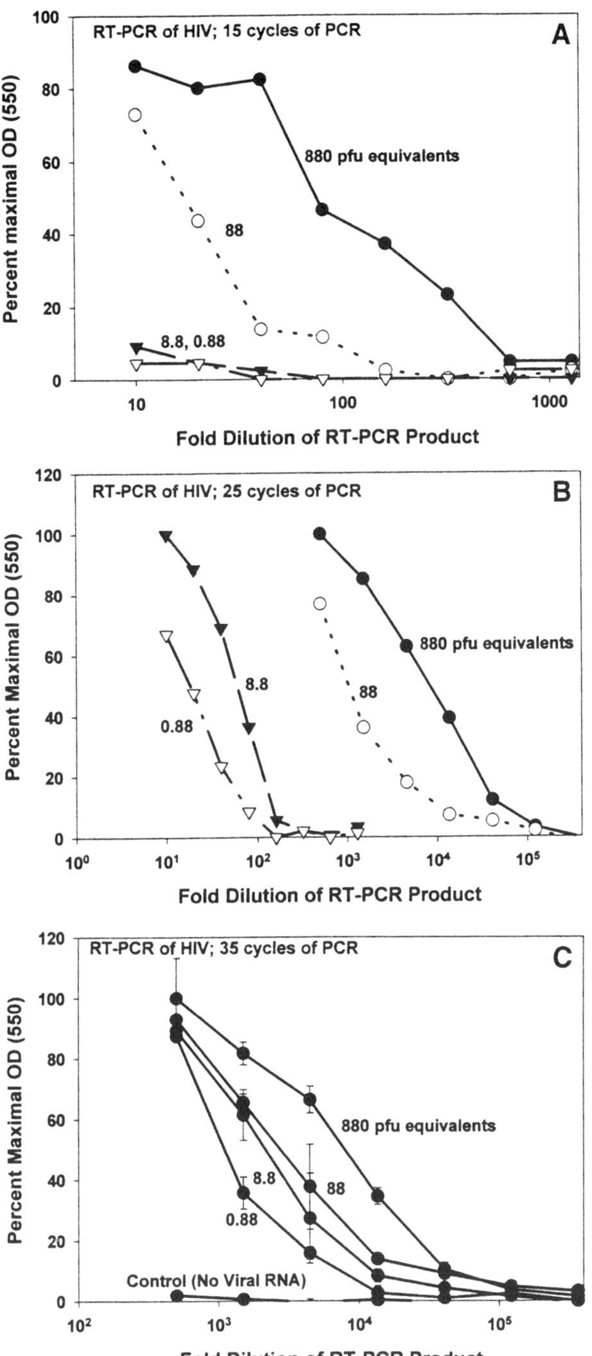



Fig. 8. Detection of RT-PCR-derived DNA from HIV-1 using different cycle numbers of PCR following RT-based replication. Primers used are 5' end-labeled with single residues of biotin, and the hybridization probe is labeled with a single residue of fluorescein.

and binding to an end-tailed probe containing multiple residues of fluorescein. This can be useful in studies in which fewer cycles of PCR are performed to generate a measurable product with fewer replication errors. It should also be noted that the use of fluoresceinated DNA in this work does not suggest that labeling with digoxigenin or other labels would not be equivalent. Our preference for this label is based on the convenience of our use of the same label for capture of immune complexes containing relevant antigens and the availability of a number of inexpensive conjugates for this purpose. It would be equally useful to capture DNA containing digoxigenin and biotin on an anti-digoxigenin matrix and use the RVV-XA-streptavidin to detect these complexes.

6. Efficiency of detection of dual-labeled products: We note that the limit of detection of RVV-XA by the current ELCA protocols is about 10 fg/mL, which is 100 attomolar (aM) or 60 molecules/μL. The concentration of RVV-XA-streptavidin used in the EDNA-ELCA protocol is 50 ng/mL (313 p$M$), or approx $3 \times 10^7$ greater than the detection limit for RVV-XA. This concentration was empirically determined to be the highest concentration that could be employed without significant nonspecific binding to the anti-fluorescein matrix. At this concentration and using the buffer conditions and incubation times described in **Subheading 3.**, the amount of RVV-XA-streptavidin that binds nonspecifically to the anti-fluorescein matrix is approx 10–30 fg/mL of RVV-XA. That results in a limit for specific detection of bound product without background of about 50–100 fg/mL of RVV-XA, or 0.5–1 fM, approx 300 molecules/mL. The limit of detection of dual-labeled DNA in this work was shown to be about 10–100 fM for DNA labeled using PCR in the presence of both primers containing fluorescein and biotinylated dUTP randomly incorporated into the product. Using end-tailed probe with multiple biotin-dUTP residues incorporated, the detection limit of the fluorescein-labeled DNA reached approx 100 f$M$. Other protocols in which less biotin was incorporated into the product had a limit of detection of approx 1 p$M$. Because RVV-XA is detectable under the conditions used in this assay at a concentration of approx 0.5–1 fM, this suggests that the multiply labeled DNA was bound to the plate and detected with an efficiency of 1/10–1/100 (molecules of RVV-XA bound/molecule of DNA) when multiple biotin and/or fluorescein residues were present in the DNA and was always less than 1/1000 when the DNA contained 1–2 biotin and few fluorescein residues.

7. We anticipate that improving the efficiency of the binding of the dual-labeled DNA to both the capture matrix and the enzyme-labeled conjugate will substantially improve the sensitivity of this assay approach. This could be accomplished either by using other protocols that incorporate multiple residues and consequently enhance avidity of association with the capture matrix and RVV-XA-streptavidin, or by using other capture and labeling strategies similar to that employed in the "branched-chain" approach used for HIV detection in the absence of PCR. In this context, it should be noted that the RVV-XA enzyme has proven to be stable at 47°C for up to 1 h incubation, so that it could be conjugated to oligonucleotide probes for hybridization-based incorporation and subsequent colorimetric detection.

## Acknowledgments

This work was supported by Grant #5-R44 CA62468-03 from the National Cancer Institute, National Institutes of Health, and by Grant #96-35201-3445 from the Food Safety and Inspection Service, U. S. Department of Agriculture.

## References


1. Levenson, C. and Chang, C.-A. (1990) Nonisotopically labeled probes and primers, in *PCR Protocols: A Guide to Methods and Applications* (Innis, M. A., Gelfand, D. H., Sninsky, J. J., and White, J. J., eds.), Academic, San Diego, CA, pp. 99–112.
2. Coutlee, F., Yang, B., Bobo, L., Mayur, K., Yolken, R., and Viscidi, R. (1990) Enzyme immunoassay for detection of hybrids between PCR-amplified HIV-1 DNA and a RNA Probe: PCR-EIA. *AIDS Res. Hum. Retrovir.* **6,** 775–783.
3. Lo, Y.-M. D., Yap, E. P. H., An, S. F., McGee, J. O. D., and Fleming, K. A. (1994) Nonisotopic probe generation by PCR, in *PCR Technology: Current Innovations* (Griffin, H. G. and Griffin, A. M., eds.), CRC, Boca Raton, FL, pp 43–57.
4. Lund A., Wasteson Y., and Olsvik O. (1991) Immunomagnetic separation and DNA hybridization for detection of enterotoxigenic *Escherichia coli* in a piglet model. *J. Clin. Microbiol.* **29,** 2259–2262.
5. Bennett, A. R., MacPhee, S., and Betts, R. P. (1995) Evaluation of methods for the isolation and detection of *Escherichia coli* O157 in minced beef. *Lett. Appl. Microbiol.* **20,** 375–379.
6. Widjojoatmodjo, M. N., Fluitt, A. D. C., Torensma, R., Verdoenk, G. P. H. T., and Verhoef, J. (1992) The magnetic immuno polymerase chain reaction assay for direct detection of salmonellae in fecal samples. *J. Clin. Microbiol.* **30,** 3195–3199.
7. Coombs, R. W., Henrard, D. R., Mehaffet, W. F., Gibson, J., Eggert, E., Quinn, T. C., and Phillips, J. (1993) Cell-free plasma HIV type 1 titer assessed by culture and immunocapture-reverse-transcription-polymerase chain reaction. *J. Clin Microbiol.* **31(8),** 1980–1986.
8. Stone, G. G., Oberst, R. D., Hays, M. P., McVey, S., and Chengappa, M. M. (1994) Detection of Salmonella serovars from clinical samples by enrichment broth cultivation-PCR procedure. *J. Clin. Microbiol.* **32,** 1742–1749.
9. Todd, J., Pachl, C., White, R., Yeghiazarian, T., Johnson, P., Taylor, B., Holodniy, M., Kern, D., Hamren, S., Chernoff, D., and Urdea, M. (1995) Performance characteristics for the quantitation of plasma HIV-1 RNA using branched DNA signal amplification technology. *J. Acquired Immune Def. Syn. Hum. Retrovirol.* **10(2),** S35–44.
10. Holodniy, M. (1994) Clinical application of reverse transcription-polymerase chain reaction for HIV infection: review. *Clinics Lab. Med.* **14,** 335–49.
11. Revets, H., Marissens, D., De Wit, S., Lacor, P., Clumeck, N., Lauwers, S., and Zissis G. (1996) Comparative evaluation of NASBA HIV-1 RNA QT, AMPLICOR-HIV monitor, and QUANTIPLEX HIV RNA assay: three methods for quantification of human immunodeficiency virus type 1 RNA in plasma. *J. Clin. Microbiol.* **34,** 1058–1064.
12. Harrigan, R. (1995) Measuring viral load in the clinical setting. *J. Acquired Immune Def. Syn. Hum. Retrovirol.* **10,** 534–540,

13. Ou, C.-Y., Kwok, S., Mitchell, S. W., Mack, D. H., Sninsky, J. J., Krebs, J. W., et al. (1988) DNA amplification for direct detection of HIV-1 in DNA of peripheral blood mononuclear cells. *Science* **239,** 295–297.

14. Doellgast, G. J. and Rothberger, H. (1985) Enzyme-linked coagulation assay I. A clot-based, solid phase assay for thrombin. *Anal. Biochem.* **147,** 529–534.

15. Doellgast, G. J. and Rothberger, H. (1986) Enzyme-linked coagulation assay II: a sensitive assay for tissue factor and factors II, VII and X. *Anal. Biochem.* **152,** 199–207.

16. Doellgast, G. J. (1987) Enzyme-linked coagulation assay III: sensitive immunoassays for clotting factors II, VII and X. *Anal. Biochem.* **162,** 102–114.

17. Doellgast, G. J., Triscott, M. X., Buss, D. H., and West, J. (1988) Extrinsic pathway enzyme-linked coagulation assay (EP-ELCA): a clot-based alternative to prothrombin time test for measurement of extrinsic pathway factors in plasma. *Clin. Chem.* **34,** 294–299.

18. Doellgast, G. J. (1987) Enzyme-linked coagulation assay IV: sensitive sandwich enzyme-linked immunosorbent assays using Russell's viper venom factor X-activator-antibody conjugates. *Anal. Biochem.* **167,** 97–105.

19. Durkee, K. H., Cheng, T. M., and Doellgast, G. J. (1990) Enzyme-linked coagulation assay V: amplified blotting assays using snake venom conjugates. *Anal. Biochem.* **184,** 375–380.

20. Doellgast, G. J., Triscott, M. X., Beard, G. A., Bottoms, J. D., Cheng, T., Roh, B. H., Roman, M. G., Hall, P. A., and Brown, J. E. (1993) Sensitive ELISA for detection of *C. botulinum* neurotoxins A, B and E using signal amplification via enzyme-linked coagulation assay. *J. Clin. Microbiol.* **31,** 2402–2409

21. Doellgast, G. J., Beard, G. A., Bottoms, J. D., Cheng, T., Roh, B. H., Roman, M. G., Hall, P. A., and Triscott, M. X. (1994) Enzyme-linked immunosorbent assay and enzyme-linked coagulation assay for detection of C. botulinum neurotoxins A, B and E and solution-phase complexes with 'dual label' antibodies. *J. Clin. Microbiol.* **32,** 105–111

22. Doellgast, G. J., Triscott, M. X., Beard, G. A., Bottoms, J. D., Anderson, P., Roh, B. H., and Brown, J. E. (1995) Amplified immunoassay of *C. botulinum* neurotoxins A, B, E and F using hyperimmune horse antisera; fluoresceinated antibody 'capture,' in *Molecular Approaches to Food Safety* (Eklund, M., Richard, J. L., and Mise, K., eds.), Alaken, Fort Collins, CO, pp. 83–98

23. Rothschild, C. B., Triscott, M. X., Bowden, D. W., and Doellgast, G. J., (1994) A microtiter plate assay using cascade amplification for detection of nonisotopically labeled DNA. *Anal. Biochem.* **225,** 64–72.

24. Davie, E. W. (1981) Blood coagulation. *Methods Enzymol.* **80,** 153–156.

25. Popovic, M, Sarngadharan, M. G., Read, E., and Gallo, R. C. (1984) Detection, isolation, and continuous production of cytopathic retroviruses (HTLV-III) from patients with AIDS and pre-AIDS. *Science* **224,** 497–500.

26. Kucera, L. S., Iyer, N., Leake, E., Raben, A., Modest, E. J., Daniel, L. W., and Piantadosi, C. (1990) Novel membrane-interactive ether lipid analogs that inhibit infectious HIV-1 production and induce defective virus formation. *AIDS Res. Hum. Retrovir.* **6,** 491–501.

# 14

# Quantitative PCR with Internal Standardization and OLA-ELISA Product Analysis for the *p53* Tumor Suppressor Gene

**Meinhard Hahn and Alfred Pingoud**

## 1. Introduction

Over the last nine years, several quantitative polymerase chain reaction (QPCR) techniques have been developed, and these are now frequently used for the quantification of DNA template copy numbers. However, only few of these PCR techniques are suitable for the precise and absolute quantification of the template copy number *(1,2)*. For this purpose, we describe here a quantitative PCR strategy that uses a known amount of an internal standard DNA that is amplified in competition with the sample template, using one common PCR primer pair and identical primer binding sites for both templates *(2–4)*. In the literature, several variants of internal control sequences were used for the purpose of standardization, e.g., (i) homologous gene sequences of closely related species; differing in few bp, slightly in length and/or absence or presence of restriction sites *(5,6)*, (ii) sample DNA derived sequences that are shortened by a deletion *(7)* or (iii) lengthened by an insertion *(8)*; (iv) competitor fragments that contain more *(9)* or (v) less *(10)* extended heterologous sequence strings; or (vi) differ only by one or two bp, thereby replacing a sample specific restriction site by unique one specific to the internal control DNA *(1,11,12)*. But only the last type of the internal control templates that differ in a negligible manner from the sample DNA sequenc is suitable for precise quantifications, as could be shown by theoretical considerations *(2)* as well as experimentally *(1)*. Even in the case of a homologous internal control sequence of identical length as the sample sequence and differing by less than 5% in sequence, the two templates are not amplified with the same efficiency and therefore do not fulfill the criteria of ideal competition *(2)*, as could be shown *(6)*.

From: *Methods in Molecular Medicine, Vol 26: Quantitative PCR Protocols*
Edited by B. Kochanowski and U. Reischl © Humana Press Inc., Totowa, NJ

Therefore, for accurate standardization, we use an internal control sequence differing by 2 bp, which carries a *Hin*dIII site instead of a sample specific *Ssp*I site (*see* **Fig. 1**). Both sequences are of identical length, carry the identical primer binding sites for QPCR-1a and QPCR-1b, differ only in two of the 342 bp, located in the center of the sequence, and are amplified with identical efficiencies *(13)* as demanded by theoretical considerations for a truly competitive PCR *(2)*. For the production of such internal-control DNAs, several mutagenesis strategies have been described, using either cloning techniques *(12)*, exclusively PCR techniques *(1,13)*, or a combination of both *(1)*. In this protocol, we describe a PCR mutagenesis procedure that does not require microbiological equipment. The product of the PCR is directly used for internal standardization without further cloning steps.

For discrimination and quantification of the competitively amplified sample and control-specific PCR products, two different techniques are possible: separation of the selectively *Hin*dIII and/or *Ssp*I restricted DNA fragments by polyacrylamide gel electrophoresis, followed by ethidium bromide staining and densitometric quantification of the DNA fragments *(1,12,15)* or an oligonucleotide ligation assay (OLA) coupled with an enzyme-linked immunosorbent assay (ELISA), called OLA-ELISA *(15)*. When applying the gel electrophoretic analysis, it is important to stress that the quantifications done by this QPCR assay system are not influenced by the heteroduplexes of sample/control-specific PCR product strands. Although quantifications by sample/internal control template systems, which possess either a sample-specific or an internal control-specific restriction site for discrimination, are impaired by heteroduplexes *(1,16)*, our double-cut system is not affected by heteroduplexes *(1,12,14)*. The gel electrophoretic assay is inexpensive but not suitable for automatization and analysis of large amounts of samples.

Therefore, we describe here in addition a microtiter plate based semi-automatic OLA-ELISA for discrimination and quantification of the competitively amplified QPCR products *(15)* which allows for analysis of many samples in parallel. In principle, the assay can be performed with a robotic work station, thereby strongly increasing the throughput using this assay. The OLA-ELISA is an inexpensive assay, because for the product discrimination step, only T4 DNA ligase, but no restriction enzymes, are needed. OLA was first developed for genotyping to discriminate allele-specific PCR products, which differ only by single-point mutations *(17–19)*, but here it is adapted for QPCR purposes *(14)*. The assay is based on the ability of two oligonucleotides to anneal immediately adjacent to each other on a complementary target DNA molecule. By the enzymatic activity of the DNA ligase, the two oligonucleotides are ligated and thereby covalently joined, provided that they are completely complementary to the ligation template. When two pairs of OLA oligonucleotides (*see* **Fig. 1**)

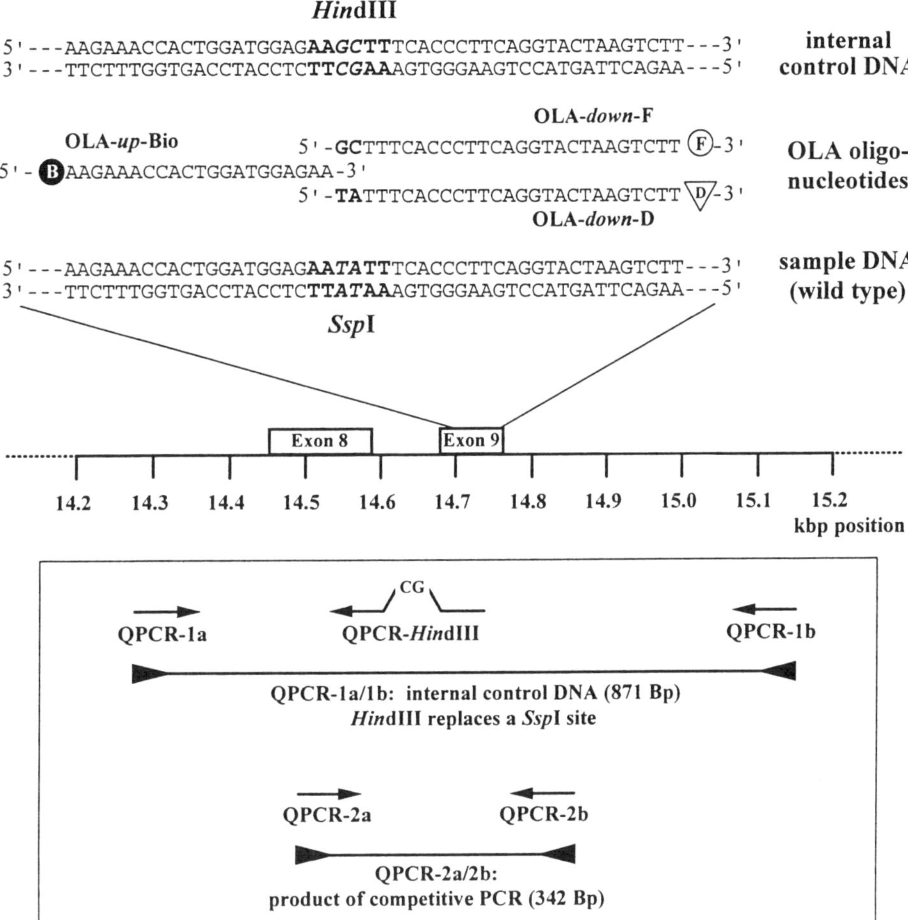



Fig. 1. Schematic presentation of the location of PCR primers (shown in the box) and of hapten and biotin-labeled OLA oligonucleotides, used for PCR product discrimination, within the human *p53* gene. In the box, the relative positions and orientations of the PCR primers and PCR products are shown. QPCR-1a/1b represents the internal control DNA template, which is generated by PCR mutagenesis and contains a new single *Hin*dIII site instead of the sample specific *Ssp*I site. QPCR-2a/2b is the product of the quantitative PCR, which is analyzed and quantified in the OLA-ELISA. The upper part shows the sequence differences between sample and control DNA in detail, differences are indicated by bold italic letters. The OLA oligonucleotides are modified by biotin (**B**), digoxigenin (**D**) or fluorescein (**F**) label. The 5' terminal bases of OLA-*down*-F and OLA-*down*-D, which are either complementary to the sample or to the internal control DNA, are highlighted by bold letters.

are used, one complementary to the sample (OLA-up-Bio and OLA-down-D), the other complementary to the internal control PCR product (OLA-up-Bio and OLA-down-F), the ligase will only join the correctly hybridized oligonucleotide pairs, which are distinguishable because of different hapten-labels (digoxigenin and fluorescein, respectively). The ligase, therefore, discriminates between the two PCR product species. The quantification of the specific ligation products is carried out in an enzyme-linked immunosorbent assay (ELISA): in the first step, the ligation products are immobilized via their biotin-label onto avidin-coated microtiter plates; in the second step—after denaturation of the immobilized double-stranded DNA fragments—the amounts of single-stranded ligation products are quantified in digoxigenin and fluorescein-specific ELISAs, respectively.

Here we present the use of the QPCR/OLA-ELISA technique for the analysis of the human tumor suppressor gene *p53*. This is the gene most often inactivated in human tumors *(20)*: very often, one allele is inactivated by deletion, which constitutes an important step in the multistep model of carcinogenesis. The activity status of *p53* is a significant prognostic factor in oncology. The presented QPCR/OLA-ELISA assay is able to detect the deletion of *p53* alleles within tumor tissue samples.

## 2. Materials

### 2.1. Oligonucleotides

1. PCR primers.

    | | |
    |---|---|
    | QPCR-1a: | 5'-CTG GCT TTG GGA CCT CTT AAC-3' |
    | QPCR-1b: | 5'-GCA GGC TAG GCT AAG CTA TGA TG-3' |
    | QPCR-*Hind*III: | 5'-TGA AGG GTG AAA GCT TCT CCA TCC AGT G-3' |
    | QPCR-2a: | 5'-CGG CGC ACA GAG GAA GAG AAT-3' |
    | QPCR-2b: | 5'-CAA ATG CCC CAA TTG CAG GTA-3'. |

    Dissolve the oligos in 10 m*M* Tris-HCl, pH 8.5, and store the 4 μ*M* solutions in aliquots at –20°C (*see* **Notes 1,2**).

2. OLA oligonucleotides:

    | | |
    |---|---|
    | OLA-up-Bio: | 5'-***BB*** AAG AAA CCA CTG GAT GGA GAA-3' |
    | OLA-down-D: | 5'-**p** TAT TTC ACC CTT CAG GTA CTA AGT CTT ***DIG***-3' |
    | OLA-down-F: | 5'-**p**-GCT TTC ACC CTT CAG GTA CTA AGT CTT ***FITC***-3'. |

The oligonucleotides carry the following modifications: ***B*** is a biotin-derivatized phosphoramidite, ***DIG*** (digoxigenin-3-*O*-methylcarbonyl-ε-aminocaproic acid-*N*-hydroxy-succinimide ester), and ***FITC*** (fluorescein isothiocyanate) are coupled with 3'-amino-modified oligonucleotides, **p** is a 5'-terminal phosphate group (*see* **step 3** in **Subheading 3.4.**). These modified oligonucleotides are available in high performance liquid chromatography (HPLC)-purified form from commercial oligonucleotide synthesis laboratories (*see* **Notes 3,4,5**). Store the oligos as 1 μ*M* solutions in 10 m*M* Tris-HCl, pH 8.5 in aliquots at –20°C.

## 2.2. PCR

1. Human genomic DNA (i) from blood lymphocytes of a healthy person and (ii) from healthy and tumorous tissue of the same patients. Isolate RNA-free DNA (i) from EDTA treated blood or (ii) from shock-frozen tissue, stored at –70°C (*see* **Note 6**). For preparation of RNA-free genomic DNA, several kits are commercially available (*see* **Note 7**). Quantify the DNA concentrations by uv spectroscopy (*see* **Note 8**).
2. Internal-control DNA. Prepare a dilution series of the internal control DNA (for its synthesis: *see* **Subheading 3.1.** and **Note 9**) in 10 m*M* Tris-HCl, pH 8.5. Vary the DNA concentrations in the range of 100 p*M* to 1 a*M* in steps of factor 10 (*see* **Note 10**). Store the solutions at –20°C in aliquots. Avoid unnecessary and repeated thaw/freeze cycles.
3. 0.2 mL Micro-Amp reaction tubes (Perkin Elmer, Weiterstadt, Germany).
4. Aerosol-resistant pipet tips, which must be sterile, DNase, and RNase free.
5. Ice or cryo box (0°C) for cooling of reaction tubes.
6. *Taq* DNA polymerase. Use a licensed high-quality polymerase, e.g., *Taq* DNA polymerase from Boehringer Mannheim (5 U/μL, recombinant, *Escherichia coli*).
7. 10X reaction buffer for *Taq* DNA polymerase: 100 m*M* Tris-HCl, 15 m*M* MgCl$_2$, 500 m*M* KCl, pH 8.3 (at 20°C).
8. dNTP-Mix: mixture of dATP, dCTP, dGTP, and dTTP, each 2 m*M*. Store the mix in aliquots at –20°C (*see* **Note 11**).
9. Thermal cycler. We recommend the GeneAmp PCR System 2400 (Perkin Elmer).
10. Commercially available kits for the isolation of PCR-generated DNA fragments from agarose gels and from aqueous solutions (*see* **Note 7**).

## 2.3. Gel Electrophoretic Analyis of PCR Products

1. Equipment for the electrophoresis of submarine agarose gels and polyacrylamide slab gels (electrophoresis chambers, suitable power supply, glass plates, spacers, and comb for gel cassettes) (*see* **Note 12**).
2. TAE buffer: 40 m*M* Tris-acetate, pH 8.0, 20 m*M* sodium acetate, 1 m*M* EDTA. Adjust the pH by acetic acid.
3. 10X TPE buffer: 800 m*M* Tris-phosphate, pH 8.0, 20 m*M* EDTA. Adjust the pH by phosphoric acid.
4. Agarose for molecular biological purposes.
5. 30% (w/v) acrylamide stock solution: dissolve 145 g acrylamide and 5.0 g *bis*-acrylamide in distilled water by extensive stirring. Fill the solution up to 500 mL, filter the solution, and store it in the dark at 4°C for not more than 2 mo (*see* **Note 13**).
6. N,N,N',N'-tetramethylethylenediamine (TEMED), store at 4°C.
7. APS: aqueous solution of 40% (w/v) ammonium peroxodisulfate (APS), store at 4°C for not longer than 4 wk.
8. Restriction endonucleases *Ssp*I and *Hin*dIII and the recommended enzyme reaction buffers of the manufacturer. Store at –20°C.
9. Incubator set at 37°C.

10. 5X gel-loading buffer: 100 m*M* EDTA, pH 8.0, 25% (w/v) ficoll 400, 0.1% (w/v) bromphenol blue, and 0.1% (w/v) xylene cyanole FF.
11. Mol wt markers for DNA fragments, e.g., commercially available 20 bp or 50 bp ladders or defined restriction digests of plasmids.
12. Ethidium bromide solution. Prepare an aqueous stock solution of 10 mg/mL, and store it in the dark at 4°C. For staining of polyacrylamide gels, prepare a dilution of 1 µg/mL ethidium bromide (*see* **Note 14**).
13. UV transilluminator with 312 nm UV light.
14. Polaroid documentation system or video documentation system (*see* **Note 15**).
15. Software for the densitometric quantification of DNA fragments (*see* **Note 15**).

## *2.4. OLA-ELISA*

1. T4 polynucleotide kinase and supplier's recommended reaction buffer.
2. 2X ligation buffer: 40% (v/v) formamide, 100 m*M* Tris-HCl, pH 7.5, 200 m*M* NaCl, 20 m*M* MgCl$_2$, 2 m*M* ATP, 10 m*M* DTE, 10 µg/mL bovine serum albumin (BSA), and 4 m*M* spermidine trihydrochloride (*see* **Note 16**). Store the buffer in aliquots of 2.0 mL at –20°C for not more than 2 mo.
3. T4 DNA ligase. Prepare a fresh dilution of 15 Weiss-U/mL using the dilution buffer recommended by the supplier.
4. NaOH. Prepare 0.1 *M* and 0.25 *M* solutions.
5. Sodium acetate. Prepare a 3 *M* sodium acetate solution, pH 6.0 adjusted with glacial acetic acid.
6. Microtiter plates. We recommend polystyrene Nunc-Immuno microtiter plates with 96 flat bottom wells and MaxiSorp surface (Nunc, Roskilde, Denmark) (*see* **Note 17**).
7. Automatic microtiter plate washer, programmed to use 100 µL wash solution per well and washing step.
8. Minishaker for microtiter plates.
9. Multichannel pipets and multipets (50–200 µL range).
10. Avidin. Prepare a solution of 14 U/mL of affinity-purified avidin (isolated from egg white) in distilled water, and store it at 4°C (*see* **Note 18**).
11. Coating buffer: 100 m*M* NaHCO$_3$, pH 9.6 adjusted with 1 *M* NaOH.
12. PBST: 4.3 m*M* Na$_2$HPO$_4$, 1.4 m*M* K$_2$HPO$_4$, 140 m*M* NaCl, 2.7 m*M* KCl, and 0.05% (v/v) Tween 20, pH 7.3 adjusted with 0.1 *M* HCl.
13. Blocking buffer or PBST-M. Dissolve 1% (w/v) skimmed milk powder in PBST buffer (*see* **Note 19**).
14. Antidigoxigenin-Fab-POD-conjugate. Dissolve the lyophilized antibody in distilled sterile water (stock solution of 150 U/mL), and store it at 4°C. 10 min before its use in the ELISA, a 10 µL-aliquot (per plate) of the stock solution is diluted 1000-fold using buffer PBST-M. Slightly agitate the dilution for 5 min.
15. Antifluorescein-Fab-POD-conjugate: carry out the same steps as described for the antidigoxigenin-Fab-POD-conjugate.
16. Staining buffer: 40 m*M* sodium acetate, 40 m*M* trisodium citrate, pH 4.4 adjusted with glacial acetic acid.

17. Staining solution. Dissolve 1 mg 3,3',5,5'-tetramethylbenzidine (TMB) in 1.0 mL dimethyle sulfoxide (DMSO) under strong agitation, which will take about 5 min. Mix this solution first with 9 mL staining buffer and then with 3 µL 30 % (w/v) hydrogen peroxide. The staining solution has to be prepared **immediately** before use (*see* **Note 20**) because it is not stable and will form the colored POD product over time.
18. Sulphuric acid. Prepare a solution of 2 $M$.
19. Microplate absorption reader to measure the ELISA absorption signals at 450 nm.

## 3. Methods
### 3.1. Generation of the Internal Control DNA

1. First perform a PCR in a final volume of 50 µL containing 1X *Taq* reaction buffer, 2 U *Taq* DNA polymerase, 200 µM of each dNTP, 0.4 µM each of the primers QPCR-1a and QPCR-1b and 100 ng genomic blood lymphocyte DNA of a healthy person (*see* **Note 21**). Use the following cycling conditions: 35 cycles of 1 min at 93°C (but 5 min in the first cycle), 1 min at 64°C, 1 min at 72°C (but 3 min in the last cycle). Isolate the sample-specific PCR product QPCR-1a1b(*Ssp*I+) (a fragment of 871 bp) using a PCR product purification kit and elute the DNA in 50 µL 10 m$M$ Tris-HCl, pH 8.5. Quantify the molar concentration of the DNA fragment by uv spectroscopy (*see* **Note 8**).
2. In this step, the DNA fragment QPCR-1a/*Hin*dIII, which will serve in the following protocol as a "mutagenesis mega-primer," is generated by PCR amplification: perform in parallel four PCRs in a final volume of 100 µL, each using the primers QPCR-1a and QPCR-*Hin*dIII and 40 ng of the template QPCR-1a1b (*Ssp*I+) (*see* **Note 21**). Apply the following cycling conditions: 30 cycles of 1 min at 93°C (but 5 min in the first cycle), 1 min at 56°C, 1 min at 72°C (but 3 min in the last cycle). Isolate and quantify the PCR product QPCR-1a/*Hin*dIII (475 bp fragment) as described above (*see* **Notes 7,23**).
3. Now generate the internal control DNA QPCR-1a1b (*Hin*dIII+) in a 50 µL PCR volume containing a minute amount of template (50 a$M$ QPCR-1a/1b (*Ssp*I+), generated in Step 1), 100 n$M$ of the "mega-primer" QPCR-1a/*Hin*dIII and 150 n$M$ of the antisense primer QPCR-1b. Apply the following cycling conditions: 26 cycles of 1 min at 93°C (but 5 min in the first cycle), 1.5 min at 62°C, 2 min at 72°C (but 5 min in the last cycle).
4. Reamplify QPCR-1a1b (*Hin*dIII+), the PCR product of the former step: perform in parallel five PCRs in a reaction volume of 100 µL each, using 2 µL of the former PCR product mixture (**step 3**) as template and the primers QPCR-1a and QPCR-1b (400 n$M$ each). Use the following cycling conditions: 14 cycles of 1 min at 93°C (5 min in the first cycle), 1 min at 62°C, 1.5 min at 72°C (5 min in the last cycle). Isolate and quantify QPCR-1a1b (*Hin*dIII+), the 871 bp PCR product, as described above (*see* **Note 7**).
5. Now QPCR-1a1b (*Hin*dIII +), the product of **step 4**, is cleaved at 37°C by 60 U of the restriction enzyme *Hin*dIII, using a reaction volume of 150 µL and the recommended enzyme buffer. After 3 h, the reaction is stopped by addition of

40 µL 5X gel loading buffer. Load the mixture and a DNA standard onto a preparative 1.2% (w/v) agarose gel using 1X TAE-buffer system *(21)*, carry out the electrophoresis overnight at a constant electric field of 1.5 V/cm. Stop the electrophoresis when the front marker bromphenol blue reaches the end of the gel. Incubate the gel for 15 min in ethidium bromide-staining solution, then destain it for 20 min in water. Transfer the gel onto an uv transilluminator and excise the gel areas that contain the two restriction fragments. Isolate the fragments by a suitable DNA preparation kit. Set up a ligation reaction of 70 µL volume containing 1 µg of the isolated restriction fragments, 0.04 Weiss-U T4 DNA ligase and 1X ligation buffer of the enzyme supplier. Incubate the reaction for 2 h at 15°C, and inactivate the enzyme by heating for 10 min at 75°C. After addition of 20 µL 5X gel-loading buffer, the mixture is loaded onto a preparative agarose gel, and electrophoresis is carried out as described above. Isolate the 871 bp ligation product QPCR-1a1b (*Hin*dIII+) from the gel matrix and quantify the yield by UV spectroscopy. Dilute the DNA in 10 m*M* Tris-HCl, pH 8.5, to get a 1-n*M*-solution. This is the initial stock solution of the internal control DNA template. Store it in aliquots at –20°C.

6. Reamplify the internal control DNA for 25 cycles (cycling conditions: *see* **step 4**): perform in parallel five 100 µL PCRs containing the primers QPCR-1a/QPCR-1b and 100 p*M* QPCR-1a1b (*Hin*dIII +) of the initial stock solution (*see* **step 5**). Isolate the PCR-product using a PCR purification kit, quantify the internal control DNA carefully by uv spectroscopy, and use it for the generation of the final dilution series of the internal control DNA (*see* **Note 10**).

## 3.2. Quantitative Competitive PCR

1. For setting up the QPCR amplification reactions of a titration series, first pipet a constant amount of sample DNA (e.g., genomic DNA of healthy or tumorous tissue) into each of the reaction tubes (e.g., 10 µL solution of 5 ng/µL genomic DNA). Now, add varying amounts (0.5–10 µL) of the different dilutions of the internal control template into the tubes (*see* **Note 24**). Perform all QPCRs in a final volume of 50 µL containing 1X *Taq* reaction buffer, 2 U *Taq* DNA polymerase, 200 µ*M* of each dNTP, and 0.4 µ*M* each of the primers QPCR-2a and QPCR-2b.

2. Apply the following cycling parameters for the QPCR amplification reactions when using the GeneAmp PCR System 2400: 35 cycles of 1 min at 93°C (but 5 min in the first cycle), 1 min at 64°C, 1 min at 72°C (but 3 min in the last cycle). The last cycle is followed by quick cooling to 4°C.

## 3.3. Gel Electrophoretic Analysis of QPCR Products

1. For discrimination of amplified sample and control DNA 7.5-µL-aliquots of each QPCR are digested by 3 U *Ssp*I or 6 U *Hin*dIII, respectively, using the recommended reaction buffers in a reaction volume of 15 µL. After 3 h, the restriction reactions are stopped by addition of 4 µL 5X gel loading buffer and stored at 4°C.

2. Prepare a 15% (w/v) polyacrylamide gel; 10 mL gel solution is needed for a gel of 90 × 85 × 1 mm: thoroughly mix 5 mL acrylamide stock solution, 1 mL 10X TPE buffer, 4 mL distilled water. Start the polymerization by addition of 20 µL APS and 20 µL TEMED, pour the solution into the assembled gel cassette, introduce the comb, and leave undisturbed for 30 min when the polymerization should be finished and the gel ready for use.

3. Mount the gel cassette into the electrophoresis chamber, fill the buffer tanks with 1X TPE buffer, remove the comb, rinse the gel slots with buffer, and apply equal amounts of the QPCR samples per slot. Usually 2 µL sample per slot of 2 mm width will be sufficient. In addition, load the DNA standard onto the gel and start the electrophoresis (*see* **Note 12**). Stop it when bromphenol blue, the front marker, migrates into the lower buffer reservoir.

4. After dismounting the gel cassette, first soak the gel for 10 min in 200 mL ethidium bromide-staining solution with gentle agitation, then destain it for 20 min in water, again with agitation. Transfer the gel onto an UV transilluminator, and document the band pattern using a polaroid or video camera.

5. Quantify the relative amounts of the DNA fragments in each gel lane using a suitable software for the densitometric analysis of gel images. Calculate the following ratios of peak areas: $S_N = S / (S + US)$ and $H_N = H / (H + UH)$ ($S_N$: relative amount of *Ssp*I fragments; S: sum of the peak areas of the two *Ssp*I fragments; US: peak area of DNA not cleaved by *Ssp*I; $H_N$: relative amount of *Hin*dIII fragments; H: sum of the peak areas of the two *Hin*dIII fragments; UH: peak area of DNA not cleaved by *Hin*dIII). Plot $S_N$ and $H_N$ against the initial molar concentrations of added internal control DNA template. The point of intersection of the *Ssp*I and *Hin*dIII specific curves indicates the point of equimolar concentrations of sample and internal control templates, therefore, the molar concentration of initial sample template copies.

## 3.4. OLA-ELISA Analysis of QPCR Products

1. For the avidin-coating of the microtiter plates, prepare an avidin dilution of 10 µg/mL in coating buffer, load 100 µL per well, and incubate the plates overnight at room temperature. These coated plates can be stored for up to 3 d at 4°C.

2. Before the plates are used in immobilization experiments, remove surplus avidin by washing each well three times with 100 µL PBST, using a microtiter plate washer if available. Subsequently, unspecific binding sites on the plates are blocked by incubating each well for 30 min with 100 µL PBST-M. After a final washing step (three times 100 µL PBST per well), the plates can be used for the immobilization of biotin-labeled molecules in the wells.

3. Prior to the OLA reaction, the oligonucleotides OLA-down-F and OLA-down-D have to be phosphorylated enzymatically at their 5'-terminus provided that this modification was not yet introduced during their chemical synthesis: prepare a 200 µL reaction mixture containing 5 µ*M* oligonucleotide, 45 U T4 polynucleotide kinase, 2 m*M* ATP and 1X kinase reaction buffer. Incubate the reaction for 60 min at 37°C and inactivate the enzyme (20 min at 65°C). Add distilled water to get a 1 µ*M* oligonucleotide solution, and store it in aliquots at –20°C.

4. For each PCR of a QPCR titration experiment, an OLA has to be done. Mix 4.0 µL of the PCR product mixture, 70 µL 2X ligation buffer, and 2.8 µL each of 1.0 µ*M* solutions of OLA-up-Bio, OLA-down-D, and OLA-down-F. Fill up to 140 µL using distilled water. Overlay the mixture with 70 µL paraffin oil, denature for 5 min at 95°C, chill it immediately to 0°C, and add 10 µL T4 DNA Ligase (15 Weiss-U/µL). Incubate the ligation mixture for 30 min at 37°C, and stop the reaction by addition of 35 µL 0.25 *M* NaOH. After neutralization with 35 µL 3 *M* sodium acetate pH 6.0 560 µL PBST are added. Immediately afterwards, carry out the ELISA.

5. For the immobilization of ligation products, transfer 80 µL of the OLA dilution into each of six wells in a row of a microtiter plate (*see* **Note 25**). Incubate the plate under agitation for 30 min using a microtiter plate minishaker. Wash the plate six times using 100 µL PBST per wash and per well. Pipet 100 µL 0.1 *M* NaOH into each well, and denature the immobilized dsDNA strands during a 10 min incubation step. Repeat the washing step for the complete removal of free oligonucleotides and denatured PCR product strands.

6. Add 100 µL of the freshly prepared antidigoxigenin-Fab-POD conjugate dilution (*see* **Note 26**) into each of three out of six wells containing the immobilized ligation products of one PCR experiment and 100 µL of the freshly prepared dilution of antifluorescein-Fab-POD conjugate into each of the other three wells. Also fill the antibody solutions into some wells that were not preincubated with OLA dilutions (blanks). Incubate the plate for 30 min with slight agitation. Afterwards, wash the plate six times using 100 µL PBST per well.

7. For the development of the colored POD product, pipet 100 µL of the freshly prepared staining solution per well. Use a multichannel pipet. After some minutes (1–10 min), when there is a distinct bluish color in the wells, the reaction is stopped by addition of 100 µL 2 *M* sulphuric acid, which changes the color from bluish to yellow. Again, use the multichannel pipet for this step. Add the acid in the same time intervals into the different rows as used for the staining solution. Gently agitate the plate containing the stopped solutions for 15 min and measure the absorption in each well at 450 nm using a microplate absorbance reader. The absorbance values should be below 1 $OD^{450\,nm}$.

8. For evaluation of the OLA-ELISA absorbance data, first subtract the mean absorbance value of the blanks (wells without ligation products) from each measured value. Then calculate the mean absorbance value of the three antifluorescein (**F**) and antidigoxigenin (**D**) ELISAs, respectively, of each OLA reaction. Using these mean values, calculate the normalized ELISA data of each OLA for antifluorescein (**F'**) and antidigoxigenin (**D'**) according to the formulas **F' = F / (F + D)** and **D' = D / (F + D)**. Plot the normalized data of each QPCR titration series against the molar concentrations of the initial added internal-control DNA template. The point of intersection of the fluorescein and digoxigenin specific curves indicates equimolar concentrations of sample/internal control template and, therefore, the molar concentration of the initial sample template copies.

## 4. Notes

1. Sequences are based on the *p53* GenBank entry X54156.
2. When designing PCR primers and OLA oligonucleotides, use commercially available computer software (e.g., OLIGO; MedProbe, Oslo, Norway). These computer programs will consider the melting temperatures and bp compositions of suitable primers and exclude self-complementarity of oligonucleotides and PCR primer pairs. In addition, they will calculate the molar extinction coefficients of oligos that you will need for the quantification of oligo solutions by UV spectroscopy, and the melting temperatures of the oligos. *See* **ref. 22** for general aspects of PCR primer design.
3. It is possible to introduce the required chemical modifications into the OLA oligonucleotides and to purify them in your own lab. For this purpose, it is an essential prerequisite that a reverse phase HPLC system is available. Suitable protocols are described in *(12,15,23)*.
4. The OLA oligonucleotide probes should be of similar length in the range of 20–30 bp and possess similar melting temperatures (in the range of 60–70°C). For this assay, it is advantageous when the oligos of the "down" oligonucleotide pair (*see* **Fig. 1**) are of identical length and vary only in their two 5'-terminal bases.
5. In the case of the oligonucleotides OLA-down-D and OLA-down-F, it is important to separate them from prematurely terminated synthesis products. The modified species have to be purified by preparative HPLC or gel electrophoresis. For OLA-up-Bio, the biotin-label can also be introduced by coupling of biotinamidocaproate N-hydroxysuccinimide ester to a 5'-amino-modified oligonucleotide.
6. When you apply this system for the quantification of the mean cellular *p53* copy number in healthy and tumorous tissue, respectively, to detect an allele deletion of the tumor suppressor *p53*, it is very important to use homogeneous and defined tissue samples that contain either tumor cells or normal cells. For this purpose, it is ideal to use microdissected tissue samples that were characterized microscopically and/or via immunohistochemical techniques.
7. For isolation of DNA fragments from agarose gels or aqueous solutions, we use the QIAEX II Gel Extraction Kit and the QIAquick PCR Purification Kit (QIAGEN, Hilden, Germany), for preparation of RNA-free DNA from blood lymphocytes or from tissue samples the QIAamp Tissue Kits.
8. Quantify the concentration of DNA solutions by UV spectroscopy. One milliliter of a solution of double-stranded DNA contains 50 μg DNA when it shows an absorbance of 1 OD at 260 nm and 10 mm optical path length. For quantification of very small volumes, cuvettes are available, that take up only 10 μL but possess a path length of 10 mm. For the quantification of oligonucleotide solutions, use the calculated molar extinction coefficients (*see* **Note 2**).
9. For generation of an internal standard DNA that only slightly differs from the sample template (e.g., in 2 bp) you can use either conventional site directed mutagenesis protocols involving laborious cloning steps *(8)* or PCR mutagenesis strategies *(1,11,13,24)*. We recommend PCR mutagenesis.

10. To prepare a dilution series of the internal-control DNA requires very careful pipeting. You should calibrate pipets gravimetrically. When preparing very diluted solutions of the internal control DNA (subpicomolar range), you should add non-specific DNA to prevent adsorption of the internal control DNA onto tube walls.

11. For preventing contaminations of PCR reactions with amplifiable DNA species, strictly separate the areas for preparation of PCR reagents and stock solutions, pipeting of PCR mixtures, and PCR product analysis.

12. We use polyacrylamide gels of the dimensions $90 \times 85 \times 1$ mm with 18 slots, having a capacity of approx 10 µL per slot. Separation of DNA fragments usually is carried out with a current of 35 mA.

13. Acrylamide and *bis*-acrylamide are highly toxic for the nervous system and carcinogenic. When working with the solid substances and its solutions, strictly obey all relevant safety instructions and regulations (e.g., see supplier's recommendations).

14. Be very careful when working with ethidium bromide (EtBr), because this substance is a carcinogen. Use gloves, do not ingest or touch EtBr or contaminate laboratory areas. Safely and correctly dispose of EtBr contaminated waste.

15. For gel documentation and quantification of relative amounts of DNA fragments, we use either a polaroid instant camera system and positive/negative ISO 80/20° instant film cassettes or a video documentation system, which creates computer files in the TIF-format. In the case of polaroid positive/negative films, we digitize the densitometric information of the negative using a Hewlett-Packard scanning system. The data are then analyzed using the software CREAM (INTAS, Göttingen, Germany).

16. The composition of the OLA buffer is optimized for this oligonucleotide probe system to get high absorbance signals, in addition to high specificity in discriminating sample and internal control-specific PCR products by the OLA. When applying this protocol for another template/oligonucleotide system, it might be necessary to vary the buffer composition slightly (e.g., varying salt or formamide concentration) to get the required specificity. In order to optimize the assay, carry out some OLAs with either sample or internal control-specific PCR products to get only one species of OLA ligation products. Using sample-specific ligation template, the optimized OLA will result in an intense anti-DIG signal and a negligible anti-FITC-signal as background signal or vice versa, respectively, for the internal control-specific ligation template.

17. Polystyrene microtiter plates with MaxiSorp surface have a high affinity for binding proteins via polar groups. Therefore, it is easy to immobilize avidin in the wells of the microtiter plates.

18. We use avidin preparations of Sigma (Deisenhofen, Germany). 14 U avidin/mL corresponds to 1 mg protein/mL.

19. PBST-M has to be freshly prepared every day and should not be stored.

20. For preparation of the staining solution, we use TMB substrate tablets (Sigma) which contain 1 mg TMB each. Be very careful when working with TMB because this substance is a carcinogen (*see* **Note 14**).

21. The amplification reaction mixture consists in all cases of 200 $\mu M$ of each dNTP, 1X *Taq* buffer, 40 U *Taq* DNA polymerase per milliliter reaction volume, 400 n$M$ of each primer, and varying amounts of template DNAs, provided that no other concentrations are quoted.

22. In most cases, genomic DNA can be used as template for the amplification of the "mega-primer," but for the *p53* system analyzed here, we got no product when using genomic blood lymphocyte DNA. When using the PCR product QPCR-1a/1b (*Ssp*I) as template, the desired mega-primer could be amplified using the primer pair QPCR-1a/QPCR-*Hin*dIII.

23. During the multiple steps of the PCR mutagenesis for generation of the internal control DNA template, you should check aliquots of each PCR by polyacrylamide gel electrophoresis (*see* **Note 12**) and verrify that the correct products were synthesized.

24. For quantification of the template copy number in a sample for which no information about the approximate concentration is available, you should perform first a "rough" titration using a dilution series of the internal control DNA in which the concentration is varied by a factor of 10 in the range of 100 p$M$ to 1 a$M$. The "rough" titration will result in values correct within a factor of three. In the second step, perform a titration series in the estimated range of concentration. This titration will result in precise sample template concentrations usually correct within a factor of 1.2. In a typical PCR of 50 μL volume containing 50 ng human genomic blood lymphocyte DNA, the concentration of a single copy gene will be about $5 \times 10^{-16}$ $M$.

25. For the quantitative evaluation of an ELISA, you will have to estimate the background absorbance of the system (microtiter plate, OLA reaction, ELISA reaction) respectively the absorbance of blank wells. Therefore, load some wells of the plate with (i) OLA reactions without ligase, (ii) 1X OLA buffer, and (iii) antibody dilutions without prior immobilization of OLA components. These wells will show the blank absorbance values of the OLA-ELISA system.

26. To estimate the optimal antibody-POD-conjugate dilutions in an ELISA, immobilize constant amounts of an OLA reaction (using an equimolar mixture of sample/internal control-specific PCR products) onto a microtiter plate and apply a dilution series of the antibody-POD-conjugates (diluted in steps of a factor of 2) and carry out the ELISA. The optimal antibody concentrations have to be high enough such that the signal will be independent of the antibody concentration, but as low as possible to reduce the assay costs.

## References

1. McCulloch, R. K., Choong, C. S., and Hurley, D. M. (1995) An evaluation of competitor type and size for use in the determination of mRNA by competitive PCR. *PCR Methods Applic.* **4,** 219–226.

2. Raeymaekers, L. (1993) Quantitative PCR: theoretical considerations with practical implications. *Anal. Biochem.* **214,** 582–585.

3. Raeymaekers, L. (1995) A commentary on the practical applications of competitive PCR. *Genome Res.* **5,** 91–94.

4. Zimmermann, K., and Mannhalter, J. W. (1996) Technical aspects of quantitative competive PCR. *BioTechniques* **21,** 268–279.

5. Garte, S. J., and Ganguly, S. (1996) Quantitative polymerase chain reaction using homologous internal standards. *Anal. Biochem.* **243,** 183–186.

6. Murata, T., Takizawa, T., Funaba, M., Fujimura, H., Murata, E., and Torii, K. (1997) Quantitation of mouse and rat β-actin mRNA by competitive polymerase chain reaction using capillary electrophoresis. *Anal. Biochem.* **244,** 172–174.

7. van Leeuwen, L., Guiffre, A. K., Sewell, W. A., and Atkinson, K. (1996) Rapid synthesis of a DNA competitor for use in quantitative polymerase chain reaction analysis using simple molecular techniques. *Anal. Biochem.* **243,** 196–198.

8. Gilliland, G., Perrin, S., Blanchard, K., and Bunn, H. F. (1990) Analysis of cytokine mRNA and DNA: Detection and quantitation by competitive polymerase chain reaction. *Proc. Natl. Acad. Sci. USA* **87,** 2725–2729.

9. Siebert, P. D. and Larrick, J. W. (1992): Competitive PCR. *Nature* **359,** 557,558.

10. Bartolin, S., Christopoulos, T. K., and Verhaegen, M. (1996) Quantitative polymerase chain reaction using a recombinant DNA internal standard and time-resolved fluorometry. *Anal. Chem.* **68,** 834–840.

11. Apostolakos, M. J., Schuermann, W. H. T., Frampton, M. W., Utell, M. J., and Willey, J. C. (1993) Measurement of gene expression by multiplex competitive polymerase chain reaction. *Anal. Biochem.* **213,** 277–284.

12. Hahn, M., Dörsam, V., Friedhoff, P., Fritz, A., and Pingoud, A. (1995) Quantitative polymerase chain reaction with enzyme-linked immunosorbent assay detection of selectively digested amplified sample and control DNA. *Anal. Biochem.* **229,** 236–248.

13. Hahn, M., Dörsam, V., and Pingoud, A. (1997) Quantitation of *p53* tumor suppressor gene copy number in tumor DNA samples by competitive PCR in an ELISA-format, in *Modern Applications of DNA Amplification Techniques: Problems and New Tools* (Lassner, D., Pustoweit, B., and Rolfs, A., eds.), Plenum, New York.

14. Friedhoff, P., Hahn, M., Wolfes, H., and Pingoud, A. (1993) Quantitative polymerase chain reaction with oligodeoxynucleotide ligation assay/enzyme-linked immunosorbent assay detection. *Anal. Biochem.* **215,** 9–16.

15. Becker-André, M. and Hahlbrock, K. (1989) Absolute mRNA quantification using the polymerase chain reaction (PCR). A novel approach by a *PCR aided transcript titration assay* (PATTY). *Nucleic Acids Res.* **17,** 9437–9446.

16. Eggerding, F. A., Iovannisci, D. M., Brinson, E., Grossman, P., and Winn-Deen, E. S. (1995) Fluorescence-based oligonucleotide ligation assay for analysis of cystic fibrosis transmembrane conductance regulator gene mutations. *Hum. Mutat.* **5,** 153–165.

17. Landegren, U., Kaiser, R., Sanders, J., and Hood, L. (1988) A ligase-mediated gene detection technique. *Science* **241,** 1077–1080.

18. Nickerson, D. A., Kaiser, R., Lappin, S., Stewart, J., Hood, L., and Landegren, U. (1990) Automated DNA diagnostics using an ELISA-based oligonucleotide ligation assay. *Proc. Natl. Acad. Sci. USA* **87,** 8923–8927.

19. Nigro, J. M., Baker, S. J., Preisinger, A. C., Jessup, J. M., Hostetter, R., Cleary, K., Bigner, S. H., Davidson, N., Baylin, S., Devilee, P., Glover, T., Collins, F. S., Weston, A., Modali, R., Harris, C. C., and Vogelstein, B. (1989) Mutations in the *p53* gene occur in diverse human tumour types. *Nature* **342,** 705–708.

20. Ausubel, F. M., Brent, R., Kingston, R. E., Moore, D. D., Seidman, J. G., Smith, J. A., and Struhl, K., eds. (1987-1995) *Current Protocols in Molecular Biology.* 3 Vols., John Wiley & Sons, New York.

21. Dieffenbach, C. W., Lowe, T. M. J., and Dveksler, G. S. (1993) General concepts for PCR primer design. *PCR Methods Appl.* **3,** S30–S37.

22. Hahn, M., Matzen, S. E., Serth, J., and Pingoud, A. (1995) Semiautomated quantitative detection of loss of heterozygosity in the tumor suppressor gene *p53*. *BioTechniques* **18,** 1040–1047.

23. Perrin, S. and Gilliland, G. (1990) Site-specific mutagenesis using asymmetric polymerase chain reaction and a single mutant primer. *Nucleic Acids Res.* **18,** 7433–7438.

24. Silver, J., Limjoco, T., and Feinstone, S. (1995) Site-specific mutagenesis using the polymerase chain reaction, in *PCR Strategies* (Innis, M. A., Gelfand, D. H., and Sninsky, J. J., eds.), Academic, London, pp. 179–188.

# 15

# Quantitative Analysis of Human DNA Sequences by PCR and Solid-Phase Minisequencing

Anu Suomalainen and Ann-Christine Syvänen

## 1. Introduction

### 1.1. General Remarks

The PCR technique provides highly specific and sensitive means for analyzing nucleic acids, but it does not allow their direct quantification. This limitation originates from the fact that the efficiency of PCR depends on the amount of template sequence present in the sample, and the amplification is exponential only at low template concentrations *(1)*. Because of this "plateau effect" of the PCR, the amount of the amplification product does not directly reflect the original amount of the template. Moreover, subtle differences in the reaction conditions, such as material from biological samples, might cause significant sample to sample variation in the final yield of the PCR product.

The problem of performing accurate quantitative PCR analyses has been addressed by two principal approaches. A quantitative PCR result can be obtained by kinetic PCR , in which the amplification process is monitored at numerous time or concentration points *(2,3)*. This approach is laborious and requires a sensitive method for detecting the PCR products at a stage during which the amplification still proceeds exponentially. The other approach, competitive PCR, utilizes an internal quantification standard sequence that is coamplified in the same reaction as the target sequence *(4–6)*. The efficiency of the amplification is affected by the sequence of the PCR primers, as well as by the size and the sequence of the template. Therefore, the internal standard should be as similar to the target sequence as possible ensuring that the ratio of the two sequences remains constant throughout the amplification. An ideal PCR quantification standard differs from the target sequence only at one nucleotide

From: *Methods in Molecular Medicine, Vol 26: Quantitative PCR Protocols*
Edited by: B. Kochanowski and U. Reischl © Humana Press Inc., Totowa, NJ

position, by which the two sequences can be identified and quantified after the amplification. Determination of the relative amounts of the PCR products that originate from the target and standard sequences allows calculation of the initial amount of the target sequence. If two target sequences are present as a mixture in a sample, it is easy and often sufficient to measure their relative amounts. To be able to determine the absolute amount of a target sequence, it is necessary to add a known amount of standard sequence to the sample before the amplification. In this case, a measure of the amount of the analyzed sample, such as the number of cells or the total amount of DNA, RNA, or protein, is needed.

We have developed a solid-phase minisequencing method for identification of point mutations or nucleotide variations in human genes *(7)*. This method is based on distinct detection of two sequences that differ from each other only at a single nucleotide, making the method an ideal tool for quantitative analysis of DNA *(8)* and RNA *(9)* sequences by competitive PCR. In the solid-phase minisequencing method, a DNA fragment containing the site of the variable nucleotide is first amplified using one biotinylated and one unbiotinylated PCR primer. The PCR product carrying a biotin residue at the 5' end of one of its strands is captured on an avidin-coated solid support and denatured. The nucleotides at the variable site in the immobilized DNA strand are then identified by two separate primer extension reactions in which a single labeled deoxynucleotide triphosphate (dNTP) is incorporated by a DNA polymerase **(Fig. 1)**. Our standard assay format utilizes [$^3$H]dNTPs as labels and streptavidin-coated microtiter plates as the solid support *(10)*. The results of the assay are numeric counts per minute (cpm)-values expressing the amount of [$^3$H]dNTP incorporated in the minisequencing reactions. The ratio between the cpm-values obtained in the minisequencing assay (R-value) reflects directly the ratio between the two sequences in the original sample **(Fig. 2)**. The method allows quantitative determination of a sequence present as a minority of less than 1% of a sample, i.e., the dynamic range for the quantitative analysis spans five orders of magnitude *(8)*. Furthermore, because the two sequences differ from each other by a single nucleotide, they are amplified with equal efficiency during PCR, and the R-value obtained is not affected by the amount of template present in the reaction **(Fig. 3)**. Consequently, the quantitative analysis can be performed irrespective of the phase of the PCR process.

## 1.2. Examples of Applications of the Solid-Phase Minisequencing Method

### 1.2.1. Determination of Allele Frequencies by Quantitative Analysis of Pooled DNA Samples

We have developed a system for forensic DNA typing in which a panel of 12 biallelic sequence polymorphisms is analyzed by the solid-phase minisequencing

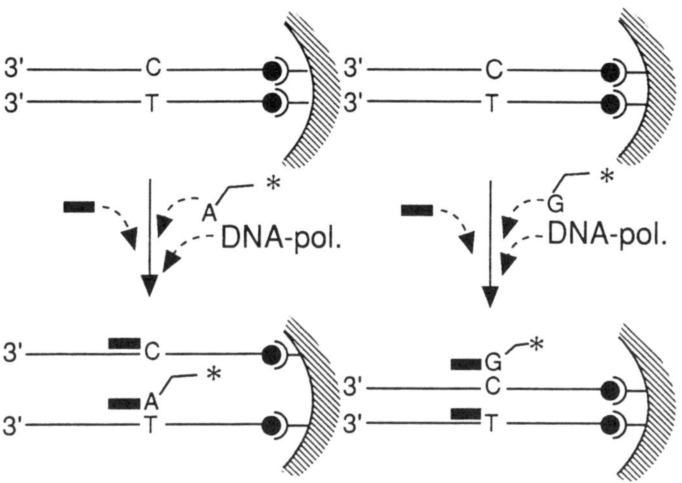



Fig. 1. Principle of the solid-phase minisequencing method. Analyses for two nucleotides performed in separate wells, are shown on the left and right. One PCR primer is biotinylated at its 5' end, resulting in a PCR product carrying biotin at the 5' end of one of its strands (filled circle). The product is captured in a streptavidin-coated microtiter well and denatured (upper part of the figure). Lower part of the figure: a detection step primer hybridizes to the single-stranded template, 3' adjacent to the variant nucleotide. The DNA polymerase extends the primer with the [$^3$H]-labeled dNTP, if it is complementary to the nucleotide present at the variable site. After washes, the sample is denatured, and the eluted radioactivity, expressing the amount of the incorporated label, is measured.

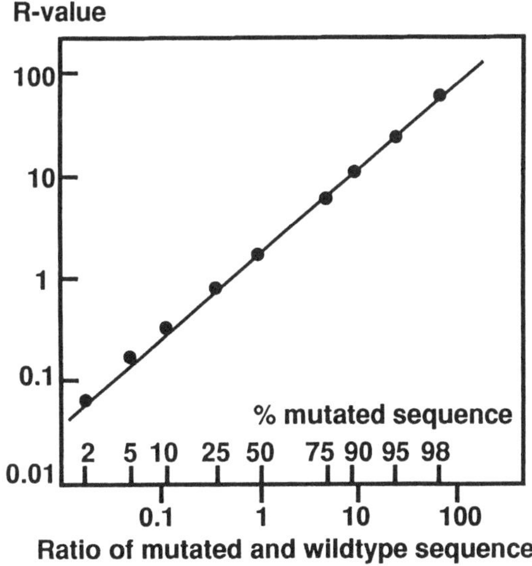



Fig. 2. Solid-phase minisequencing standard curve prepared by analyzing mixtures of two 63-mer oligonucleotides differing from each other at one nucleotide in the mitochondrial tRNA$^{Leu(UUR)}$ gene (12). The Ccpm/Tcpm ratio obtained in the minisequencing reactions is plotted as a function of the original ratio between two oligonucleotides in the mixtures.

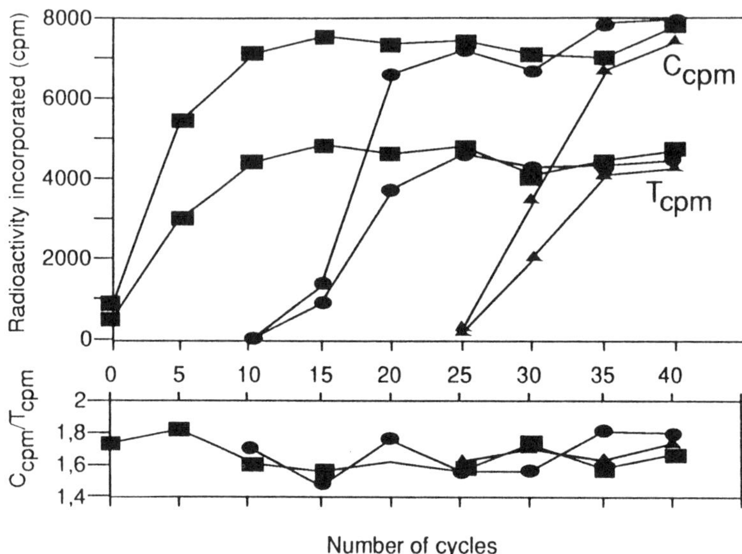



Fig. 3. Result of the solid-phase minisequencing assay obtained at different PCR cycles and amounts of template. Mixtures of equal amounts ($10^3$, $10^7$, or $10^{11}$ molecules) of the same oligonucleotides as in Fig. 2 were analyzed. The upper panel shows the cpm-values obtained in the minisequencing assay at different PCR cycles , and the lower panel shows the corresponding Ccpm/Tcpm ratios. Models of template: ■, $10^{11}$; ●, $10^7$; ▲, $10^3$.

method *(10)*. The statistical interpretation of the forensic and paternity testing results requires information on the allele frequencies of the analyzed markers in each particular population. To obtain rapidly this information in the Finnish population, we utilized the quantitative nature of the solid-phase minisequencing method to determine the allele frequencies of the polymorphic markers by analyzing pooled DNA samples derived from hundreds of individuals. The ratio between the two sequences at each polymorphic locus in the pooled DNA samples is equivalent to the allele frequencies in the population. **Table 1** shows the results from the analysis of allele frequencies of a polymorphism in the PROS1 gene on chromosome 13. A good correlation between the allele frequencies determined from the pooled samples and those determined from about 100 alleles individually were observed *(10)*. Analogously, we have determined the carrier frequency of the recessively inherited aspartylglucosaminuria in Finland by determining the frequency of the mutant allele by quantitative analysis of the pooled DNA samples *(8)*.

**Table 1
Example of the Determination of the Allele Frequencies
of a Polymorphism in the PROS1 Gene by Quantitative Analysis
of Pooled DNA Samples[a]**

| Sample[a] | [3H]dNTP incorporated (cpm) | | R-values | Allele frequency[c] | |
|---|---|---|---|---|---|
| | A-reaction[b] | G-reaction[b] | Acpm/Gcpm | A-allele | G-allele |
| Pool 1390 | 2750 | 1280 | 2.14 | 0.59 | 0.41 |
| Pool 860 | 2240 | 1048 | 2.15 | 0.59 | 0.41 |
| Pool 920 | 2510 | 1190 | 2.11 | 0.58 | 0.42 |
| Control (AA) | 2100 | 52 | 40 | / | / |
| Control (GG) | 96 | 1930 | 0.050 | / | / |
| Control (AG) | 2480 | 1660 | 1.49 | / | / |
| No DNA | 64 | 39 | / | / | / |

[a]The figure gives the number of individuals in each pool. AA, GG and AG indicate the genotypes of the individual used as control. Mean values of five (pools) or two (individual controls) parallel assays are given.

[b]The specific activities of [3H]dATP and [3H]dGTP were 58 and 32 Ci/mmol, respectively.

[c]The allele frequencies determined from 50 individual samples were 0.61 (A) and 0.39 (G) *(10)*.

## 1.2.2. Detection of Heteroplasmic Point Mutations of the Mitochondrial DNA

Disease-causing point mutations of mitochondrial DNA (mtDNA) are most often heteroplasmic (i.e., the tissues of the patients contain both mutant and normal mtDNA). The solid-phase minisequencing method is particularly useful for detecting heteroplasmic mtDNA mutations, allowing both identification and quantification of the mutation in the same assay. Using this method, we have observed a correlation between the degree of heteroplasmy and the severity and age of onset of the disease in two mitochondrial disorders associated with mtDNA point mutations *(11,12)*.

## 1.2.3. Determination of Gene Copy Numbers

We have developed an alternative method for the widely used fluorescence *in situ* hybridization techniques to determine the copy number of human genes. The copy number of aspartylglucosaminidase, a marker gene located on chromosome 4q, was determined by solid-phase minisequencing in the samples of three patients with either deletions or duplications involving the distal region of chromosome 4q. A known amount of DNA from a patient homozygous for a mutation in the marker gene was mixed with the DNA samples to be analyzed, to serve as an internal standard, and the relative amount of the normal sequence was determined by the solid-phase minisequencing method *(13)* (*see* **Table 2**).

**Table 2
Determination of the Copy Number
of the Aspartylglucosaminidase Gene *(13)***

| Karyotype of sample | $C_{cpm}/(C_{cpm} + G_{cpm})$ | Deduced AGA gene copy number |
|---|---|---|
| 46,XY,–4,+der(4) t(4;12) (q31.3;p12.2)mat | 0.28–0.33[a] | 1 |
| 46,XX,del(4)(q33) | 0.26–0.32[a] | 1 |
| 46,XX,–21,+der(21) t(4;21) (q28;p13)mat | 0.63–0.70[a] | 3 |
| Controls[b] | 0.50–0.54 | 2 |

[a]Range of variation of five parallel assays.
[b]DNA from individuals heterozygous for the target nucleotide in the aspartylglucosaminidase gene.

This application demonstrates the suitability of the method for determining monosomies, trisomies, and loss of heterozygosity, provided that a DNA standard containing a suitable polymorphism is available.

### 1.2.4. Identification of Mixed Samples

The R-values obtained by the solid-phase minisequencing method, when individual genomic DNA samples are analyzed for a variable nucleotide, fall into three distinct categories that unequivocally define the genotype of the sample. R-values falling outside these three categories that normally differ from each other by a factor of ten suggest the presence of contaminating DNA in the sample *(10)*. The ability of the method to identify a mixed sample is an advantage in forensic analyses, wherein stain samples may contain DNA from several individuals as well as in prenatal diagnosis, where placental biopsy samples may contain contaminating maternal DNA.

## 2. Materials

### 2.1. Equipment

1. Programmable heat block and facilities to avoid contamination in PCR.
2. Microtiter plates with streptavidin-coated wells (e.g., Combiplate 8, Labsystems, Finland) (*see* **Note 1**).
3. Multichannel pipet and microtiter plate washer (optional).
4. Shaker at 37°C.

5. Water bath or incubator at 50°C
6. Liquid scintillation counter.

## 2.2 Reagents

All the reagents should be of standard molecular biology grade. Use sterile distilled or deionized water. Store reagents for PCR at –20°C:

1. Thermostable DNA polymerase. We use *Thermus aquaticus* (5 U/µL, Promega Corp.) or *Thermus brockianus* (Dynazyme™ II, 2 U/µL, Finnzymes) DNA polymerase (*see* **Note 2**).
2. 10X concentrated DNA polymerase buffer: 500 m$M$ Tris-HCl, pH 8.8, 150 m$M$ (NH$_4$)$_2$SO$_4$, 15 m$M$ MgCl$_2$, 1% (v/v) Triton X-100, 0.1% (w/v) gelatin.
3. dNTP mixture: 2 m$M$ dATP, 2 m$M$ dCTP, 2 m$M$ dGTP, and 2 m$M$ dTTP.

   Reagents for the minisequencing analysis:

4. PBS/Tween: 20 m$M$ sodium phosphate buffer, pH 7.5, and 0.1% (v/v) Tween 20. Store at 4°C. 50 mL is enough for several full plate analyses.
5. TENT (washing solution): 40 m$M$ Tris-HCl, pH 8.8, 1 m$M$ EDTA, 50 m$M$ NaCl, and 0.1% (v/v) Tween-20. Store at 4°C. Prepare 1–2 L at a time, which is enough for several full-plate analyses.
6. 50 m$M$ NaOH (make fresh every 4 wk), store at room temperature (~20°C). Prepare 50 mL.
7. [$^3$H]-Labeled deoxynucleotides (dNTPs): dATP to detect a T at the variant site, dCTP to detect a G and so forth. (Amersham; [$^3$H]dATP, TRK 625; dCTP, TRK 576; dGTP, TRK 627; dTTP, TRK 633), store at –20°C (*see* **Note 3**).
8. Scintillation reagent (for example Hi-Safe II, Wallac).

## 2.3. Primer Design

9. PCR primers: Biotinylate one of the PCR primers at its 5' end during the synthesis using a biotin-phosphoramidite reagent (for example Amersham or Perkin Elmer/ABI) (*see* **Note 4**).
10. If oligonucleotides are used as quantification standards (*see* **Subheading 2.4.**), the length of oligonucleotides that can be synthesized with acceptable yields sets an upper limit for the length of the PCR product to about 80–100 bp.
11. The detection step primer for the minisequencing analysis is an oligonucleotide complementary to the biotinylated strand of the PCR product, designed to hybridize with its 3' end immediately adjacent to the variant nucleotide to be detected (*see* **Fig. 1**) The detection step primer for our standard protocol is a 20-mer. The primer should be at least five nucleotides nested in relation to the unbiotinylated PCR primer.

## 2.4. Quantification Standards

To quantify accurately a sequence in a sample that contains only one sequence type, a standard should be designed to differ from the target sequence at the nucleotide to be detected in the minisequencing reaction (*see* **Fig. 1**). To construct a standard curve, a second standard identical to the target sequence is required (*see* **Subheading 3.3.**). Oligonucleotide standards can be synthesized using a DNA synthesizer (*see* **Note 5**), or PCR products or cloned DNA fragments can be used. Measure the molecular concentrations of the standards. The optimal amount of the standard added to a sample depends on the abundance of the target sequence in the original sample. The ratio of the target to the standard sequence should preferably be between 0.1 and 10. If no estimate of the amount of the target sequence is available, it may be necessary to initially titrate the optimal amount of the standard in the analysis, for example $10^2$, $10^4$, $10^6$, and $10^8$ molecules. For accurate quantification, standards representing both sequence variants should be available, and analysis of mixtures of known amounts of the two standards should be analyzed to construct a standard curve, as demonstrated in **Figs. 1** and **2,** and **Table 3**.

## 3. Methods

### 3.1. PCR for Solid-Phase Minisequencing Analysis

The PCR follows the routine protocols, except that the amount of the biotin-labeled primer should be reduced not to exceed the biotin-binding capacity of the microtiter well (*see* **Note 1**). For a 50 µL PCR reaction, we use 10 pmol of biotin-labeled primer and 50 pmol of the unbiotinylated primer. The PCR should be optimized (i.e., the annealing temperature and the amount of the template) to be efficient and specific. To be able to use [$^3$H]dNTPs, which have low specific activities, for the minisequencing analysis, 1/10 of the PCR product should produce a single visible band after agarose gel electrophoresis and staining with ethidium bromide.

### 3.2. Solid-Phase Minisequencing Analysis

1. Affinity capture: Transfer 10 µL aliquots of the PCR product and 40 µL of PBS/ Tween to two streptavidin-coated microtiter wells (*see* **Note 6**). Include as negative controls two wells without PCR product. Seal the wells with a sticker, and incubate the plate at 37°C for 1.5 h with gentle shaking.
2. Discard the liquid from the wells, and tap the wells dry against a tissue paper.
3. Wash the wells three times at room temperature by adding 200 µL of TENT to each well, discard the washing solution, and empty the wells thoroughly between the washing steps (*see* **Note 7**).
4. Denature the captured PCR product by adding 100 µL of 50 m*M* NaOH to each well, followed by incubation at room temperature for 3 min. Discard the NaOH, and wash the wells as in **step 3**.
5. For each DNA fragment to be analyzed prepare two 50 µL mixtures of nucleotide-specific minisequencing solution, one for detection of the wild-type and one for the mutant nucleotide, by mixing 5 µL of 10X DNA polymerase buffer, 10 pmol of detection step primer (for example 2 µL of 5 µ*M* primer), 0.2 µCi (usually 0.2 µL)

**Table 3**
**Example of the Result of a Solid-Phase Minisequencing Analysis of Mixtures of Two 63-mer Oligonucleotides Differing from Each Other at One Nucleotide in the Mitochondrial tRNA$^{Leu(UUR)}$ gene *(12)***

| Ratio wild-type/ mutated sequence | Tcpm (wt oligo)[a] | Ccpm (mut oligo)[b] | Ccpm/Tcpm |
|---|---|---|---|
| Wild-type | 3110 | 44 | 0.014 |
| 50:1 | 3640 | 190 | 0.05 |
| 20:1 | 2780 | 420 | 0.15 |
| 10:1 | 2830 | 730 | 0.26 |
| 4:1 | 2520 | 1690 | 0.67 |
| 1:2 | 1650 | 2810 | 1.7 |
| 1:4 | 790 | 3630 | 4.6 |
| 1:10 | 350 | 3790 | 10.8 |
| 1:20 | 210 | 4760 | 22.7 |
| 1:50 | 120 | 4800 | 40.0 |
| Mutant | 43 | 4580 | 106.5 |
| $H_2O$ | 41 | 23 | / |

[a]The specific activities of the [$^3$H]dNTPs: dTTP 126 Ci / mmol, dCTP 67 Ci / mmol.
[b]In this case, two [$^3$H]dCTPs were incorporated into the mutant sequence.

of one [$^3$H]dNTP, 0.1 U of DNA polymerase, and $H_2O$ to a total volume of 50 µL. It is obviously convenient to prepare master mixes for the desired number of analyses with each nucleotide (*see* **Note 8**).

6. Add 50 µL of one nucleotide-specific mixture to each well, and incubate the plate at 50°C for 10 min (*see* **Note 9**).
7. Discard the contents of the wells, and wash them as in **step 3**.
8. Release the detection step primer from the template by adding 60 µL 50 m$M$ NaOH and incubate for 3 min at room temperature.
9. Transfer the NaOH solution containing the eluted primer to the scintillation vials, add scintillation reagent, and measure the radioactivity, i.e., the amount of incorporated label, in a liquid scintillation counter (*see* **Note 10**).
10. The result is obtained as cpm-values. The cpm-value of each reaction expresses the amount of the incorporated [$^3$H]dNTP. Calculate the ratio (R) (*see* **Table 3** and **Note 11**) :

R = (cpm incorporated in the reaction detecting the wild-type (target) sequence)/
(cpm incorporated in the reaction detecting the mutated (standard) sequence

### 3.3. Preparation of the Standard Curve

Mix the wild-type and mutated standard sequences in known proportions, for example 1:50, 1:20, 1:10, 1:4, 1:2, 4:1, 10:1, 20:1, and 50:1. Plot the resulting R-values on a log–log scale as a function of the ratio between the sequences present in the original mixture, which should result in a linear standard curve (*see* **Table 2** and **Fig. 2**). This curve can then be utilized for the analysis of the actual samples with an unknown amount of target sequence either to determine the relative or the absolute amount of the target.

## 4. Notes

1. The binding capacity of the streptavidin-coated microtiter well (Labsystems) is 2–5 pmol of biotinylated oligonucleotide. If a higher binding capacity is desired, avidin-coated polystyrene beads (Fluoricon, 0.99 μm, IDEXX Corp., Portland ME) or streptavidin-coated magnetic polystyrene beads (Dynabeads M-280) can be used *(14)*.
2. The use of a thermostable DNA polymerase in the single-nucleotide primer extension reaction is advantageous, since a high temperature favorable for the simultaneous primer annealing reaction can be used.
3. Although the [$^3$H]dNTPs are weak β-emitters, their half lives are long (13 yr), and the necessary precautions for working with [$^3$H] should be taken. Also, dNTPs or dideoxy-nucleotides labeled with other isotopes ([$^{35}$S] or[$^{32}$P], ref *7*) or with fluorophores *(15)* can be used.
4. The efficiency of the 5'-biotinylation of an oligonucleotide on a DNA synthesizer is most often 80–90%. The biotin-labeled oligonucleotides can be purified from the unbiotinylated ones either by high-performance liquid chromatography *(16)*, polyacrylamide gel electrophoresis *(17)*, or by disposable ion exchange columns manufactured for this purpose (Perkin-Elmer/ABI). If the biotin-labeled primer is used without purification, the success of the biotinylation can be confirmed after the PCR by affinity capture of an aliquot of the biotinylated PCR product on an avidin-matrix with high biotin-binding capacity (*see* **Note 1**). Analyze the supernatant after the capturing reaction by agarose gel electrophoresis. If the biotinylation has been efficient, no product, or a faint product of significantly lower intensity than the unbound PCR product, is observed in the supernatant.
5. For use as quantification standards, full-length oligonucleotides should be purified from prematurely terminated ones by high-performance liquid chromatography *(16)* or by polyacrylamide gel electrophoresis *(17)*. The molecular concentration of the purified full-length standard DNA can then be accurately determined.
6. Each nucleotide to be detected at the variant site is analyzed in a separate well. Thus, at least two wells are needed per PCR product. For quantitative applications, we carry out two parallel assays for each nucleotide, i.e., four wells per PCR product.
7. The washing can be performed utilizing an automated microtiter plate washer, or by manually pipeting the washing solution to the wells, discarding the liquid and tapping the plate against tissue paper. It is important for the specificity of the minisequencing reaction to thoroughly empty the wells between the washing steps to remove completely all dNTPs from the PCR. The presence of other dNTPs

than the intended [$^3$H]dNTP during the minisequencing reaction will cause unspecific extension of the detection step primer.

8. The minisequencing reaction mixture can be stored at room temperature for 1–2 h. It is convenient to prepare it during incubation in **step 2**.

9. The conditions for hybridizing the detection step primer are not stringent, and the temperature of 50°C can be applied to analysis of most PCR products irrespective of the sequence of the detection step primer. However, if the primer is considerably shorter than a 20-mer, or it is very AT-rich (melting temperature close to 50°C), lower temperatures for the primer annealing may be required.

10. Streptavidin-coated microtiter plates made of scintillating polystyrene are available (ScintiStrips, Wallac, Finland). When these plates are used, the final washing, denaturation, and transfer of the eluted detection primer to scintillation vials can be omitted, but a scintillation counter for microtiter plates is needed *(18)*.

11. The ratio between the cpm-values for the two nucleotides reflects the ratio between the two sequences in the original sample. The R-value is affected by the specific activities of the [$^3$H]dNTPs used, and if either the wild-type or the mutant sequence allows the detection step primer to be extended by more than one [$^3$H]dNTP, this will obviously also affect the R-value. Both of these factors can easily be corrected when calculating the ratio between the two sequences. Alternatively, a standard curve can be used to correct for these factors (*see* **Subheading 3.3.**).

# References


1. Syvänen, A. C., Bengtström, M., Tenhunen, J., and Söderlund, H. (1988) Quantification of polymerase chain reaction products by affinity-based hybrid collection. *Nucleic Acids Res.* **16,** 11,327–11,338.
2. Murphy, L. D., Herzog, C. E., Rudick, J. B., Fojo, A. T., and Bates, S. E. (1990) Use of polymerase chain reaction in the quantitation of mdr-1 gene expression, *Biochemistry* **29,** 10,351–10,356.
3. Noonan, K. E., Beck, C., Holzmayer, T. A., Chin, J. E., Wunder, J. S., Andrulis, I. L., Gazdar, A. F., Willman, C. L., Griffith, B., von Hoff, D. D., and Roninson I. B. (1990) Quantitative analysis of MDR1 (multidrug resistance) gene expression in human tumors by polymerase chain reaction. *Proc. Natl. Acad. Sci. USA* **87,** 7160–7164.
4. Chelly, J., Kaplan, J.-C., Maire, P., Gautron, S., and Kahn, A. (1988) Transcription of the dystrophin gene in human muscle and non-muscle tissues. *Nature* **333,** 858–860.
5. Wang, A. M., Doyle, M. V., and Mark, D. F. (1989) Quantitation of mRNA by the polymerase chain reaction. *Proc. Natl. Acad. Sci. USA* **86,** 9717–9721.
6. Gilliland, G., Perrin, S., Blanchard, K., and Bunn, H. F. (1990) Analysis of cytokine mRNA and DNA: detection and quantitation by competitive polymerase chain reaction. *Proc. Natl. Acad. Sci. USA* **87,** 2725–2729.
7. Syvänen, A.-C., Aalto-Setälä, K., Harju, L., Kontula, K., and Söderlund, H. (1990) A primer-guided nucleotide incorporation assay in the genotyping of apolipoprotein E. *Genomics* **8,** 684–692.

8. Syvänen, A.-C., Ikonen, E., Manninen, T., Bengtström, M., Söderlund, H., Aula, P., and Peltonen, L. (1992) Convenient and quantitative detection of the frequency of a mutant allele using solid-phase minisequencing: Application to aspartylglucosaminuria in Finland. *Genomics* **12**, 590–595.

9. Ikonen, E., Manninen, T., Peltonen, L., and Syvänen, A.-C. (1992) Quantitative determination of rare mRNA species by PCR and solid-phase minisequencing. *PCR Methods Appl.* **1**, 234–240.

10. Syvänen, A.-C., Sajantila, A., and Lukka, M. (1993) Identification of individuals by analysis of biallelic DNA markers, using PCR and solid-phase minisequencing. *Am. J. Hum. Genet.* **52**, 46–59.

11. Suomalainen, A., Kollmann, P., Octave, J.-N., Söderlund, H., and Syvänen, A.-C. (1993) Quantification of mitochondrial DNA carrying the tRNA8344Lys point mutation in myoclonus epilepsy and ragged-red-fiber disease, *Eur. J. Hum. Genet.* **1**, 88–95.

12. Suomalainen, A., Majander, A., Pihko, H., Peltonen, L., and Syvänen, A.-C. (1993) Quantification of tRNA3243Leu point mutation of mitochondrial DNA in MELAS patients and its effects on mitochondrial transcription. *Hum. Mol. Genet.* **2**, 525–534.

13. Laan, M., Grön-Virta, K., Salo, A., Aula, P., Peltonen, L., Palotie, A., and Syvänen, A.-C. (1995) Solid-phase minisequencing confirmed by FISH analysis in determination of gene copy number. *Hum. Genet.* **96**, 275–280.

14. Syvänen, A.-C., and Söderlund, H. (1993) Quantification of polymerase chain reaction products by affinity-based collection. *Meth. Enzymol.* **218**, 474–490.

15. Pastinen, T., Partanen, J., and Syvänen, A.-C. (1996) Multiplex, fluorescent solid-phase minisequencing for efficient screening of DNA sequence variation. *Clin. Chem.* **42**, 1391–1397.

16. Bengtström, M., Jungell-Nortamo, A., and Syvänen, A.-C. (1990) Biotinylation of oligonucleotides using a water soluble biotin ester. *Nucleotides* **9**, 123–127.

17. Wu, R., Wu, N.-H., Hanna, Z., Georges, F., and Narang, S. (1984) *Oligonucleotide Synthesis: A Practical Approach* (Gait, M. J., ed.), IRL Press, Oxford, UK.

18. Ihalainen, J., Siitari, H., Laine, S., Syvänen, A.-C., and Palotie, A. (1994) Towards automatic detection of point mutations: use of scintillating microplates in solid-phase minisequencing. *BioTechniques* **16**, 938–943.

# 16

# High Resolution PCR Quantitation by AmpliSensor Assay

**Chang-Ning Wang**

## 1. Introduction

A real-time kinetic tool for polymerase chain reaction (PCR) quantitation, AmpliSensor assay *(1)* quantifies PCR product by relating the rate of an amplification reaction through the progressive depletion of a rate-limiting primer. AmpliSensor assay invokes a two-step amplification scheme:

1. An initial amplification with one primer in excess in order to overproduce one strand of the target, and
2. A subsequent hemi-nested amplification for fluorescence signal detection.

The assay monitors the energy-transfer-based fluorescence of an AmpliSensor primer duplex introduced as rate-limiting primer during the late-log phase of an amplification and reports the rate of dissociation of the primer duplex as it is integrated into the amplification product.

AmpliSensor is a signal-generating duplex labeled to each strand with a fluorescence energy-transfer donor and an energy-transfer acceptor, respectively. The donor and acceptor fluorophores are so configured that maximum fluorescence will issue when the duplex remains hybridized. As the primer component of AmpliSensor is engaged in elongation, the duplex dissociates, causing disruption of the energy-transfer process. The extent of disruption and the accompanying signal drop correlates directly to the event of strand duplication and can be measured in a cycle-by-cycle manner. Assuming the duplication is target-specific, the rate of the target amplification can then be assessed progressively by normalizing the amount of amplified product to the amount of residual AmpliSensor primer duplex. Therefore, the rate difference between

From: *Methods in Molecular Medicine, Vol 26: Quantitative PCR Protocols*
Edited by: B. Kochanowski and U. Reischl © Humana Press Inc., Totowa, NJ

individual reaction can be resolved and the initial target dosage of each sample can be determined accurately by simple standard-curve extrapolation.

Unlike other detection methods, AmpliSensor assay monitors the perturbation of a rate-limiting primer rather than the accumulation of amplified product. The linearity belonging to the exponential phase of a PCR growth curve can thus be extended into the post-exponential phase mathematically. The advantage of the assay lies in its ability to take into account of the rate difference between individual amplification reactions, making high-resolution quantitation possible. To prevent false negatives, the assay employs internal mimic target as a control or the user may simply check the remaining polymerase activity at the end of reaction. The assay has been applied to areas such as mRNA quantitation *(2)*, viral load study *(3)*, gene stability analysis *(4)*, detection of food contamination *(5)* and screening of genetic disorders *(1)*.

## *1.1. Primer Design*

A single-stranded target-specific primer can be converted into an AmpliSensor primer duplex through a simple ligation reaction. After ligation, the primer duplex can be used in a hemi-nested PCR reaction to monitor the extent of amplification. A system primer ready signal tagged oligonucleotide (PReSTO) provides an effective means for such a conversion. In the PReSTO system, the AmpliSensor is 11 bp long signal-generating oligonucleotide duplex that can be readily ligated to a 5'C ending oligonucleotide primer. The ligation is mediated by a two-nucleotide sticky end and thus imposes minimum sequence requirement on the target-specific primer.

There are three primers involved in an AmpliSensor assay (*see* **Fig. 1**). In order to have an optimum amplification, these primers must be carefully selected to satisfy the following criteria:

1. The amplicon of the initial amplification should be between 100 and 200 bp;
2. The amplicon of the hemi-nested amplification should be between 50 and 80 bp;
3. The hemi-nested primer should end with a C at its 5' termini for effective ligation;
4. The optimum size for the excess primer is 22–25 mers, 20–22 mers for the limiting primer, and 15–20 mers for the hemi-nested primer;
5. The GC content of each primer should not exceed 70%.

## 2. Materials

Items 1–8 are available from AcuGen/Biotronics Inc. (Lowell, MA).

1. PReSTO AmpliSensor Kit (cat. no. 10-1788-10).
2. 96-Well thermoplate temperature cycler (cat. no. 90-1030-T).
3. Fluorescence thermoplate reader (cat. no. 90-1000-A).

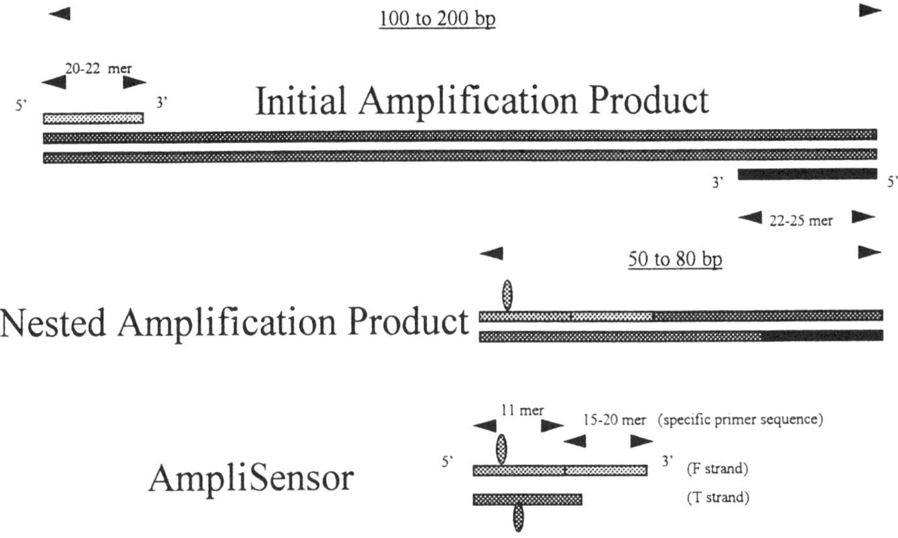


Fig. 1. Scheme of the AmpliSensor assay.

4. Data-acquisition software (cat. no. 90-1000-S).
5. Electronic multi-dispenser (cat. no. 90-1040-P).
6. 96-Well thermoplate (cat. no. 95-1000-30).
7. Thermoplate cover (cat. no. 95-1000-35).
8. Mineral/dispensing oil (cat. no. 95-1000-20).
9. Taq DNA Polymerase (e.g., Perkin Elmer, Norwalk, CT).
10. 10X dNTP solution containing 1 m$M$ each of the four deoxyribonucleotides.
11. 10X PCR buffer. Prepare a solution comprising of 500 m$M$ Tris-HCl, pH 8.8, 40 m$M$ MgCl$_2$, 400 m$M$ KCl, 50 m$M$ NH$_4$Cl, 10 m$M$ dithiothreitol (DTT) and 1% Triton X-100.
12. 10X Excess primer (40 ng/μL).
13. 10X Limiting primer (8 ng/μL).
14. Hemi-nested primer (500 ng/μL).
15. 10X Inhibition test probe (50 ng/μL).
16. TE buffer: 10 m$M$ Tris-HCl, 1 m$M$ ethylenediaminetetriaacetic acid (EDTA), pH 8.0.
17. 10X Kinasing buffer: (700 m$M$ Tris-HCl, pH 7.6, 100 m$M$ MgCl$_2$, 5 m$M$ DTT, 10 m$M$ ATP.
18. Personal computer (IBM Compatible, Windows 95, 16 MB RAM, SVGA).
19. MicroPipet and corresponding pipet tips.
21. 1.5-mL Reaction tubes.
22. Deionized or distilled water

# 3. Methods

## 3.1. Kinasing and Ligation

1. To a 0.6 mL microtube, add the following reagents at room temperature: 2 µL 10X kinasing buffer, 2 µL 10 m*M* ATP, 2 µL T4 polynucleotide kinase (10 U/µL), 6 µg hemi-nested primer (15 mer). Add ddH$_2$O to a final volume of 20 µL.
2. Incubate at 38°C for 4 h followed by 90°C for 5 min to inactivate the enzyme. Concentration accuracy is crucial to the success of the assay. For primer of different size, adjust its dosage according to the formula Q = 6 µg X (primer length/15 nucleotides).
3. Add 5 µL of the kinased primer to the PReSTO ligation mix (*see* **Note 1**).
4. Vortex vigorously for 30 s until the mix turns opaque.
5. Incubate the reaction at 30°C for 3 h. At the end of incubation, the reaction mix will form two phases with the ligated product segregated to the lower phase.
6. Remove 30 µL of the top phase by pipeting under UV illumination to avoid uptaking the condensed fluorescent interphase.
7. Add 30 µL 1X PCR buffer to dilute the ligation product and label it as 20X AmpliSensor primer stock. Store the stock in freezer.

## 3.2. Initial Amplification

The amplification should be carried out in a 96-well thermoplate. The assay is a two-step reaction. In the first step, PCR is carried out to its late-log phase before the limiting primer approaches depletion. In the second step, AmpliSensor primer duplex, whose primer has the same polarity as the limiting primer is added to prime a hemi-nested amplification in concert with the excess primer.

1. Set up an assay template. Include in the template a Negative Control, an Apex, and at least four positive standards. The Apex is a constant signal reference to which all the readings will be normalized. To ensure proper quantitative analysis, at least four positive standards should be included in each assay.
2. To a 1.5-mL tube add the following reagent to make the PCR Master Mix: 150 µL 10X PCR buffer, 50 µL 10X dNTP mix, 150 µL 10X excess primer, 150 µL 10X limiting primer. Add H$_2$O to 1000 µL.
   The reaction mix is intended for a full plate of 96 assays and can be scaled down according to the need. Optimum concentration for the limiting primer is 0.8 to 1.0 ng per µL reaction and 4.0 to 6.0 ng for the excess primer.
3. Mix 100 U of Taq DNA polymerase to the Master Mix and aliquot 10 µL to each reaction except the Apex well. Overlay each well with 10 µL of mineral oil.
4. Add 5 µL of specimen to each reaction according to the template layout. All the specimen should be prepared properly for PCR reaction and denatured at 100°C for 5 min prior to the assay. Duplicate the reaction for each specimen to ensure assay consistency. The negative control is preferably from samples of known negative status. The absolute quantity of positive standards should be confirmed

by an independent method. Set the upper limit of quantitation usually as one of the positive standard.

5. Subject the assays to thermal cycling. A typical cycling profile is 95°C for 25 s, 60°C for 25 s and 72°C for 30 s, and finished with an additional incubation at 72°C for 30 s and 20°C for 1 min. The total sum of cycle numbers $n$ for this step is determined by the upper quantitation limit Qp of the assay. Once the Qp is set, then $n$ is equal to [(12– log Qp)/0.3]–2.

## 3.3. AmpliSensor Assay and Data Acquisition

1. To a 1.5-mL tube add the following to make the 1X AmpliSensor mix for 100 reactions: 25 µL 20X AmpliSensor primer stock, 50 µL 10X PCR buffer. Add $H_2O$ to 500 µL.
2. Aliquot 5 µL 1X AmpliSensor mix to each reaction.
3. Run the assay for two more cycles to equilibrate the signal. Lower the annealing temperature according to the length of the hemi-nested primer. Typical annealing temperature for a 15 mer is 55°C. As a rule of thumb, increase the annealing temperature by 1°C for each base increment.
4. In a bottom-reading fluorimeter, take the 625 nm fluorescence reading of the reactions directly from the thermoplate at a excitation wavelength of 485 nm. Save the data as initial reading. Ensure the thermoplate is registered firmly in the reader every time before taking the reading. The initial reading forms the base to which all the subsequent readings at later cycles will be compared.
5. Continue the thermal cycling and acquire the data for every two or three cycles. Save the data as assay reading (*see* **Note 2**).

## 3.4. Inhibition Test

1. To a 1.5-mL tube add the following to make the 1X inhibition test mix for 100 reactions: 50 µL 10X inhibition test probe, 50 µL 10X PCR buffer. Add $H_2O$ to 500 µL. The inhibition test probe is a complement of the hemi-nested primer and will hybridize to the AmpliSensor primer duplex.
2. Add 5 µL of 1X inhibition test mix to each reaction and run three more thermal cycles. In the absence of inhibition, the remaining AmpliSensor primer duplex in the test reaction will be dissociated as the inhibition-test probe elongates, causing a complete drop of fluorescence.
3. Take the readings and save it as polymerization activity check (PAC) reading. After the reading data is reduced, its omega value will reach 100 in the absence of inhibition activity.

## 3.5. Data Processing and Interpretation

A typical amplification reaction can be described by a first order differential equation $dP/dn = kP(M-P)$, where $P$ is the amount of amplification product at cycle $n$ and $M$ represents the maximum value of $P$. It is obvious from this equation that the growth rate shrinks as $P$ approaches $M$, rendering postexponential quantitative analysis unreliable. Thus, for high-resolution quantitation, it is essential to standardize the growth rate between reactions. Solving the differential equation for $P$, we obtain $P_n = P_m (1 + e)^{n-m} (M - P_n)/(M - P_m)$, where $Pn$ is the target dosage at cycle $n$ and e is the cycle-independent amplification efficiency. The equation can be reduced further by assuming the target dosage $P_0$ at time zero is insignificant in comparison to M. Taking logarithm on both sides of the equation after amassing $P_n$, we obtain $\ln P_0 = \ln M - n \ln(1 + e) + ?_n$, where $W_n = \ln\{P_n /(M - P_n)\}$. Accordingly, the initial target dosage of an unknown sample can be derived by extrapolating a standard dosage curve relating the cycle number n at $W = 0$ and initial target dosage $P_0$. Empirically, the W value can be determined from the readings collected from an Ampli-Sensor assay based on following equations.

$$W_{i,j} = 50\{1 + \text{Log}[(N_{apex,j} - N_{i,j})/ N_{i,j}]\}$$
$$N_{i,j} = (R_{i,j}/Q_i) - S$$
$$Q_i = R_{i,base}/R_{apex,base}$$
$$R_{i,j} = \text{Reading of sample i at cycle j}$$

$S$ is a constant representing the leakage signals of the donor and acceptor fluorophores.

## 4. Notes

1. The PReSTO ligation mix consists of 1 µg of PReSTO AmpliSensor and 10 U of T4 ligase in 45 µL quantitative ligation buffer. There are four classes of PReSTO AmpliSensor labeled A, T, G, and C, according to the nucleotide identity of its 5' termini at the ligation sticky end. Each class of the PReSTO AmpliSensor will ligate only to primers with corresponding base at its 5' penultimate site. The mix is sufficient for ligation of 1.5 µg equivalent of a 15 mer oligonucleotide. The donor and acceptor fluorophores of PReSTO AmpliSensor are fluorescein and Texas red, respectively.
2. A minimum of five sets of data are required for the quantitative analysis. Stop the data acquisition when the required assay cutoff has been reached or the reading of the negative control starts to decrease relative to the Apex. To interpret the data, the reading has to be first converted into an index value termed omega, which can be used to gauge the amount of amplification product accumulated at each assay cycle. The algorithm of data manipulation and reduction will be discussed in **Subheading 3.5.**

## References


1. Wang, C. N., Wu, K. Y., and Wang, H.-T. (1995) Quantitative PCR using the AmpliSensor assay, in *PCR Primer: A Laboratory Manual* (Diefennbach, C. W. and Dveksler, G. S., eds.), Cold Spring Harbor Laboratory, Cold Spring Harbor, NY, pp. 193–202.
2. Yokoi, H., Nonoguchi, K., Kishishita, M., Iwai, M., Higashitsuji, H., and Fujita, J. (1996) Use of AmpliSensor to quantitate gene expression in small amount of

samples: comparison with the quantitative RT-PCR method using CCD imaging system. *Clin. Pathol.* **44,** 847–852.

3. Zhang, F. C., Wu, W. F., and Dong, H. C. (1997) Quantitative PCR for diagnosis of serum HBV DNA and its clinical application. *J. Infect. Dis.* (Chinese) **15(1),** 25–28.

4. Chiang, P.-W., Song, W. L., Wu, K. Y., Korenberg, J. R., Fogel, E. J., Van Keuren, H. L., Lashkari, D., and Kurnit, D. M. (1996) Use of a fluorescence-PCR reaction to detect genomic sequence copy number and transcriptional abundance. *Genome Methods* **6,** 1013–1026.

5. Chen, S., Yee, A., Griffiths, M., Wu, K. Y., Wang, C. N., Rahn, K., and De Grandis, S. A. (1997) A rapid, sensitive and automated method for detection of salmonella species in foods using AG-9600 AmpliSensor analyzer. *J. Appl. Microbiol.* **83,** 314–321.

# 17

# Construction of Polycompetitors
# for Competitive PCR

## David B. Corry and Richard M. Locksley

## 1. Introduction

Many different protocols are now available for competitive polymerase chain reaction (PCR) and most rely on the use of a mimic or competitor that serves as a reference for quantitation *(1–4)*. The success (or failure) of all these protocols is critically dependent on the design, construction, and utilization of these constructs. This protocol provides detailed instructions for developing individual mimics, or competitors, for use in competitive PCR reactions. Individual competitors can be joined together in logical order in one plasmid, producing a single reagent, or polycompetitor, with multiple specificity. Although the protocol has been used successfully in producing cytokine polycompetitors, for both human and mouse *(5)*, it should work well for almost any molecule of interest, provided sequence information is available. If a polycompetitor is to be synthesized, careful planning is especially required for a trouble-free outcome. Detailed restriction-endonuclease maps of the cloning vectors and PCR products to be cloned must be used in the design of primers and to plan appropriate strategies for incorporation of individual competitor constructs. Although many different cloning vectors may be used, in order not to be too general, this protocol provides detailed information using a commercially available vector, pGEM 11Z, and steps used in the construction of a specific polycompetitor, the human polycompetitor for T-cell cytokines, pDC10. The general principles, however, are applicable to the construction of polycompetitors for any genes, using many different commercially available vectors.

From: *Methods in Molecular Medicine, Vol 26: Quantitative PCR Protocols*
Edited by: B. Kochanowski and U. Reischl © Humana Press Inc., Totowa, NJ

## 2. Materials

1. Oligonucleotide primers: These are synthesized as 22–28 residue oligonucleotides encompassing a sequence of approximately 300–400 bp within the protein-coding region of the mRNA of interest. These constraints allow for high-stringency annealing temperatures during PCR, maximizing specificity, and for the tandem alignment of a fairly large number of competitors without overwhelming the vector. It is essential that the primers be designed so that the encompassed sequence includes a reliable restriction-enzyme site that is not present in the cloning vector, and for which the corresponding enzyme is readily available. This site will be used in the subsequent synthesis of the competitor. Restriction sites used in the synthesis of pDC10, that are not present in pGEM 11Z, the cloning vector, are listed in **Table 1**. The reader is referred to another source for a general discussion of the design of primers for PCR (*6*; *see* **Note 2**).
2. Tissue for RNA extraction: Tissues to be used for RNA extraction, cDNA synthesis, and PCR must be carefully selected to include the transcribed gene(s) of interest. Appropriate tissues may include biopsy or whole-organ specimens, either fresh or frozen, and cell cultures, either in suspension or adherent. A protocol for RNA extraction from each of these tissues is included in **Subheading 3.**
3. RNAzol B (Tel-Test, Inc., Friendswood, TX): Although there are several commercially available products for extraction of total RNA, our laboratory has had the most consistently favorable results with this product.
4. Chloroform, high-pressure liquid chromatography (HPLC)-grade.
5. Isopropanol, HPLC grade.
6. Ethanol, HPLC-grade.
7. Agarose: Both high- and low-gelling temperature agarose are used in the following protocols. Only high-grade, nuclease-free preparations should be used.
8. Diethylpyrocarbonate (DEPC)-treated water. All aqueous solutions used while working with RNA should be prepared with RNase-free, DEPC-treated water. This solution is easily prepared by adding 1 mL of DEPC per 500 mL of distilled, deionized water, shaking vigorously until the DEPC completely disperses, and autoclaving the solution.
9. Water baths set at 37°, 42°, and 65°C.
10. RNase-free glassware and pipets .
11. RNase-free glycogen or tRNA.
12. Random hexamer primers.
13. 5X reverse transcription buffer.
14. Dithiothreitol (DTT).
15. SuperScript II reverse transcriptase (Gibco-BRL, Gaithersburg, MD).
16. RNase inhibitor.
17. Deoxynucleotide triphosphate mix (dNTP; dATP+dTTP+dCTP+dGTP, each at 20 m*M*).
18. 10 m*M* Tris-HCl buffer, pH 8.0.
19. Thermal cycler for PCR.

**Table 1**
**Summary of the Essential Steps in the Construction
of the Human T-Cell Cytokine Polycompetitor for Competitive PCR, pDC10[a]**

| Synthesis sequence | Starting vector | Vector digested with | Competitor plasmids | Competitors digested with |
| --- | --- | --- | --- | --- |
| 1 | pGEM-11Z | *Sal*I, *Bam*HI | pTNF-α | *Xho*I, *Sac*I |
|   |   |   | pIL-12 | *Bam*HI, *Sac*I |
| 2 | pTNF-α/IL-12 | *Bam*HI, *Xba*I | pIFN | *Eco*RI, BamHI |
|   |   |   | pIL-4 | *Eco*RI, *Xba*I |
| 3 | pTNF-α/IL-12/IFN/IL-4 | *Xba*I, *Not*I | pIL-5 | *Eco*RI, *Xba*I |
|   |   |   | pIL-10 | *Eco*RI, *Not*I |
| 4 | pTNF-α/IL-12/IFN/IL-4/IL-5/IL-10 | *Not*I, *Nsi*I | pIL-2 | *Eco*RI, *Not*I |
|   |   |   | pIL-13 | *Eco*RI, *Nsi*I |
| 5 | pTNF-α/IL-12/IFN/IL-4/IL-5/IL-10/IL-2/IL-13 | *Nsi*I, *Hind*III | pHPRT | *Eco*RI, *Nsi*I |
|   |   |   | pLT | *Eco*RI, *Hind*III |

[a]For each step in the synthesis sequence, a starting vector was digested with the restriction endonucleases indicated. In parallel, singlet competitor plasmids were digested with the indicated restriction endonucleases, liberating the competitors from the vector, and the fragments were gel purified along with the digested starting vector. The two competitors were incorporated into the starting vector in a single trimolecular ligation, forming the starting vector for the next step, until all 10 constructs were incorporated into the same molecule.

20. Benchtop vortexer.
21. Cloning vectors: "T-tailed" cloning vectors are available commercially but are easily prepared at far reduced cost (*see* **Subheading 3.4.**) (*7*; *see* **Note 1**).
22. Genomic DNA: Unsheared genomic DNA is prepared by precipitating DNA from proteinase K-digested tissues with an equal volume of isopropanol. The DNA is recovered using a pipet tip and resolubilized in a minimum volume of tris-buffered water, pH 8.0 *(9)*.
23. Restriction enzymes: The choice of restriction enzymes will vary with each application. Careful planning during the design of primers will prevent requiring the use of expensive and difficult to acquire enzymes during later cloning steps.
24. Ethidium bromide.
25. 1 Kb DNA ladder (Gibco-BRL).
26. 10X PCR buffer.
27. *Taq* DNA polymerase (Perkin- Elmer, Foster, CA).
28. Scalpels for cutting agarose.
29. Calf intestine alkaline phosphatase (CIP).
30. Dephosphorylation buffer.

## 3. Methods

### *3.1. RNA Extraction*

Select the appropriate tissue or cellular source of mRNA representing the gene(s) of interest. Cell cultures should be viable at the time of RNA extraction, whereas tissues may be used either fresh or after freezing at $-80°C$. Tissues should be frozen immediately following procurement, before significant necrosis has occurred (*see* **Note 3**).

1. For adherent cell cultures, aspirate supernatant and immediately add RNAzol B, approximately 1 mL/10 $cm^2$ surface area. Cells should not be allowed to desiccate prior to the addition of RNAzol B. Gently aspirate and reapply RNAzol mixture to flask until cells are completely disrupted and RNA is solubilized.
2. For cells in suspension, place in a centrifuge tube and pellet gently, at 100*g* for 5 min.
3. Remove supernatant, being careful not to disrupt the soft cell pellet, leaving approx 30 μL behind.
4. Gently resuspend cells in this volume by tapping the bottom of the tube. For more adherent cell suspensions, especially those containing macrophages, light vortexing is required to completely dissociate cells.
5. Add 1 mL RNAzol to resuspended cells while vortexing, until thoroughly mixed.
6. RNA extraction from fresh tissues is most efficiently accomplished by homogenizing samples in a tissue homogenizer in the presence of RNAzol B (1 mL/50 mg tissue). To extract RNA from frozen samples, tissues are wrapped with several layers of aluminum foil and placed in liquid nitrogen for 5 min. The tissues are then pulverized to a fine powder by striking the foil several times with a

hammer. While still frozen, the powder is transferred to RNAzol B (maximum 50 mg tissue per 1 mL RNAzol B) and vortexed for several s at room temperature. If large clumps of undispersed tissue remain, the solution may be sonicated using a sonicator tip for 2-5 s, only until clumps disperse.

7. After solubilizing tissues or cells in RNAzol, add HPLC-grade chloroform (200 μL/mL RNAzol) and shake the samples vigorously for 20 s, until a fine emulsion forms (*see* **Note 4**).
8. Keep samples on ice for 5 min and centrifuge for 15 min at 16,000*g*. The resulting solution consists of an upper aqueous phase and a lower, organic phase. A fine layer consisting of precipitated protein, lipid, and DNA is often visible between the two liquid phases.
9. Carefully aspirate and save the upper, aqueous phase, being careful not to contaminate it with any precipitated or organic phase material.
10. Add an equal volume of HPLC-grade isopropanol and mix thoroughly. Incubate on ice for 20 min. and centrifuge at 16,000*g* for 15 min (*see* **Note 5**).
11. Carefully wash the RNA pellet with 70% ethanol-DEPC water twice by adding 500 μL to the pellets and gently rocking the liquid over the pellet several times and discarding. Centrifuge at 16,000*g* for 1 min after each wash.
12. Resuspend RNA in a minimal volume of DEPC-treated water, usually 5–30 μL, to achieve an RNA concentration of 0.5–3 mg/mL. Store samples frozen at −80°C.

## 3.2. cDNA Synthesis

1. Prepare a mixture of 1–2 μg total RNA, 1 μg random hexamer primers , 1 U RNase inhibitor and DEPC-treated water in a total volume of 10 μL (*see* **Note 6**).
2. Heat mixture to 65°C for 5 min and place immediately on ice.
3. Add to this mixture a cocktail containing 5 μL 5X reverse transcription buffer, 2.5 μL of 100 m*M* DTT (add only if not present in buffer), 0.625 μL of 20 m*M* deoxynucleotide triphosphate mix, SuperScript II reverse transcriptase (200 U/μg RNA), 1 U RNase inhibitor and 5.4 μL DEPC-treated water.
4. Mix well and incubate at 42°C for 30–45 min.
5. Enzyme activity is neutralized by heating to 65°C for 10 min.
6. Dilute cDNA to a final volume of 80 μL with tris buffer, pH 8.0.

## 3.3. Polymerase Chain Reaction

1. To 5 μL of cDNA, add 94.5 μL of PCR cocktail containing 10 μL 10X PCR buffer, 2 μL each of 20 μ*M* oligonucleotide primers, 1 μL of 20 m*M* dNTP, and 79.5 μL deionized, distilled water.
2. If required for the machine design, add one drop of high-grade mineral oil to each tube, cover tightly, and vortex until well mixed.
3. Centrifuge the PCR mixture at 16,000*g* for 10 s to separate the oil from the aqueous solution.
4. Place PCR tubes into the thermal cycler and begin heating to 94°C.

5. "Hot start" the reaction by adding 0.5 U Taq polymerase, using a micropipet, after the reaction mixture has reached 90°C. Care should be taken to inject the polymerase into the aqueous mixture, and not into the oil (*see* **Note 8**).
6. Continue heating the mixture according to the following general protocol: denature at 94°C for 30 s; anneal at 60°C for 20 s; extension at 72°C for 40 s The annealing temperature should be altered according to the recommended temperatures in **Table 2**.
7. Continue amplification for 30–45 cycles, depending on relative transcript abundance. Only reactions that yield sharp, single bands of the appropriate size should be used for subsequent cloning steps.

## 3.4. TA Cloning

The first seven steps describe the synthesis of a T-tailed vector from pGEM-11Z.

1. Digest 1 μg of pGEM- 11Z vector to completion using *Hin*cII restriction endonuclease in 10 μL total volume (*see* **Notes 7** and **9**).
2. Transfer the entire aliquot to a PCR tube. Add 40 μL of PCR cocktail (*see* **Subheading 3.3., step 1**) containing 2 m*M* dTTP alone in place of the dNTP mixture.
3. Heat the mixture in a thermal cycler to 72°C, then add 0.5 U Taq polymerase. Continue heating at 72°C for 2 h.
4. Purify the T-tailed product by adding an equal volume of cold, HPLC-grade isopropanol and allowing the DNA to precipitate over 20 min while on ice.
5. Pellet the DNA by centrifuging at 16,000*g* for 15 min in a microcentrifuge tube.
6. Wash the DNA pellet twice with 70% cold ethanol.
7. After removing all ethanol, resuspend the pellet in 30 μL of 20 m*M* tris buffer, pH 8.0 (*see* **Note 10**). The T-tailed vector is now suitable for subsequent cloning steps.
8. Perform standard ligation reactions in which the T-tailed vector and PCR product of interest are mixed in approximately 1:2 to 1:3 molar ratios. Alternatively, vector and PCR product may be mixed empirically in a series of volume ratios (i.e., 1 μL vector: 1 μL PCR product; 1 μL vector: 5 μL PCR product, and so on) with one ratio likely to favor ligation of the desired product (*see* **Note 11**).
9. Transform and select appropriate mutants (*see* **Note 12**). After screening, the resulting product, $V_{PCR}$ , is suitable as the starting vector for single competitor synthesis (*see* **Note 13**).

## 3.5. Single competitor synthesis.

1. Digest both $V_{PCR}$ (1 μg) and genomic DNA (2 μg) with the same restriction enzyme, the recognition site of which is unique to the PCR product (*see* **Table 2**), in total volumes of 10 μL and 20 μL, respectively. Digestions should be allowed to proceed to completion (at least one hour incubation).
2. Dephosphorylate the vector by adding 2 U CIP, 2 μL of 10X dephosphorylation buffer and sufficient water to 20 μL total volume. Incubate for 1 h at 37°C (*see* **Note 14**).

**Table 2**
**Primer Sequences, Spacer Ligation Sites and Sizes of the Wild-Type and Competitor cDNA's (CcDNA) Contained in pDC10[a]**

| Gene | Primer sequences | Spacer site | WT cDNA(bp) | CcDNA(bp) | $T_a$(°C) |
|---|---|---|---|---|---|
| IL-2 | 5'-ACTCACCAGGATGCTCACAT<br>3'-AGGTAATCCATCTGTTCAGA | Afl II | 266 | ~390 | 50 |
| IL-4 | 5'-TCCACGGACACAAGTGCGATATCACCT<br>3'-GCTTGTGCCTGTGGAACTGCTGTGCAG | Pst I | 240 | ~325 | 67–72 |
| IL-5 | 5'-CTTGGAGCTGCCTACGTGTATGC<br>3'-CCACTCGGTGTTCATTACACC | Sty I | 357 | ~485 | 60 |
| IL-10 | 5'-ATGCCCCAAGCTGAGAACCAAGACCCAGAC<br>3'-TCTCAAGGGGCTGGGTCAGCTATCCCA | Cla I | 352 | ~430 | 65–72 |
| IL-12 | 5'-CCAAGAACTTGCAGCTGAAG<br>3'-TGGGTCTATTCCGTTGTGTC | Pst I | 355 | ~451 | 55 |
| IL-13 | 5'-CCCAGAACCAGAAGGCTCCGCTCTG<br>3'-GTTGAACCGTCCCTCGCGAA | Pst I | 278 | ~370 | 55–60 |
| HPRT | 5'-CCTGCTGGATTACATCAAAGCACTG<br>3'-TCCAACACTTCGTGGGGTCCT | Sty I | 289 | ~355 | 60 |
| IFN-γ | 5'-AGTTATATCTTGGCTTTTCA<br>3'-ACCGAATAATTAGTCAGCTT | Afl II | 356 | ~440 | 50 |
| TNFα | 5'-GAGTGATCGGCCCCAGAGG<br>3'-TGCGGCTGATGGTGTGGGTG | Sty 1 | 315 | ~415 | 60 |
| LT | 5'-CCTCACACCTTCAGCTGCCC<br>3'-GAGAAACCATCCTGGAGGAA | Nsi I | 162 | ~300 | 55 |

[a]Note that primers are written in the 5' to 3' direction and should be synthesized precisely as written. $T_a$, suggested annealing temperature; Wt, wild type, Cc competitor; LT = lymphotoxin (formerly TNF-β).

3. Electrophorese the digested vector and genomic DNA in the same 1% low-gelling temperature agarose. The genomic DNA is run adjacent to a low molecular weight DNA ladder, such as the 1 Kb DNA ladder.

4. Remove the vector band, and a thin band (approx 2–4 mm wide) from the genomic digest that corresponds to a molecular size of 75–100 bp, using a scalpel and place in separate microcentifuge tubes

5. Heat the tubes to 65°C for 15 min, until all agarose has melted.

6. Freeze the melted agarose at –80°C for a minimum of 20 min.

7. Allow the tubes to thaw, and centrifuge at 16,000*g* for 15 min.

8. A fluid layer will be noted above agarose that has precipitated out of solution. The fluid is enriched with DNA and is suitable for cloning (*see* **Note 15**)

9. Ligate the digested, CIP-treated vector and genomic fragments as in **Subheading 3.4., step 8**, and transform. Again, accurate quantitation of the molecular species prior to ligation is not necessary to achieve successful cloning.

10. Following transformation, colonies are screened as in **Subheading 3.4., step 9**. Most ligations will yield numerous colonies, most of which will be competitors, i.e., products that are 75–100 bp larger than wild-type products. The actual sizes of PCR products and the corresponding competitors contained in pDC10 are listed in **Table 2** (*see* **Notes 16** and **17**).

11. The aforementioned steps are repeated until individual competitors have been synthesized for all the desired PCR products (*see* **Note 17**).

## *3.6. Polycompetitor Construction*

The following synthesis algorithm summarizes the 5 principal steps used to incorporate all 10 competitor constructs in tandem into pGEM- 11Z.

1. Digest pGEM-11Z with *Bam*HI and *Sal*I. In parallel, digest the plasmids containing the TNF-α and IL-12 competitors (pTNF-α and pIL-12) with *Sac*I and *Xho*I, and *Bam*HI and *Sac*I, respectively. All digestions should be allowed to proceed to completion (*see* **Note 18**).

2. Purify the digestion products as in **Subheading 3.5., steps 3–8** before using 1% low-gelling temperature agarose.

3. Mix the linearized pGEM vector with the two liberated competitors and perform standard ligation and transformation reactions, as during the original competitor synthesis. Multiple vector-insert ratios (by volume) should be used to ensure retrieval of the desired recombinant product (*see* **Note 19**). The resulting plasmid (pTNF-α/IL-12) should be propagated through competent bacteria and approximately 5–10 μg of purified plasmid prepared for subsequent cloning steps.

4. Screen the resulting transformant by both restriction enzyme digestion and PCR, using gene- specific primers, to verify that both constructs are present in the correct position and efficiently amplify (*see* **Note 12**).

5. Digest the resulting product (pTNF-α/IL-12) with *Bam*HI and *Xba*I. In parallel, digest two new competitor vectors, pIFN and pIL-4, with *Eco*RI and *Bam*HI, and *Eco*RI and *Xba*I, respectively.

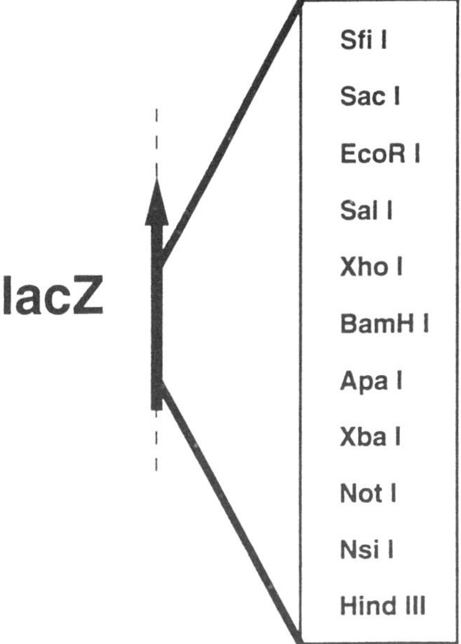


Fig. 1. Multiple cloning site contained within the lacZ gene of pGEM-11Z. Individual sites are referred to throughout **Methods**.

6. Gel purify and isolate the single competitor constructs and the digested vector, pTNF-α/IL-12, and ligate the products together as above.
7. Repeat the same steps, utilizing sequentially the other endonuclease sites contained in pGEM-11Z, until all single competitors are incorporated into the same vector. These steps are summarized in **Table 1**.
8. As with the intermediate steps, the final construct should be tested for efficient and specific amplification of all the incorporated products. Following verification, large amounts of the vector are prepared, carefully purified, and frozen in small dilute aliquots for single use in competitive PCR reactions.

## 4. Notes

1. In the general protocol described here, the pGEM-11Z multi-cloning vector (Promega, Inc., Madison, WI) is used, but many different vectors may be applied to particular cloning projects. pGEM-11Z contains a multiple cloning site into which PCR products may be inserted (**Fig. 1**). The lacZ gene imparts a blue color to bacterial colonies transformed with the native vector. Mutants in which an insert has been cloned into the multiple cloning site, disrupting the lacZ gene, can be identified by their white appearance.
2. Primer Design: An additional consideration in the initial design of primers for competitor synthesis is the other restriction-endonuclease sites encompassed

other than the site chosen for competitor cloning. These sites could interfere with subsequent steps during the synthesis of a polycompetitor. Although simply changing the order in which single competitors are incorporated into one vector may circumvent this problem, by avoiding potentially troublesome cloning sites from the beginning, maximum flexibility during later subcloning is assured.

3. In addition to using autoclaved glassware and DEPC-treated water, it is suggested that the work bench and all materials and instruments potentially coming contact with RNA be cleaned by wiping with a 10% bleach solution. Working within a laminar flow hood will prevent airborne RNase contamination of RNA solutions.

4. HPLC-grade reagents are required for consistent, reliable results with these protocols. We have noticed on occasion, using reagents of lesser grade, that unusual precipitates form during RNA extraction, which results in degradation, or complete loss of RNA.

5. RNase-free glycogen or transfer RNA (tRNA) may be added at this point (2 µg/mL) to facilitate yield of total RNA, particularly for samples with very low RNA abundance.

6. cDNA synthesis: This protocol is sufficient for high-quality cDNA synthesis from most RNA species. Occasional RNAs do not efficiently reverse transcribe using this protocol, however. For these instances, commercial cDNA synthesis kits are recommended. These kits contain proprietary reagents that improve cDNA synthesis efficiency.

7. *Hinc*II is an isoschizomer of *Sal*I and cleaves bluntly at this site. Blunt ends are preferred for accepting dATP overhangs.

8. Hot start. All PCR reactions are performed by adding *Taq* DNA polymerase last, after the reaction mixture has been placed in the thermal cycler and heated beyond 90°C. This extra step insures maximum reliability and reproducibility of PCR.

9. *Taq* DNA polymerase or another thermostable polymerase that adds dATP at the 3' end of DNA templates should be used in this step. Some thermostable polymerases are relatively unreliable with regard to A- and T-tailing, or do not have this capability at all, and should not be used.

10. Nucleic-acid purification: Other methods for nucleic-acid purification may be used for this and subsequent steps. The described methods are preferred because of their simplicity, ease of application, low cost, and reliability.

11. Ligation reactions. Ligation reactions involving these molecular species are most efficient when the insert exceeds the vector by a two- to threefold molar ratio *(9)*. Although relatively accurate estimates of the concentrations of these molecules can be determined using quantitating gels (incorporating markers of known concentration and size), ultraviolet spectrophotometry, and other methods, consistent, reliable results can be obtained using the empirical method described, and is far easier. In addition, we have noted that ligation reactions involving these molecular species are most efficiently performed at 4°C, for a minimum of 8 h.

12. Screening for appropriate mutants. Because of the vector design, "blue-white selection" can be used to initially identify desirable products. Desirable products

(those in which the full-length PCR product have been incorporated into the vector; $V_{PCR}$) are easily verified as follows:

a. The exact length PCR product should be cleanly and efficiently amplified by PCR directly from the purified plasmid;

b. Restriction-enzyme digestion of $V_{PCR}$ at sites adjacent to the site of insertion should yield a product identical in size to the wild-type PCR product;

c. Digestion of the PCR product using enzymes unique to it should yield fragments of the expected sizes.

13. Difficulty securing appropriate transformants. The most likely difficulty to be encountered in creating a competitor is inability to efficiently generate appropriate $V_{PCR}$. Assuming that a trivial error has not occurred, the most likely cause is a poorly A-tailed PCR product. This can be easily remedied by simply adding fresh *Taq* DNA polymerase and 2 m$M$ dATP to the PCR product and heating for an additional 2 hours at 72°C.

14. To prevent ligation of the vector onto itself at the subsequent step, the digested vector may be dephosphorylated using CIP immediately following digestion. The restriction enzyme should be heat-inactivated (if possible), after which 1–2 U CIP and appropriate buffer and water diluent are added directly to the reaction. There is no need to extract the DNA through phenol-chloroform or otherwise purify the vector between these steps.

15. Again, no additional DNA purification steps are necessary. Because only a single recombinant species is required following subsequent ligation steps, the relative inefficiency of the method is not a hindrance, as multiple recombinants (typically 10–100) are generated, when performed correctly.

16. Although a range of sizes will be noted among the competitor products, most will be larger by exactly the size of the genomic fragments prior to ligating into the vector. These larger products should be discarded in favor of the smaller competitors.

17. Before finally deciding on a particular competitor, it is best to mix candidate competitors with wild-type product in equal amounts and performing PCR to ensure reliable concomitant amplification. Only competitors that give clean, single bands, accompanied by single wild-type bands, should be accepted. Occasional competitors produce troublesome extra bands following amplification that confound data interpretation. This likely represents heteroduplex formation and, although often remediable, these competitors should be discarded in favor of others that do not possess this attribute *(10)*.

18. The first digestion will linearize the native pGEM and render it suitable for simultaneously accepting the two competitors. The other digestions will liberate the competitors from their vectors and provide them with the necessary complementary ends for the desired recombination. Note that dephosphorylation of the vector is not necessary as it is unable to religate on itself because of the noncomplimentary ends.

19. The desired product is pGEM-11Z containing the TNF-α and IL-12 competitor constructs between the original *Sal*I and *Bam*HI sites (pTNF-α/IL-12). Note that

the two competitors join together at their shared *Sal*I sites, while their remaining *Bam*HI and *Xho*I sites ligate to the vector at its *Bam*HI and *Sal*I sites, respectively. The unusual *Xho*I-*Sal*I ligation is complementary, but results in a new sequence not recognized by either enzyme.

## References


1. Kozbor, D., Hyjek, E., Wiaderkiewicz, R., Wang, Z., Wang, M., and Loh, E. (1993) Competitor mRNA fragments for quantitation of cytokine specific transcripts in cell lysates. *Mol. Immunol.* **30**, 1–7.
2. Platzer, C., Richter, G., Uberla, K., Muller, W., Blocker, H., Diamantstein, T., and Blankenstein, T. (1992) Analysis of cytokine mRNA levels in interleukin-4-transgenic mice by quantitative polymerase chain reaction. *Eur. J. Immunol.* **22**, 1179–1184.
3. Wang, A. M., Doyle, M. V., and Mark, D. F. (1989) Quantitation of mRNA by the polymerase chain. *Proc. Natl. Acad. Sci. USA* **86**, 9717–9721.
4. Carding, S. R., Lu, D., and Bottomly, K. (1992) A polymerase chain reaction assay for the detection and quantitation of cytokine gene expression in small numbers of cells. *J. Immunol. Methods* **151**, 277–287.
5. Reiner, S. L., Zheng, S., Corry, D. B., and Locksley, R. M. (1993) Constructing polycompetitor cDNAs for quantitative PCR. J. Immunol. Methods **165**, 37–46.
6. Innis, M. A. and Gelfand, D. H. (1990) Optimization of PCRs., in *PCR Protocols: A Guide to Methods and Applications* (Innis, M. A., Gelfand, D. H., Sninsky, J. J., and White, T. J., eds.), Academic, San Diego, CA, pp. 3–6.
7. Marchuk, D., Drumm, M., Saulino, A., and Collins, F. S. (1991) Construction of T-vectors, a rapid and general system for direct cloning of unmodified PCR products. *Nucleic Acids Res.* **19**, 1154.
8. Laird, P. W., Zijderveld, A., Linders, K., Rudnicki, M. A., Jaenisch, R., and Berns, A. (1991) Simplified mammalian DNA isolation procedure. *Nucleic Acids Res.* **19**, 4293.
9. Dugaiczyk, A., Boyer, H. W., and Goodman, H. M. (1975) Ligation of *Eco*RI endonuclease-generated DNA fragments into linear and circular structures. *J. Mol. Biol.* **96**, 171–184.
10. Henley, W. N., Schuebel, K. E., and Nielsen, D. A. (1996) Limitations imposed by heteroduplex formation on quantitative RT-PCR. Biochem. *Biophys. Res. Commun.* **226**, 113–117.

# 18

# Tailed RT-PCR for the Quantitation of Chloramphenicol Acetyl Transferase (CAT)mRNA

Marlyse C. Knuchel and Aftab A. Ansari

## 1. Introduction

Reporter gene plasmids have been used extensively to monitor gene expression and elucidate intracellular pathways *(1–4)*. They have been particularly useful in understanding the architecture of promoter regions and the interactions between promoter elements and cellular or viral regulatory factors *(5–9)*. The conventional strategy has been to transfect host cells transiently with a plasmid bearing the sequences of interest linked to a chloramphenicol acetyl transferase (CAT) reporter gene. Subsequently, CAT activity is measured as a readout by thin-layer chromatography (TLC) or the levels of CAT protein are determined using an enzyme-linked immunosorbent assay (ELISA). However, most transfections–whether stable or transient–result in low levels of CAT gene expression, as long as no activation signal is provided *(10–12)*. Although this is an ideal situation to study gene activation pathways, it is poorly suited to monitor gene repression or negative regulatory mechanisms. To overcome this problem, investigators use cell-activating agents, such as phorbolester and phytohemagglutinin, or transfect a second plasmid expressing a transactivator (e.g., viral transactivator). However, use of such agents might interfere with the pathway(s) being studied and could provide erroneous results. Also, current assays such as TLC or CAT ELISA are often not able to quantitate such low levels of expression, thus hampering studies aimed at dissecting the down modulation of gene expression *(13–15)*.

To address this issue, our laboratory has developed a sensitive quantitative RT-PCR protocol that allows for direct monitoring of CATmRNA expression in stably transfected and cloned cell lines. Because no intron is present in the CAT reporter gene, a technique was derived to distinguish CAT cDNA from

From: *Methods in Molecular Medicine, Vol 26: Quantitative PCR Protocols*
Edited by B. Kochanowski and U. Reischl © Humana Press Inc., Totowa, NJ

CAT DNA. This was accomplished by using a tailed reverse transcription (RT) primer whose 5' end (27 bp) is not complementary to the CAT gene *(16)*. Following reverse transcription, the generated cDNA is amplified using a CAT specific sense primer and an antisense primer specific for the 5' tail of the RT primer (**Fig. 1**). The use of a 5'-tailed RT primer eliminates the need for a DNA digestion step, which is often a source of problems, because DNase is rarely totally free of RNase. In addition, this protocol neither requires limiting dilution analysis of the samples nor uses competitive PCR templates; therefore it is less labor-intensive and less prone to sample contamination *(17–21)*. The tailed RT-PCR presented here is not only complementary to existing methods, because it is designed to monitor CAT at the transcriptional level, but it is also reasonable to state that the described tailed RT-PCR assay is far more sensitive and thus able to define more rigorously changes in CAT gene expression *(22–23)*.

Although we describe a protocol specific for quantitation of CAT mRNA, the same principle is applicable for all unspliced genes, as long as their constitutive expression is constant. Thus, this assay is easy to adapt for the specific needs of a variety of in vitro experiments.

## 2. Materials
### 2.1. General Supplies

1. Cell incubator.
2. Tabletop centrifuge.
3. Template Tamer (Coy Laboratory Products, Grass Lake, MI).
4. Microman pipet with capillaries and pistons (Gilson, Middletown, WI).
5. Water bath set at 45°C.
6. Thermocycler.
7. Heat block.
8. Spectrophotometer.
9. Phosphorimager (Molecular Dynamics, Sunnyvale, CA) or densitometer.
10. UV Stratalinker 2400 (Stratagene, La Jolla, CA) or oven.
11. Incubator with shaker set at 42°C.

### 2.2. Cell Culture Medium

Most cell lines will grow in RPMI 1640 (GIBCO, Grand Island, NY) supplemented with 10–20% heat-inactivated (56°C for 30 min) fetal bovine serum (FBS), 50 µg/mL gentamicin and 2 m*M* L-glutamine (both from GIBCO). In this protocol, an Epstein-Barr virus (EBV) immortalized and cloned cell line derived from a sooty mangabey (an African nonhuman primate species) was stably transfected with the pMSG-CAT vector (Pharmacia, Piscataway, NJ) and was termed FEc-CAT (*see* **Notes 1** and **2**).

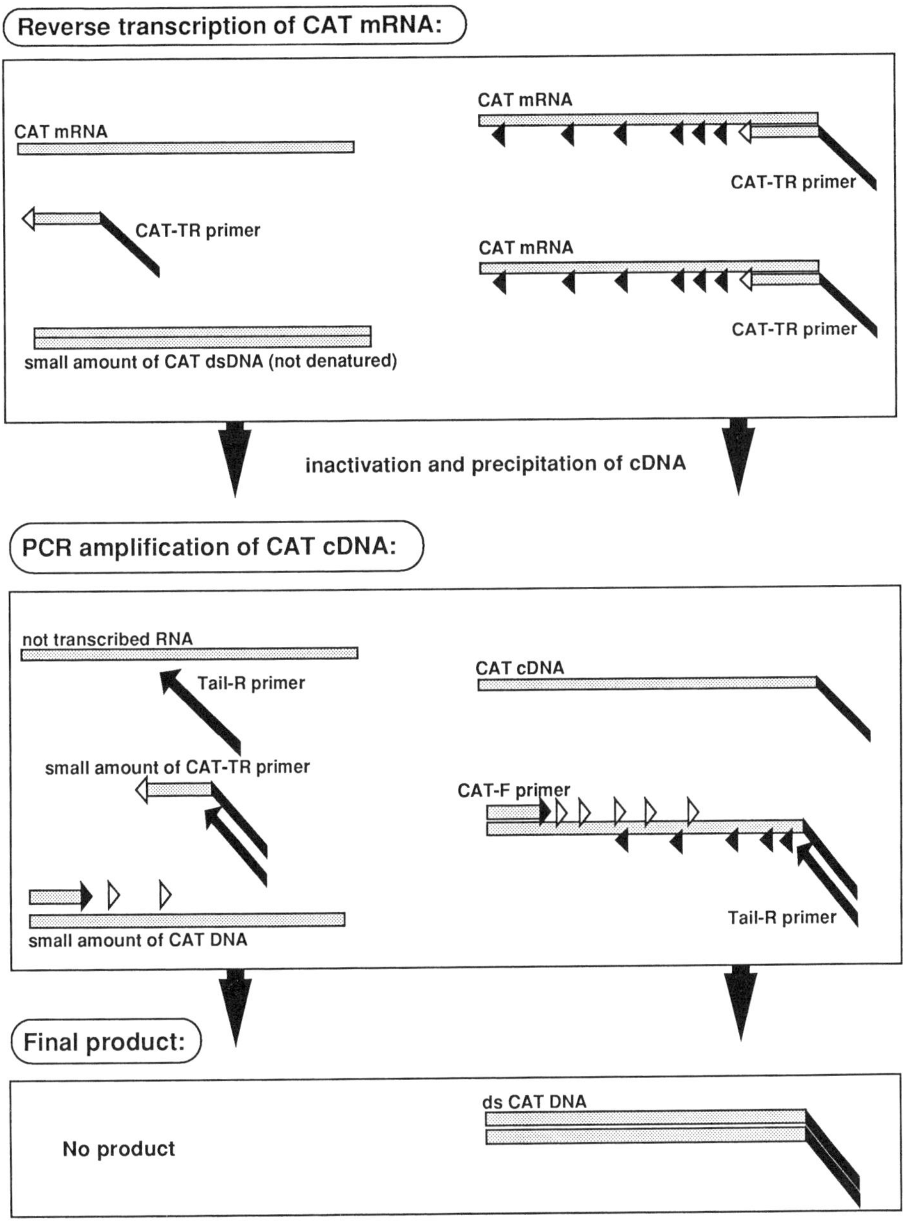

Fig. 1. Flow diagram of the RT-PCR assay for the quantitation of CATmRNA from cell lines stably transfected with a CAT-expressing plasmid. Products that are not amplified are shown on the left; specific amplification is depicted on the right.

### 2.3. Cell Harvest

Phosphate-buffered saline (PBS): dissolve 8 g of NaCl, 0.2 g of KCl, 1.44 g of $Na_2HPO_4$, and 0.24 g of $KH_2PO_4$ in 800 mL of distilled water. Adjust the pH to 7.4 with HCl and add water to 1 L. Autoclave and keep at 4°C. Make up a large batch, because PBS will be needed in large quantities to wash the cells.

### 2.4. RNA Extraction

1. RNAzol B (Tel-Test, Friendswood, TX): keep at 4°C protected from light. It is stable for about 6 mo.
2. Diethyl pyrocarbonate (DEPC) treated water: add 0.1% DEPC to the water, incubate overnight at 37°C and autoclave.
3. Chloroform, as well as isopropanol and ethanol should be of high purity and should be used for RNA work only (*see* **Notes 3–5**).

### 2.5. RT Reaction

1. 5X RT buffer (Promega, Madison, WI).
2. AMV reverse transcriptase (Promega).
3. Stock of 10 m*M* deoxynucleotide triphosphates (USB, Cleveland, OH) in DEPC-treated water.
4. Tailed RT primer (CAT-TR) diluted at 100 µ*M* in DEPC water: 5'-*CATCGATGACAAGCTTAGGTATCGATA*CCATTCATCCGCTTATTATC – 3', the 3' end of this nucleotide oligomer is homologous to the CAT gene (20 bases) and its 5' end (27 bases, in italic) is unrelated to the CAT gene. The resulting cDNA will be 741 bp long (*see* **Note 6**).

### 2.6. PCR Reagents

1. 10X PCR buffer: 100 m*M* Tris-HCl, pH 9.5, 500 m*M* KCl, 20 m*M* $MgCl_2$, and 0.1% (w/v) gelatin. Autoclave, aliquot, and store at –20°C. The high gelatin concentration limits the stability of this buffer to about 6 mo.
2. Stock of 1.25 m*M* deoxynucleotide triphosphates (USB).
3. *Taq* DNA polymerase (Perkin Elmer Cetus, Norwalk, CT).
4. 100 µ*M* stock of each oligomer. Sense CAT primer (CAT-F): 5'-CTAAAATGGAGAAAAAAATCACTGG-3'. Antisense primer (TAIL-R): 5'-CATCGATGACAAGCTTAGGTATCGATA- 3'. This 27 oligonucleotide oligomer is complementary to the tail of the CAT-TR primer. Mineral oil, autoclave (*see* **Notes 6–8**).

### 2.7. Generation of Standards

1. pMSG-CAT vector (Pharmacia).
2. 100 µ*M* stock of sense CAT primer (CAT-F) and antisense primer (CAT-R2): 5'-TAACACGCCACATCTTGCGAATATA- 3'.

## 2.8. Generation of Positive Controls

pMSG-CAT vector. 100 $\mu M$ stock of primer CAT-F and antisense primer CAT-R1: 5'-CCATTCATCCGCTTATTATCACTTA -3'.

## 2.9. DNA Visualization and Probing

1. 1% agarose gel in 1X TBE with 0.5 µg/mL ethidium bromide. 10X TBE: 108 g Tris Base, 55 g boric acid, 40 mL 0.5 $M$ EDTA (pH 8.0), add water to 1 L.
2. Oligonucleotide probe (CAT-P): 5'-GCTGAACGGTCTGGTTATAGGT ACATTGAGCAACTGACTGAAATGCCTCA-3'.
3. ProbeQuant G50 micro columns (Pharmacia).
4. T4 polynucleotide kinase (PNK) with 10X buffer (New England Biolabs, Beverly, MA).
5. Hybond-N membrane (Amersham, Arlington Heights, MD).
6. 10% sodium dodecyl sulfate (SDS) stock solution: dissolve 100 g of electro-phoresis-grade SDS in 900 mL water. Heat to 70°C, adjust the pH to 7.2 and the volume to 1 L with water.
7. 20X SSC: dissolve 175.3 g of NaCl and 88.2 g of sodium citrate in 800 mL water. Adjust the pH to 7.0 and the volume to 1 L.
8. Blocking solution: 20 m$M$ Tris-HCl (pH 7.5), 0.9 $M$ NaCl, 0.1% SDS, 6.0 m$M$ EDTA and 0.25% nonfat dry milk.
9. Wash solution I: 2X SSC/0.5% SDS.
10. Wash solution II: 0.1X SSC/0.5% SDS.
11. 6X DNA loading buffer: 4 g sucrose and 25 mg bromophenol blue in 10 mL water (*see* **Notes 9** and **10**).

# 3. Methods

## 3.1. RNA Extraction

1. Cell harvest. Collect the cells with a 1 mL pipet, and transfer to a 10 mL centri-fuge tube. Wash the cells twice with 10 mL of sterile PBS. Resuspend in 1 mL of PBS and transfer to a 1.5 mL Eppendorf tube, count the cells, and spin in a microfuge for 2 min (450$g$). Discard the supernatant, and resuspend the pellet well by vortexing (*see* **Notes 11** and **12**).
2. Add RNAzol B at 200 µL/$10^6$ cells. (*see* **Note 13**).
3. Vortex for 15 s.
4. Let sit on ice for 10 min.
5. Add 1/10 volume of chloroform, and vortex vigorously for 30 s.
6. Place on ice until the two phases are well separated (10–15 min).
7. Centrifuge for 15 min at 4°C (14,000$g$) and collect the aqueous (top clear) phase in a fresh tube. (*see* **Note 14**).
8. Add 0.6 volume of isopropanol. Precipitate at 4°C for 1 h or up to overnight. (*see* **Notes 15** and **16**).
9. Centrifuge the tubes for 20 min at 4°C.
10. Wash the pellet twice with 200 µL cold 70% ethanol in DEPC water by centrifug-ing the tubes 5 min at 4°C. (*see* **Note 17**).

11. Remove the ethanol, and let the pellet dry under the hood with the lid open for 5 min (*see* **Note 18**).
12. Add 35.5 μL DEPC water and 10 U RNasin per $10^6$ cells (*see* **Note 19**).

## *3.2. Reverse Transcription*

1. Per $10^6$ cells

| | Add | Final concentration |
|---|---|---|
| 5X RT buffer | 10 μL | 1 X |
| Primer | 1 μL | 1 μ*M* |
| dNTPs | 2.5 μL | 500 μ*M* |
| AMV-RT | 1 μL | 1 U |

These reagents can be made up as a master mix, and then distributed to each tube for a final volume of 50 μL.
2. Quick spin the tubes to ensure an even distribution of the reagents.
3. Incubate for 2 h at 45°C.
4. Stop the reaction by placing the tubes at 95°C for 10 min
5. Add 0.6 volume of isopropanol and precipitate at 4°C for 1 h.
6. Spin down the cDNA for 20 min at 4°C.
7. Wash twice with 70% ethanol (spin 5 min at 4°C).
8. Dry the pellet using a speed vac for 5 min.
9. Add 71.6 μL water/$10^6$ cells (*see* **Note 20**).

## *3.3. cDNA Amplification*

1. Per $10^6$ cells

| | Add | Final concentration |
|---|---|---|
| 10X PCR buffer | 10 μL | 1X |
| dNTPs | 16 μL | 200 μ*M* |
| Sense primer | 1 μL | 1 μ*M* |
| Antisense primer | 1 μL | 1 μ*M* |
| Taq polymerase | 0.4 μL | 2.5 U |

Make a master PCR mix and distribute to PCR tubes. In one tube, add 71.6 μL water (without any DNA) to control for contaminants from the master mix.
2. Add the cDNA and two drops of oil.
3. Spin quickly in a microcentrifuge.
4. Place in the thermocycler and cycle at: 1X (94°C, 5 min) 35x [94°C, 1 min; 55°C, 1 min; 72°C, 1 min] 1X (72°C, 10 min) and maintain at 4°C until use.

## *3.4. Standards*

The standards serve the purpose of representing the linear range of the assay. To define this linear range, a limiting dilution of RNA from the FEc-CAT cell line is prepared and amplified by RT-PCR using the CAT-F and Tail-R primers. However, to avoid RNA stability problems, we chose to amplify a large batch of DNA, three dilutions of which equal the optical densities encompassing the linear range defined with the FEc CAT RNA dilution. These standards are amplified from the pMSG-CAT plasmid with the CAT-F and CAT-R2 primers, diluted to match the optical densities of three points on the linear range, aliquoted, and frozen to allow for a standard regression curve for each subsequent CAT assay using this cell line.

## 3.5. Positive Controls

Together with the cDNA samples, a positive control is amplified in duplicate using 1.25 fg of pMSG-CAT plasmid and the primer CAT-F and CAT-R1. The resulting band densities should be equivalent to $144 \pm 10$ copies of CATmRNA ($1.8 \times 10^{-10}$ pmol) on the standard graph and serves as an internal control for the PCR amplification.

## 3.6. Electrophoresis and Radioactive Probing

1. Pour a 1% agarose gel containing ethidium bromide. Load the DNA-free negative control, the standards, the positive controls and 25 μL of the samples with 5 μL of 6x DNA loading buffer.
2. Run the gel in 1X TBE for 1 h at 120 V.
3. Control the reaction under UV lights, and record the results (black and white photography).
4. Transfer onto Hybond-N membrane with 20X SSC, overnight *(19)*.
5. Crosslink the blot twice using a UV Stratalinker or bake the blot 30 min at 80°C.
6. Prehybridize for 1 h at 42°C in blocking solution.
7. Add 10–20 μL of the $^{32}$P-labeled oligonucleotide probe (CAT-P) for overnight at 42°C. To make the probe, mix 5 μL of $^{32}$P-γ ATP, 200 ng primer, 3 μL 10X PNK buffer, 3 μL PNK, and 18 μL water, incubate for 30 min at 37°C and purify using a ProbeQuant G50 micro column.
8. Wash the blot with wash solution I and II for 30 min each at 42°C.
9. Expose to a Kodak XAR2 film overnight at −70°C or to a phosphorimager plate for 2 h.

## 3.7. Quantitation

The use of stably transfected and cloned cell lines allows for a constant production of CATmRNA, which is required for reproducible quantitation of this RT-PCR assay. Under these conditions, RNA extracted from a constant amount of cells should result in equivalent amounts of amplified DNA after RT-PCR amplification. For quantitation of the samples, a standard curve is drawn for each experiment using the standards representing the linear range. The assay is considered valid only when the mean value of the positive control falls within 134 and 154 CATmRNA copies (**Fig. 2**). Once this is controlled, the samples are normalized against the mean value of the positive control and quantitated using the standard regression curve. If their values are not within the linear range, the starting RNA amount needs to be adjusted, and the experiment repeated.

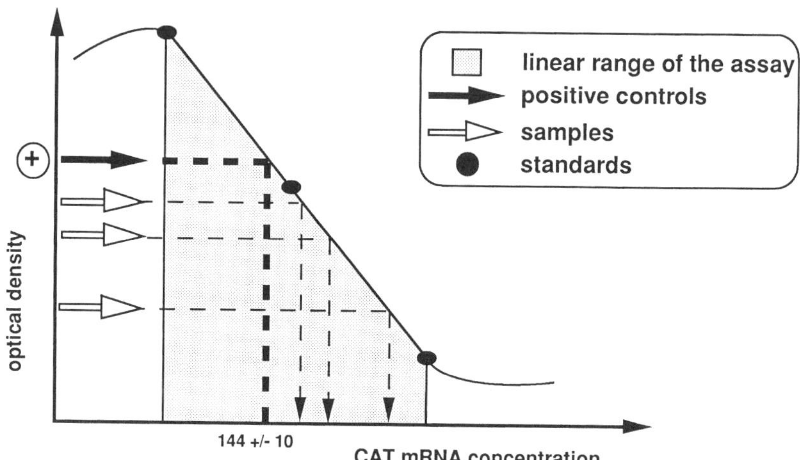


Fig. 2. Quantitation of the tailed CAT RT-PCR. To validate the assay (inter-assay reproducibility), the positive controls must result in 144 ± 10 copies of CAT mRNA. The samples concentration is then assessed by normalizing the values to the positive controls and plotting them on the standard curve.

## 4. Notes

1. FBS can be purchased from a variety of sources. It is advantageous to compare different brands; plan to purchase the selected lot as a large batch in order to keep the conditions similar throughout the experiments. For some experiments, it is preferential not to have any FBS. In this case, it is recommended to use special serum-free medium such as AIM-V (GIBCO) for lymphocytes. However, it is better to use regular medium for the maintenance of the cells.

2. Cloned and stably transfected cell lines provide a constant expression of the CAT gene, which is necessary for defining the linear range of the assay. However, it is very important that the cells are in the logarithmic growth phase and that the viability is greater than 95% when used. Also, the cells should be counted at the last wash step in a volume of 1 mL for greater accuracy.

3. Until cDNA is generated, it is important to use only buffers made up with DEPC treated water and to keep all reagents at 4°C. RNase contamination often occurs through contact with the investigator's hand or dust. It is therefore recommended to wear gloves at all times and to open the tubes using a micro tube opener (Robbins Scientific Corp., Sunnyvale, CA). Also, the hood in which the samples are handled should be treated with UV light for 30 min before using it.

4. A new reagent from Tel-Test, the RNA STAT-60 reagent, specifically enriches mRNA and should be tested as an alternative to RNAzol B in individual protocols.

5. DEPC is suspected to be carcinogenic and should be handled with great care. However, DEPC will be inactivated by autoclaving; it is necessary to aliquot the DEPC water in order to avoid new RNase contamination.

6. All RT- or PCR-reagents should be aliquoted and kept at –20°C. PCR buffers should be made up using endotoxin free water (GIBCO). The use of high quality *Taq* DNA polymerase is highly recommended. Also, it is always a good idea to double check the concentration of the oligomers.

7. The master mix for the RT reaction or the PCR reaction should be made up in a Template Tamer set up specifically for this purpose. Also, the cDNA or amplified DNA should be kept physically away from the rooms where the RT reaction or PCR reaction has been set up.

8. Autoclave mineral oil for 20 min on a liquid cycle. The oil will be cloudy after autoclaving and will need several days to clear.

9. Ethidium bromide (EtBr) is a powerful mutagen. Gloves should be worn when working with solutions that contain this dye. The stock solution (10 mg/mL in water) can be kept at room temperature and is very stable. EtBr should be added to the agarose when the temperature is greater than 60°C. Contaminated solutions (electrophoresis buffer) can be cleared from EtBr by adding activated charcoal and stirring the solution overnight.

10. SDS is also called sodium lauryl sulfate.

11. It is very important to completely loosen up the pellet, because it will not dissolve in RNAzol B.

12. All microfuge centrifugation steps in this protocol are perfomed at $14,000g$ (= maximum speed) except for pelleting cells.

13. From this step on, it is necessary to avoid any contamination. Pipets, tips, water, tubes, racks, and so forth should be treated with UV light (UV Stratalinker) before using them.

14. The numbers of cells will vary from experiment to experiment even when the same amount of cells has been seeded. To work with a constant amount of RNA, extract the RNA with RNAzol B at 200 $\mu$L/$10^6$ cells, and collect a constant volume of supernatant after centrifugation. Any contamination from the white interface should be avoided.

15. For practical reasons and also because the RNA recovery is more consistent, we prefer to precipitate the RNA overnight. However, we recommend to keep the precipitation time constant from one experiment to the other.

16. Isopropanol should always be kept at room temperature. In contrast to other precipitation methods, isopropanol will precipitate only large DNA fragments. In this protocol, the tailed RT-primer will not be precipitated (or only in insignificant amounts), which is important, because this primer could interfere with the PCR amplification step.

17. The pellet is generally not visible. It is therefore recommended to mark the tubes in order to know where the pellet is supposed to be.

18. Do not use a speed vac to dry RNA. A dry pellet is very hard to resuspend in water; a small amount of ethanol will not inhibit the RT reaction.
19. Although it is theoretically possible to freeze the reaction at this point, we do not recommend it. In our hands, it is best to continue directly with the RT step and the PCR amplification.
20. Because each enhancer has a different pattern of nuclear binding sites, and each cell type has distinct pathways of interaction with these sites, it is reasonable to assume that the dynamics of the constitutive replication of the CAT constructs will be different depending on the cell type and promoter used. Therefore, it is necessary to determine the amount of cellular RNA needed and the linear range of the assay for each particular cell line or CAT construct.

## References


1. Gorman, C. M., Moffat, L. F., and Howard, B. H. (1982) Recombinant genomes which express chloramphenicol acetyl transferase in mammalian cells. *Mol. Cell. Biol.* **2,** 1044–1051.
2. Swingler, S., Easton, A., and Morris, A. (1992) Cytokine augmentation of HIV-1 LTR-driven gene expression in neural cells. *AIDS Res. and Human Retroviruses* **8(4),** 487–493.
3. Verdin, E., Becker, N., Bex, F., Droogmans, L., and Burny, A. (1990) Identification and characterization of an enhancer in the coding region of the genome of human immunodeficiency virus type 1. *Proc. Natl. Acad. Sci. USA* **87,** 4874–4878.
4. Rosen, C. A., Sodroski, J. G., and Hascltine, W. A. (1985) The location of cis-acting regulatory sequences in the human T cell lymphotropic virus type III (HTLV-III/ LAV) long terminal repeat. *Cell* **41,** 813–823.
5. Sodroski, J. G., Rosen, C., Wong-Staal, F., Salahuddin, S. K., Popovic, M., Aryas, S., Gallo, R. C., and Haseltine, W. A.(1985) Trans-acting transcriptional activation of human T-cell leukemia virus type III long terminal repeat. *Science* **227,** 171–173.
6. Toyama, R., Bende, S. M., and Dahr, R. (1992) Transcriptional activity of the human immunodeficiency virus-1 promotor in fission yeast Schizosacharomyces pombe. *Nucleic Acids Res.* **20(10),** 2591–2596.
7. Spandidos, D. A., Zoumpourlis, V., Kotsinas, A., Tsiriyotis, C., and Sekeris, C. E. (1990) Response of human immunodeficiency virus long terminal repeat to growth factors and hormones. *Anticancer Res.* **10,** 1241–1246.
8. Feuchter, A. and Mager, D. (1990) Functional heterogeneity of a large family of human LTR-like promoters and enhancers. *Nucleic Acids Res.* **18(5),** 1261–1270.
9. Golub, E. I., Gongrong, L., and Volsky, D. J. (1990) Differences in the basal activity of the long terminal repeat determine different replicative capacities of two closely related human immunodeficiency virus type 1 isolates. *J. Virol.* **64(8),** 3654–3660.
10. Markovitz, D. M., Hannibal, M., Perez, V. L., Gauntt, C., Folks, T. M., and Nabel, G. J. (1990) Differential regulation of human immunodeficiency viruses (HIVs): A specific regulatory element in HIV-2 respond to stimulation of the T-cell antigen receptor. *Proc. Natl. Acad. Sci. USA* **87,** 9098–9102.

11. Bielinska, A., Krasnow, S., and Nabel, G. J. (1989) NF-kB-mediated activation of the human immunodeficiency virus enhancer: Site of transcriptional initiation is dependent of the TATA box. *J. Virol.* **63(9),** 4097–4100.

12. Tong-Starksen, S. E., Luciw, P. A., and Peterlin, B. M. (1989) Signaling through T lymphocyte surface proteins, TCR/CD3 and CD28, activates the HIV-1 long terminal repeat. *J. Immunol.* **142(2),** 702–707.

13. Powell, J. D., Yehuda-Cohen, T., Villinger, F., McClure, H. M., Sell, K. W., and Ansari, A. A. (1990) Inhibition of SIV/SMM replication in vitro by CD8+ cells from SIV/SMM infected seropositive clinically asymptomatic sooty mangabeys. *J. Med. Primatol.* **19,** 239–249.

14. Powell, J. D., Bednarik, D. P., Yehuda-Cohen, T., Villinger, F., Folks, T. M., and Ansari, A. A. (1991) Regulation of immune activation/retroviral replication by CD8+ T cells. Ann. *N.Y. Acad. Sci.* **636,** 360–362.

15. Powell, J. D., Bednarik, D. P., Folks, T. M., Yehuda-Cohen, T., Villinger, F., Sell, K. W., and Ansari, A. A. (1993) Inhibition of cellular activation of retroviral replication by CD8$^+$ T cells derived from nonhuman primates. *Clin. Exp. Immunol.* **91(3),** 473–481.

16. Shuldiner, A. R., Tanner, K., Moore, C. A., and Roth, J. (1991) RNA template-specific PCR: an improved method that dramatically reduces false positives in RT-PCR. *BioTechniques* **11(6),** 760–763.

17. Holodniy, M., Katzenstein, D. A., Sengupta, S., Wang, A., Casipit, C., Schwartz, D. H., Konrad, M., Groves, E., and Merigan, T. C. (1991) Detection and quantitation of human immunodeficiency virus RNA in patient serum by use of the polymerase chain reaction. *J. Infect Dis.* **163,** 862–866.

18. Ballagi-Pordany, A., Ballagi-Pordany, A., and Funa, K. (1991) Quantitative determination of mRNA phenotypes by the polymerase chain reaction. *Analyt. Biochem.* **196,** 89–94.

19. Li, B., Sehajpal, P. K., Khanna, A., Vlassara, H., Cerami, A., Stenzel, K. H., and Suthanthiran, M. (1991) Differential regulation of transforming growth factor β and interleukin 2 genes in human T cells: demonstration by usage of novel competitor DNA constructs in the quantitative polymerase chain reaction. *J. Exp. Med.* **174,** 1259–1262.

20. Piatak, M., Jr., Saag, M. S., Yang, L. C., Clark, S. J., Kappes, J. C., Luk, K. C., Hahn, B. H., Shaw, G. M., and Lifson, J. D. (1993) High levels of HIV-1 in plasma during all stages of infection determined by competitive PCR. *Science* **259,** 1749–1754.

21. Sambrook J., Fritsch, E. F., and Maniatis, T. (1989) *Molecular Cloning.* Cold Spring Harbor Laboratory Press, Cold Spring Harbor, NY.

22. Knuchel, M., Bednarik, D. P., Chikkala, N., Villinger, F., Folks, T. M., and Ansari, A. A. (1994) Development of a novel quantitative assay for the measurement of chloramphenicol acetyl transferase (CAT)mRNA. *J. Virol. Methods* **48,** 325–338.

23. Knuchel, M., Bednarik, D. P., Chikkala, N., and Ansari, A. A. (1994) Biphasic in vitro regulation of retroviral replication by CD8+ cells from nonhuman primates. *JAIDS* **7,** 438–446.

# 19

# A Stochastic PCR Approach
# for RNA Quantification in Multiple Samples

Adrian Puntschart and Michael Vogt

## 1. Introduction

When studying the effect of various treatments on gene expression in humans, one occasionally is faced with the problem of detecting small changes in transcript levels in minute tissue samples. In addition, interindividual variations can be quite large and may even be the major source of variation *(1)*. Therefore, numerous samples usually have to be analyzed to detect such small variations in gene expression.

The limited amount of starting material favors the use of the polymerase chain reaction (PCR). It offers unsurpassed sensitivity by way of the exponential amplification of a specific target DNA. However, this nonlinear nature of the amplification process renders quantification of DNA or cDNA difficult. Small differences in the amplification conditions from reaction to reaction can have dramatic effects on amplification efficiencies and therefore on the amount of product accumulated after a limited number of cycles *(2)*. Such variations are inevitable in experimental practice, they originate from small differences in the composition of different samples or from minute variations in pipetting steps. Numerous methods have been developed to control for this inescapable variation in efficiency, the most popular being competitive PCR *(2)*, regression analysis of PCR kinetics *(3)*, or the use of internal *(4)*/external *(5)* standards (*see* **6** and **7** for review). These methods are mostly labor-intensive and time-consuming. This is because of the need for construction of template-related standards, preparation of dilution series for each sample and multiple repetitions of PCR reactions. Despite these attempts, the "truly quantitative" PCR has still proven elusive *(8)*. Even tightly controlled PCR reactions will always show some variations in their reaction efficiencies. A single PCR run must

From: *Methods in Molecular Medicine, Vol 26: Quantitative PCR Protocols*
Edited by: B. Kochanowski and U. Reischl © Humana Press Inc., Totowa, NJ

therefore be considered as an independent outcome of a stochastic process with small, but inevitable differences in reaction efficiencies *(6,9)*. We reasoned that it is possible to treat the error generated in such a stochastic process with a statistical quantitation approach, i.e., to measure each sample repeatedly, without trying to control each processing step for high precision. We developed the rationale for this approach when studying the effect of exercise training on skeletal muscles in humans, where we became aware that the major source of variation between two groups may actually be interindividual variations in the training response. This fact some time ago gave rise to the notion, "Do more, less well" *(2)*, meaning that one should measure more samples (individuals) several times rather than trying to measure few samples only once which would demand to control the precision of every processing step. It has led to the development of a PCR quantitation approach, which minimizes work and still reaches an acceptable level of precision.

The main features of the approach are:

1. All samples are amplified in parallel without rigorously controlling for the efficiency during PCR. The reactions are stopped in the exponential phase of amplification. The specific PCR products accumulated are determined by an enzyme-linked immunosorbent assay (ELISA) assay and compared to a Reference Sample amplified in parallel. Because the initial template concentration of the reference sample is known, the template concentration in the samples can be calculated. PCR reactions are repeated three to four times in order to account for stochastic variations in the reaction efficiencies.

2. The statistical nature of the approach necessitates that a minimum number of samples are analyzed (at least 4–5) in each group to be compared (a prerequisite that has to be fulfilled in any study where heterogeneous populations are compared). Overall, only about three to four reactions are performed with each sample, which is the lower range for most other PCR quantification procedures.

3. Samples have to be standardized for the amount of tissue used for RNA extraction for the efficiency of the reverse-transcription step. The amount of RNA used for reverse transcription is often used to standardize reverse transcriptase-polymerase chain reaction (RT-PCR) *(4,5,10–12)*. However, with the limited amounts of starting material that necessitate the use of PCR, this is usually not possible. Housekeeping genes have also been used to account for differences in input RNA *(13–15)*. In our original study, we determined the volume of the tissue used and added a constant amount of an unrelated, synthetic RNA (chicken myosin light chain) to account for variations in the recovery of extracted and reverse-transcribed total RNA *(16)*. Alternatively, we have used 28S ribosomal RNA *(17)* to standardize different human samples, which has proven to be easier. The type of standardization used will depend on the question addressed. A perfect standard useful for all applications probably does not exist.

4. To make the results comparable between experiments, we also amplify a sample in which the amount of template is known (a so-called reference sample, RS). In this way, cDNA as well as genomic DNA can be quantitated in absolute terms. The reference samples are prepared by PCR amplification as well. Therefore, the availability of cDNAs is not restrictive and no cloning steps are needed.

5. To control for systemic differences among samples caused by the presence of inhibitors or stimulators *(18)*, two samples are mixed with each other (half the volume of the separate reactions for each) and amplified in parallel with the separate samples. The amount of PCR product obtained in the control mixtures should therefore be the mean of the separately amplified samples. The presence of inhibitors or stimulators is thought to affect significantly these amplifications. But so far we never have detected a difference in amplification rate caused by such inhibitors or stimulators.

6. It is a prerequisite for this type of quantitative PCR to be stopped in the exponential phase. This can be determined easily. As we have shown earlier *(16)*, the first 4–5 cycles that yield clearly visible bands on ethidium bromide gels are still in the exponential phase.

7. The precision of our approach was assessed for a narrow linear range *(16)*. Differences of about 30% between groups should be discernible. This is comparable to or better than previously published methods *(10,13,19)*.

We believe that our PCR quantitation approach is highly versatile, because the availability of cDNAs is not restricting and no cloning steps are necessary. Any RNA or DNA whose sequence or part of it is known can be quantitated. The approach is especially well-suited in cases where the analysis of multiple RNA species in numerous samples is required. The following sections are arranged according to the sequential steps that have to be followed after having optimized a particular PCR reaction.

## 2. Materials

1. Water (PCR only): ddH$_2$O is autoclaved and stored at 4°C.
2. TE-Solution (PCR only): 10 m$M$ Tris-HCl, pH 7.4, 1 m$M$ ethylenediaminetetraacetic acid (EDTA), pH 8.0.
3. Phenol/chloroform (1:1 v/v): Stored at 4°C.
4. Ethanol: Stored at 4°C.
5. 75% Ethanol: Stored at 4°C.
6. 3 $M$ NaAcetate, pH 5.2: Stored at room temperature.
7. Yeast carrier tRNA (Boehringer Mannheim, Mannheim, Germany): Used as a carrier for the dilutions of the reference samples. Stored as 10 µg/µL solution at –20°C.
8. Spectrophotometer (Beckmann): To determine the cDNA concentration of the PCR product that are used for Reference Sample preparation.
9. Thermostable DNA-polymerase: 2 U/µL DynaZyme (Finnzymes Oy, Finland).

10. 10X Buffer (DynaZyme): With or without $Mg^{2+}$, supplied by the manufacturer. Repeated freezing and thawing sometimes results in reduced PCR efficiency. Therefore, fresh buffer is aliquoted and stored at $-20°C$. Thawed aliquots are stored at $4°C$ for further use.

11. 50 m$M$ $MgCl_2$ solution: purchased by the supplier of the 10X PCR buffer or self-made. Self-made buffer is autoclaved and stored at $4°C$.

12. Primers (MWG Biotech, Ebersberg, Germany): about 25–30 nucleotides in length. In our experience, there are differences in the properties between primers from different purchasers (e.g., Primer-Dimer formation). TE is used to dissolve primers to a storing concentration of about 100 µ$M$. Primers are stored at $-20°C$.

    PCR primer selection was done manually according standard recommendations *(20)* or alternatively by using Primer Analysis Software OLIGO⁻ 5.0, which always resulted in efficient PCR-experiments. It may be of importance to note that our quantitation approach was established using primers of 25–30 bp in length. The primers were selected to give PCR products of about 200–400 base pairs in length.

13. PCR DIG labeling mix (Boehringer Mannheim): Including: 2 m$M$ dATP, dCTP, dGTP each, 1.9 m$M$ dTTP, and 0.1 m$M$ digoxigenin-11-dUTP.

14. Thermocycler: UNO PCR-Cycler (Biometra, Biomedizinische Analytik GmbH, Germany). A cycler with a heated lid that makes use of mineral oil superfluous, containing a module for 96 0.2-mL tubes. Other cyclers with 0.5-mL tubes work equally well.

15. Thermo-Fast-Plates of 96 0.2-mL tubes and strips of eight 0.2-mL caps (Biometra, Biomedizinische Analytik GmbH). In case of other thermocyclers, ordinary (non-thin-walled) 0.5-mL tubes are used.

16. Aluminium racks: Helpful in holding the Thermo-Fast-Plate or the tubes on ice during preparation of the PCR-experiment. These racks were made in our laboratory. The dimensions of these racks are $11.5 \times 9$ cm with $8 \times 12$ holes for 0.2-mL tubes and $13.5 \times 12.5$ cm with $8 \times 10$ holes for 0.5-mL tubes.

17. Standard agarose-gel electrophoresis equipment, including ultraviolet (UV)-transilluminator.

18. 4% NuSieve Agarose (FMC) in 1X TBE containing 0.5 µg/mL ethidium-bromide.

19. 10X TBE (1 L): Tris(hydroxymethyl)aminomethan (108 g), borate (55 g), 0.5 $M$ EDTA, pH 8.0 (40 mL).

20. Molecular weight marker: e.g., pBR322 $\times$ *Hpa*II.

21. Streptavidin (Boehringer Mannheim): 1 mg/mL stock solution is prepared in 1X PBS and stored at $4°C$.

22. 1 L 1X PBS: 8 g NaCl, 0.2 g KCl, 1.44 g $Na_2HPO_4$. Adjust to pH 7.4 with HCl.

23. 1 L 20X SSC: 175.3 g NaCl, 88.2 g NaCitrate. Adjust to pH 7.0 with 10 $N$ NaOH.

24. Acid-Hybridization-Solution: 12 mL 20X SSC, 1 $M$ Tris, 400 µL pH 7.4, 800 µL 0.5 $M$ EDTA, 1.33 mL 6 $M$ HCl, 26.8 mL dd$H_2O$. 40 mL of the hybridization solution is prepared at once. This solution is stored at $4°C$ and can be used up to 2 mo.

25. TBS-T solution (per 1 L): 2.42 g Tris(hydroxymethyl)aminomethan, 8 g NaCl. Adjust to pH 7.5, then add 1 mL Tween-20.
26. Anti-digoxigenin-AP Fab fragments (Boehringer Mannheim, cat. no. 1093274). Fab fragments from anti-digoxigenin antibody conjugated with alkaline phosphatase (AP). Fresh dilutions of 1:1000 in TBS-T are prepared for each ELISA experiment.
27. Substrate solution (5 mL prepared fresh for each ELISA experiment): 0.478 mL Diethanolamin, 50 µL 50 m$M$ MgCl$_2$, 68 µL 6 $M$ HCl, 4.404 mL H$_2$O. Vortex mixture vigorously, pH should be about 9.8, then add 20 mg 4-Nitrophenylphosphate (4-NPP, Boehringer Mannheim).
28. Bio-Oligos (MWG Biotech, Germany; Intron, Switzerland). 5' Biotin-labeled oligo probes of about 20 nucleotides in length are selected as nested oligos within the sequence of the amplified PCR-product by using Primer Analysis Software OLIGO™ 5.0. Supplied oligos are diluted with TE to a stock concentration of 10 nmol/mL and stored at –20°C.
29. COSTAR E.I.A/R.I.A Microtiter plates (BioRAD, cat. no. 224-0096). The source of the microtiter plates is crucial, because there are great differences in the binding capacities between different microtiter plates.
30. Microplate Reader (Model 450, BioRAD).

## 3. Methods

### 3.1. Preparation of RS

Total RNA is prepared by the acid phenol method of Chomozynski and Sacchi *(21)* (*see* **Note 1**). cDNA was prepared by reverse transcription of total RNA with Superscript™ RNase H⁻ (Gibco-BRL, Gaithersburg, MD) according to the manufacturer's instructions using random hexamer primers (Boehringer Mannheim). The cDNA is diluted to 200 µL with TE and stored at –20°C.

1. Run about 200–300 µL PCR to a product band clearly visible on an EtBr gel or pool previous PCR samples. For PCR primer-, Nucleotide- and Mg$^{2+}$-concentrations were adjusted according to standard recommendations *(20)*. The concentrations usually used are 0.2 µ$M$ for each primer, 0.5–4 m$M$ for Mg$^{2+}$ and 0.04 m$M$ for each nucleotide. Nucleotide concentrations lower than 0.02 m$M$ resulted in inefficient reactions, while no substantial differences in efficiency were observed with higher concentrations.
2. Typical three step PCR cycles are 95°C/10 min denaturation, 60–65°C/60–90 min annealing and 72°C/10 min elongation. For higher annealing temperatures, two-step PCR cycles are adequate: 95°C/10 s denaturation and 70°C/90 s annealing.
3. Phenol extract by adding an equal volume of Phenol/CHCl$_3$ (1:1 v/v), vortex vigorously, centrifuge for 5 min.
4. Ethanol precipitate aqueous phase by adding 1/10 vol 3 $M$ NaAcetate, pH 5.2, and 2.5 vol EtOH$_{abs}$. Incubate on ice for 15–30 min. Centrifuge for 15 min at 4°C (*see* **Note 2**).

5. Air dry pellet until no liquid droplets are visible. The pellet might not be visible.
6. Resuspend pellet in 300 μL TE. Determine the DNA concentration without dilution at 260/280 nm. Add tRNA (10 μg/mL) as carrier to make RS stable against degradation.
7. Dilute the DNA with TE/tRNA to a concentration that is about 100–1000 times higher than the expected concentration of the sequence to be analyzed. This concentrated RS (RS-X$_{conc}$ of a particular sequence X) is stored at –20°C. We usually prepare total RNA from about 1–5 mg human skeletal muscle tissue and dilute the concentrated RS to about 1–100 pg/μL.
8. Make serial 1:3 dilutions of RS-X$_{conc}$ in TE/tRNA. Amplify these dilutions in parallel with a selection of the samples to be analyzed (*see* **Fig. 1**). The dilution that gives a signal similar in strength to the samples is used as RS-X and 2–3 100 μL aliquots are stored at –20°C.

## 3.2. Determination of Cycle Numbers

1. Prepare a PCR run with all samples and the dilution series of the corresponding RS.
2. Run the PCR for about 15–25 cycles.
3. Analyze 5 μL PCR product on a 4% NuSieve Agarose gel.
4. If necessary, run additional PCR cycles until clear specific bands appear (*see* **Notes 3** and **4**).

## 3.3. PCR-Experiment

1. **Figure 1** should help to clarify the setup for one PCR run. All work is done on ice. Two microliters of each cDNA sample including the respective RS are pipetted into separate reaction tubes. To control for systemic differences among samples possibly caused by the presence of inhibitors or stimulators (*see* **Introduction**), aliquots from two samples are mixed together (1 μL each). A negative control without template is also amplified to subtract background staining and to control for an eventual PCR contamination. Mix Controls are performed for only one RNA (*see* **Notes 5–7**).
2. Prepare a master mix containing the following components indicated in final concentrations: 1X 10X buffer (with or without Mg$^{2+}$), supplied with the enzyme; 0.5–3.5 m*M* MgCl$_2$; 0.2–0.4 μ*M* primers each (mix of 5'- and 3'-primer); 0.04 m*M* nucleotides: PCR DIG-Labeling-Mix (Boehringer Mannheim); and 20 U/mL thermostable DNA-polymerase.
3. Add 38 μL of the master-mix to each tube prepared in step one.
4. Transfer tubes to the preheated PCR cycler.
5. Incubate at 95°C for at least 2 min (*see* **Note 8**).
6. Run the PCR-experiment for the appropriate cycles (*see* **Subheading 3.2.**).
7. Analyze 5 μL of each sample on a 4% NuSieve Agarose gel. Agarose-gel electrophoresis is used to test if a PCR run was successful. When processing large numbers of samples, select a few to analyze for PCR product (*see* **Note 9**).
8. For further work, PCR products can be stored at 4 °C for several days.

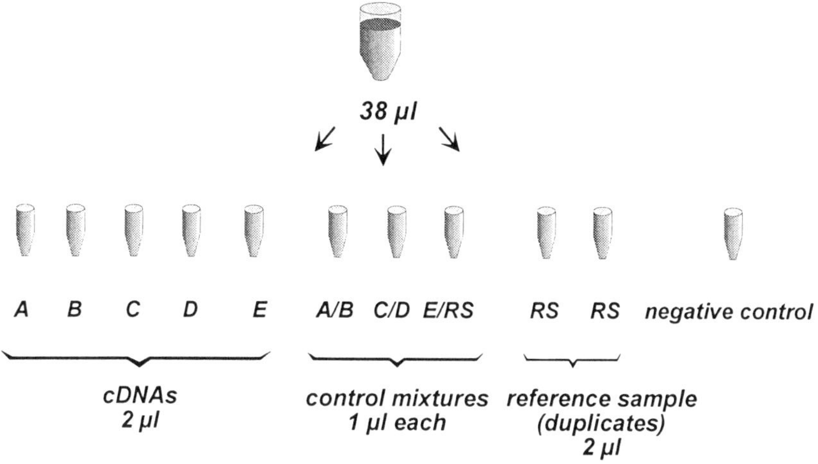


Fig. 1. Quantitative PCR setup.

## 3.4. ELISA Quantification of PCR Products

1. Coat a microtiter plate with 100 µL streptavidin solution: (10 µg/mL in PBS)/ well. Incubate overnight at 4°C or for 6 h at room temperature (*see* **Note 10**).
2. Transfer 10 µL of each PCR sample to a separate microtiter plate (the preparation plate). Add 40 µL of alkaline Bio-Oligo Solution (0.1 µ*M* Biotin-labeled oligo probe, 0.125 *M* NaOH) to each well. Incubate at room temperature for 10 min to denature the dsDNA (*see* **Note 11**).
3. Add 50 µL acid-hybridization solution (*see* **Materials**) to each sample. Incubate at room temperature for 60–120 min.
4. Wash the streptavidin-coated microtiter plate three times with 100 µL TBS-T/well.
5. Transfer 50 µL of the DNA/Bio-Oligo hybrid from the preparation plate to the streptavidin-coated microtiter plate. Incubate at room temperature for 30 min.
6. Wash the plate three times with 100 µL TBS-T/well.
7. Add 50 µL of diluted anti-digoxigenin-AP (*see* **Materials**)/well and incubate at room temperature for 30 min.
8. Wash the plate three times with 100 µL TBS-T/well.
9. Prepare 5 mL of fresh substrate solution (*see* **Materials** and **Note 12**). Add 50 µL substrate solution/well. Take care to prevent the formation of small bubbles.
10. Measure the degree of yellow staining with microplate reader at 405 nm wavelength. Determine the absorption 30–120 min after addition of the substrate solution (*see* **Note 13**).

## 3.5. Calculations

1. PCR/ELISA experiments are repeated three to four times for a particular sequence X.

2. The concentration of a sequence X in a sample A ($X_A$) is calculated as follows:

    a. Subtract the background absorbance (= absorbance of the negative control $Abs_{neg.control}$) from the absorbance of each sample. The corrected absorbance of each sample (e.g., $Abs_A$ - $Abs_{neg.control}$) is then divided by the corrected absorbance of the corresponding RS. This yields a relative amount of each sample ($X_{rel\_A}$):

    $$X_{rel\_A} = \frac{Abs_A - Abs_{neg.control}}{Abs_{RS\_X} - Abs_{neg.control}}$$

    b. $X_A$ is calculated by multiplication of the relative amount $< X_{rel\_A} >$ (mean of all PCR experiments) with the concentration of the corresponding $RS_X$ used for PCR.

    $$X_A = < X_{rel\_A} > * RS_X$$

3. To normalize each sample we used 28S rRNA, which is also measured three to four times by PCR. Use the aforementioned calculation method for $X_A$ to determine the concentration of 28S rRNA ([28S]). Divide $X_A$ by $X_{28S}$ and finally calculate the mean value of all samples.

## 4. Notes

1. When preparing total RNA from skeletal muscle tissue some modifications turned out to be necessary in order to eliminate the mitochondrial DNA, which exists in several thousand copies per nuclear genome *(17)*. Therefore, an additional DNase digestion step with concomitant acid phenol extraction and ethanol precipitation has to be performed.

2. Precipitation of small PCR products (<200 bp) can be difficult. Addition of $Mg^{2+}$ may be advisable *(22)*.

3. The presented quantitative PCR method without internal standards requires that the PCR reaction is stopped while it is still in the exponential phase (plateau effect). For this reason, it is important to adjust the appropriate cycle number for each transcript. As we showed earlier *(16)*, the first 4–5 cycles that yield clearly visible bands on ethidium bromide gels are still in the exponential phase. The number of cycles depends on the abundance of the target species and the efficiency of the PCR reaction. When the concentration of a specific sequence differs by more than one order of magnitude between samples, aliquots of these samples may have to be diluted further in TE/tRNA.

4. We are performing PCR with cycle numbers ranging from 16 (28S rRNA, whose cDNA was further diluted 1:10 before amplification because of its high celluar content) to 37 (for some immediate early genes correspondending to a concentration of some attograms/ng 28S).

5. To test for DNA contamination, dilute 2 µL of your RNA preparation (before reverse transcription) in 18 µL TE. Amplify these -RT samples in parallel with the cDNA samples. No bands should be visible for the -RT samples after performing 5 cycles more than for the corresponding cDNA samples.

6. Some precautions before starting PCR are recommended: Always wear gloves. Use pipets, water, TE- and $Mg^{2+}$- solution, which are solely used for pre-PCR work. When following these precautions, no problems with contaminations have occurred in our laboratory so far.

7. In the protocol described here we determine the amount of PCR product by an ELISA assay. The PCR products can also be radioactively labeled. In that case, use 20 $\mu M$ each dNTP and 0.08 $\mu$Ci/$\mu$L ($\alpha$-$^{32}$P)dTTP. PCR products were then separated by polyacrylamide-gel electrophoresis. The gels were dried and exposed to X-ray films. The corresponding bands were cut out and the radioactivity was determined by liquid scintillation counting.

8. In certain cases, longer denaturation (up to 5 min) might sometimes be helpful to prevent primer dimerization.

9. If mineral-oil overlay must be used, transfer the aqueous phase of the PCR samples into new tubes to remove sufficiently the mineral oil for the next steps.

10. Make sure that the wells of the plate do not dry out during processing. If necessary, cover plates temporarily (e.g., with Parafilm).

11. For denaturation of PCR products, we use an alkaline-denaturing procedure. Heat denaturation (95°C for 5 min) yields similar results.

12. It is important to add the 4-Nitrophenylphosphate just before incubation of the substrate in order to prevent unspecific staining. Substrate is turned over by AP to give a yellow staining product.

13. In our experience staining should be visible within 15 min. On the BioRad reader, absorbance from Optical density (OD) = 0.2–2.0 is in a linear range. Color reaction is linear within this range at least between 30 and 120 min after substrate incubation.

## Acknowledgments

We thank Drs. R. Billeter and H. Hoppeler for helpful discussions and for reading the manuscript. This work was supported by Grant 31-28821.90 and 3100-42449.94 of the Swiss National Science Foundation, as well as by Eidgenössische Sportkommission, Institute of Sports Sciences, Magglingen (Switzerland).

## References

1. Gundersen, H. J. G. and Osterby, R. (1980) Optimizing sampling efficiency of stereological studies in biology: or 'Do more less well!' *J. Microscopy* **121,** 65–73.

2. Gilliland, G., Perin, S., Blanchard, K., and Bunn, H. F. (1990) Analysis of cytokine mRNA and DNA: Detection and quantitation by competitive polymerase chain reaction. *Proc. Natl. Acad. Sci. USA* **87,** 2725–2729.

3. Wiesner, R. J., Ruegg, J. C., and Morano, I. (1992) Counting target molecules by exponential polymerase chain reaction: Copy number of mitochondrial DNA in rat tissues. *Biochem. Biophys. Res. Comm.* **183,** 553–559.

4. Wang, A., Doyle, M. V., and Mark, D. F. (1989) Quantitation of mRNA by the polymerase chain reaction. *Proc. Natl. Acad. Sci. USA* **86,** 9717–9721.

5. Chelly, A., Montarras, D., Pinset, C., Berwald-Netter, Y., and Kaplan, I.-C. (1990) Quantitative estimation of minor mRNAs by cDNA-polymerase chain reaction: application to dystrophin mRNA in cultured myogenic and brain cells. *Eur. J. Biochem.* **187,** 691–698.

6. Ferre, F. (1993) Quantitative or semi-quantitative PCR: reality versus myth. *PCR Methods Applic.* **2,** 1–9.

7. Clementi, M., Menzo, S., Bagnarelli, P., Manzin, A., Valenza, A., and Varaldo, P. E. (1993) Quantitative PCR and RT-PCR in virology. *PCR Methods Applic.* **2,** 191–196.

8. Reischl, U. and Kochanowski, B. (1995) Quantitative PCR. *Mol. Biotech.* **3,** 55–71.

9. Nedelman, J., Haegerty, P., and Lawrence, C. (1992) Quantitative PCR with internal controls. *Comput. Applic. Biosci.* **8,** 65–70.

10. Becker-Andre, M. and Hahlbrok, K. (1989) Absolute mRNA quantification using the polymerase chain reaction (PCR): a novel approach by a PCR aided transcript titration assay (PATTY). *Nucleic Acid Res.* **17,** 9437–9446.

11. Murphy, L. D., Herzog, C. E., Rudick, J. B., Fojo, A. T., and Bates, S. E. (1990) Use of the polymerase chain reaction in the quantitation of mdr–1 gene expression. *Biochemistry* **29,** 10,351–10,356.

12. Noonan,K. E., Beck, C., Holzmayer, T. A., Chin, J. E., Wunder, J. S., Andrulis, I. l., Gazdar, A. F., Willman, C. L., Griffith, B., Von Hoff, D. D., and Roninson, I. B. (1990) Quantitative analysis of MDR1 (multidrug resistance) gene expression in human tumors by polymerase chain reaction. *Proc. Natl. Acad. Sci. USA* **87,** 7160–7164.

13. Hoof, T., Riordan, J. R., and Tuemmler, B. (1991) Quantitation of mRNA by the kinetic polymerase chain reaction assay: a tool for monitoring P-glycoprotein gene expression. *Anal. Biochem.* **196,** 161–169.

14. Chelly, J., Kaplan, J.-C., Maire, S., Gautron, S., and Kahn, A. (1988) Transcription of the dystrophin gene in human muscle and non-muscle tissues. *Nature* **333,** 858–860.

15. Rappole, D. A., Wang, A., Mark, D., and Werb, Z. (1989) Novel method for studying mRNA phenotypes in single or small numbers of cells. *J. Cell. Biochem.* **39,** 1–11.

16. Puntschart, A., Jostarndt, K., Hoppeler, H., and Billeter, R. (1994) An efficient polymerase chain reaction approach for the quantitation of multiple RNAs in human tissue samples. *PCR Methods Applic.* **3,** 232–238.

17. Puntschart, A., Claassen, H., Jostarndt, K., Hoppeler, H., and Billeter, R. (1995) mRNAs of enzymes involved in energy metabolism and mtDNA are increased in endurance-trained athletes. *Am. J. Physiol.* **269,** C619–C625.

18. Yang, B., Yolken, R., and Viscidi, R. (1993) Quantitative polymerase chain reaction by monitoring enzymatic activity of DNA polymerase. *Anal. Biochem.* **208,** 110–116.

19. Ito, H., Miller, S. C., Akimoto, H., Torti, S. V., Taylor, A., Billingham, M. E., and Torti, F. M. (1991) Evaluation of mRNA levels by the polymerase chain reaction in small cardiac tissue samples. *J. Mol. Cell. Cardiol.* **23,** 1117–1125.
20. Dieffenbach, C. W. and Dveksler, G. S. (1995) *PCR Primer: A Laboratory Manual.* Cold Spring Harbor Laboratory, Cold Spring Harbor, NY.
21. Chomczynski, P. and Sacchi, N. (1987) Single-step method of RNA isolation by acid guanidinium thiocyanate-phenol-chloroform extraction. *Anal. Biochem.* **162,** 156–159.
22. Sambrook, J., Fritsch, E. F., and Maniatis, T. (eds.) (1989) *Molecular Cloning. A Laboratory Manual* (2nd ed.), Cold Spring Harbor Laboratory, Cold Spring Harbor, NY.

# 20

# Quantitation of mRNA Species by RT-PCR on Total mRNA Population with Nonradioactive Probes

Sabine Herblot, Benoît Rousseau, and Jacques Bonnet

## 1. Introduction

Quantitative polymerase chain reaction (PCR) is aimed to determine the absolute or relative amounts of RNA or DNA sequences in a given sample. There are two facts limiting the convenience of this approach. First, in most cases, only one or two sequences are amplified in a given round of amplification. If a family of sequences are to be quantitated, as many amplification reactions are necessary. However, it has been shown that complex populations could be amplified in a sequential independent way *(1–3)*. A major concern about the amplification of whole populations are the biases for or against some sequences. In fact, it appears that these biases are not important and that the amplified populations are quite representative of the original mixture of sequences *(1,4)*. This makes possible a score of PCR applications such as differential display analysis *(5)* or representational difference analysis *(6)*, which are aimed to detect qualitative and quantitative differences between sequences present in genomes or messenger RNA (mRNA) populations. This also implies that it is possible to measure the amount of numerous sequences in the amplicons.

Second, to be valid, PCR measures need to be corrected from sample to sample variation and to be compared to standards. With population amplification, these steps are simplified, because on the one hand, all the sequences are amplified with the same primers; on the other hand, the total amount of the amplicons can be used as a reference. Indeed, the fact that most—if it not all—sequences in the population are amplified results in the sequences are competing for the amplification system at the end of the amplification reaction. As a consequence, the ratio between the amounts of a given sequence in the amplicons and the total amount of the amplicons is representative of the initial proportion of the sequence of interest.

From: *Methods in Molecular Medicine, Vol. 26: Quantitative PCR Protocols*
Edited by: B. Kochanowski and U. Reischl ©Humana Press Inc., Totowa, NJ

Consequently, this system of quantitation has the advantage in that it allows the quantitation of multiple sequences from a single sample and a single experiment of amplification without the need of standardization for each sequence. It has the correlative inconvenience of not being as sensitive as classic quantitative PCR.

## 2. Materials

1. Primers UPdN$_6$: The UPdN$_6$ (Universal primer [dN]$_6$) oligonucleotide, used in cDNA synthesis, is a mixture of all the sequences GCCGGAGCTGCAGAATTCNNNNNN, where N is either A, C, G, or T. Consequently, this primer carries first a 3' degenerated hexanucleotide sequence and a 5' defined sequence. The hexanucleotide is used to randomly prime the reverse transcription, allowing to anchor the defined sequence for the subsequent PCR.
2. UP primer: The UP oligonucleotide sequence GCCGGAGCTGCAGAATTC is the defined 5' end of the above oligonucleotide.
3. NIII(dT)$_{15}$VV: For the cDNA synthesis from total RNA, we use this modified oligo(dT) carrying first two degenerated nucleosides at the 3' end (either A, C, or G) to avoid reverse transcription of a long poly(A) stretch and second, a defined sequence in 5' : CGG GAA TTCGCTCGACATGTTTTTTTTTTTTTTTVV. The (dT)VV is used to prime the reverse transcription from the beginning of the poly(A) tract and allows the anchoring of the defined sequence for the subsequent PCR.
4. NIII primer: The NIII primer sequence CGGGAATTCGCTCGACATG is the 5' defined end of the above oligonucleotide.
5. 5X Reverse transcriptase buffer: 250 m*M* Tris-HCl, pH 8.3, 375 m*M* KCl, 50 m*M* dithiothreitol, 2.5 m*M* spermidine
6. Enzymes: We use the M-MLV reverse transcriptase for the first strand cDNA synthesis and the Klenow large fragment enzyme for the second strand cDNA synthesis. For PCR amplification, we use the Goldstar *Taq* DNA polymerase (Eurogentec, Angers, France) (*see* **Note 1**).
7. dNTP mix: dATP, dCTP, dGTP and dTTP tri sodium salts, 10 m*M* each.
8. RNAse inhibitor.
9. RNAse-free water: Millipore water is treated with 0.05 % of DEPC, at 37°C for 12 h, then autoclaved at 120°C for 20 min.
10. Sephacryl S400 HR spin column: Any DNA purification system can be used to eliminate the oligonucleotides, the enzyme, and the salts from the cDNA. We prefer the Sephacryl spin column procedure because of its speed and efficiency.
11. Nylon membrane: The DNA dot blots are performed on a positively charged Nylon membrane (*see* **Note 2**).
12. Probe labeling and detection kits: The probes can be synthesized and digoxigenin-labeled by a variety of methods and commercial kits (random priming, nick translation, PCR) (*see* **Note 3**). We prefer the labeling by PCR amplification when it is possible, because one PCR reaction provides enough probe for many hybridizations. The chemiluminescent detection is performed with an alkaline phosphatase-antidigoxigenin Fab fragment and the CSPD substrate. Labeling and detection kits are available from Boehringer Mannheim (Meylan, France).

13. Thermocycler: A programmable thermocycler is used for PCR amplification (Polylabo, Strasbourg, France).

14. Dot blot apparatus: For instance, the hybridot apparatus (from Biorad, Ivry su Seine, France) connected to a vacuum pump.

15. Hybridization oven: Hybridization and stringent washes are performed in sealed bottles that are constantly rotated.

16. Water bath: One boiling water bath is required for probe denaturation.

17. Roller or bidimensional agitator for detection steps.

18. UV-crosslinker or transilluminator, providing a 254 nm wavelength light.

19. Hybridization buffer: For 100 mL of hybridization buffer, dissolve 10 g of SDS (sodium dodecyl sulfate) in 60 mL of millipore water (heat to 60–65°C), add 25 mL of a 1 $M$ phosphate buffer stock solution pH 7.5, and 0.2 mL of a 0.5 $M$ EDTA stock solution. Complete the volume to 100 mL with millipore water. For a 100 cm$^2$ membrane, 40 mL of hybridization buffer are necessary.

20. Wash buffer: Mix 20 mL of a 1 $M$ phosphate buffer pH 7.5 stock solution, 2 mL of a 0.5 $M$ EDTA stock solution and 100 mL of a 10 % SDS stock solution and complete the volume to 1 L with millipore water. For a 100 cm$^2$ membrane, 90 mL of wash buffer are necessary.

21. Wash buffer (B1): Dissolve 11.6 g of maleic acid in 800 mL of millipore water and adjust the pH to 8 with NaOH pellets. Add 8.7 g of NaCl, and adjust the final volume to 1 L with millipore water. Sterilize by autoclaving at 120°C for 20 min. Extemporaneously add 0,2 % (v/v) of Tween 20. For a 100 cm$^2$ membrane, about 150 mL of B1 buffer are necessary.

22. Saturation buffer (B2): Make 1 % (w/v) of Blocking Reagent (Boehringer Mannheim) in B1 buffer by dissolving 1 g of Blocking Reagent in 100 mL of B1 buffer. Sterilize by autoclaving at 120°C for 20 min. Store at 4°C or –20°C. For a 100 cm$^2$ membrane, 25 mL of B2 buffer are necessary. (*see* **Note 4**).

23. Detection buffer (B3): Mix 25 mL of a 1 $M$ Tris HCl pH 9.5 stock solution, 10 mL of a 5 $M$ NaCl stock solution, and 0.5 mL of a 1 $M$ MgCl stock solution. Sterilize by autoclaving at 120°C for 20 min. For a 100 cm$^2$ membrane, 15 mL of B3 buffer are necessary.

24. Thin plastic sheets (e.g., cut from bags) for autoclaving or transparent for overhead projector.

25. Dehybridizing solution 1: 0.1% SDS (50 mL/100 cm$^2$).

26. Dehybridizing solution 2: 0.1% SDS, 50 m$M$ NaOH (50 mL/100 cm$^2$).

27. Photographic film and dark room: Classic X-Ray films are used.

28. Microinformatic hardware and softwares. The hybridization signals are recorded with a Macintosh microcomputer equipped with a standard video camera (for instance 500 × 582 pixels, 0.1 lux). Analyses are performed with NIH-Image program and Microsoft-Excel (*see* **Note 5**).

## 3. Methods

The protocol is outlined in the flowchart shown in **Fig 1**. If several samples are to be run, it is important to amplify them in parallel with the same master mix and

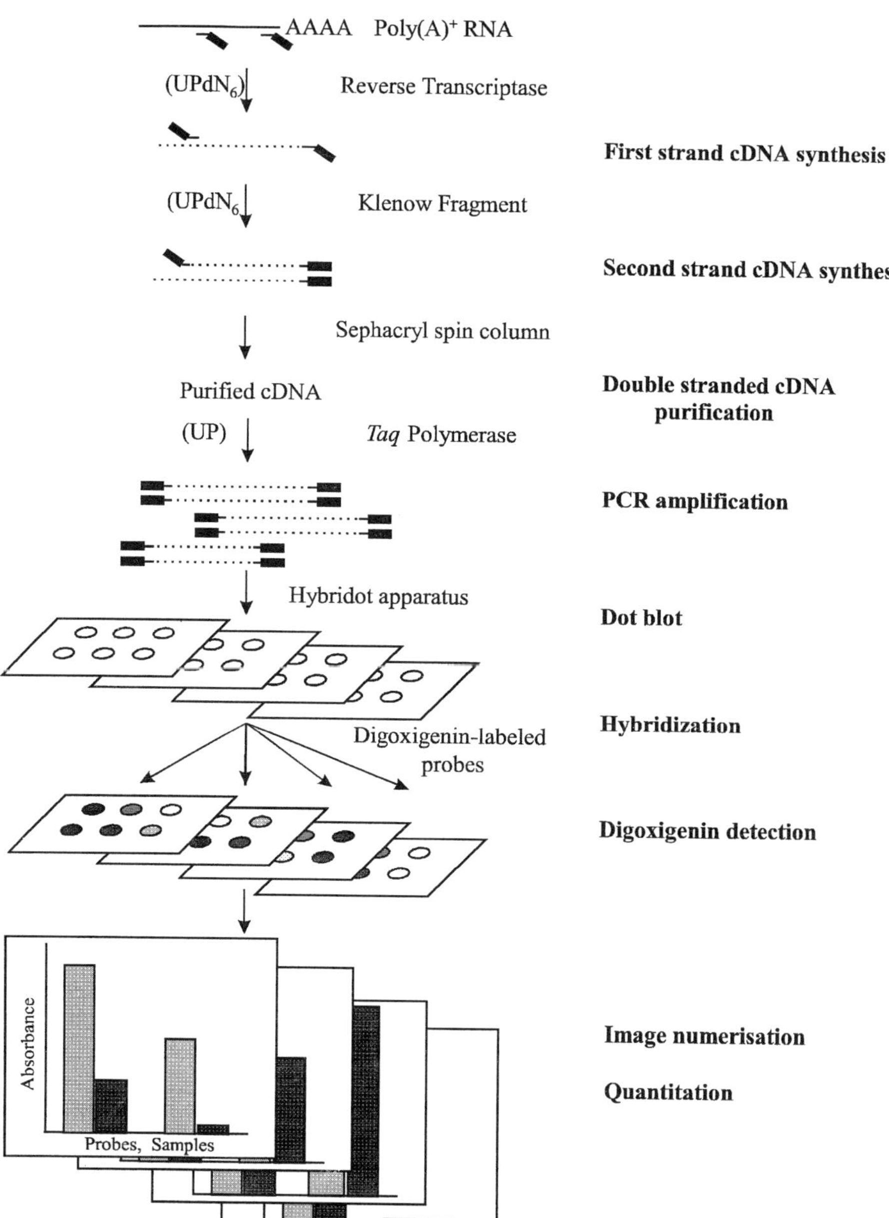


Fig. 1. Outline of the procedure: Double-stranded cDNA is synthesized from poly (A)+ RNA and then amplified. Amplicons are dotted and hybridized with specific probes. Finally, the hybridization signals are quantitated by densitometry.

to dot them on the same membrane and hybridize them simultaneously. Indeed, this is a multiple-step procedure, and each step can influence the final result.

## 3.1. cDNA Synthesis and Amplification from Purified mRNA

### 3.1.1. cDNA Synthesis

When using purified mRNA, we use a technique based on the procedure of Froussard *(7)*. mRNA can be prepared by a variety of methods, usually by capturing the poly(A)$^+$ molecules with an oligo(dT) attached to latex or magnetic beads or on an oligo(dT) column (*see* **Note 6**). The amount of mRNA can be determined by the absorbance at 260 nm, but very often there is not enough RNA to be measured by spectrophotometry. In that case, the amount can be evaluated after dotting of an aliquot on a nylon membrane and staining with colloidal gold (Genogold, Tebu, Le Perray, France), according to the manufacturer's protocol.

1. Mix about 10–100 ng of mRNA, 100 ng of UPdN$_6$ primer with RNAse free water to obtain a final volume of 7 μL.
2. Heat the mixture to 70°C for 10 min (*see* **Note 7**) and chill on ice.
3. Add 2.5 μL of 5X reverse transcriptase buffer, (generally provided with the enzyme), 0.6 μL of dNTP mix, 100 U of M-MLV reverse transcriptase and 1 U of RNAse inhibitor. Final volume : 12.5 μL.
4. Incubate the mixture at 37°C for 2 h.
5. Heat to 99°C for 2 min, and cool rapidly on ice (*see* **Note 8**).
6. Add 5 U of Large Fragment of Klenow enzyme, 5 μL of 10X DNA polymerase buffer (generally provided with the enzyme), 0.6 μL of dNTP mix, and adjust the final volume to 50 μL with millipore water. Incubate the mixture at 37°C, for 30 min, then heat at 70°C, for 10 min to inactivate the enzyme (*see* **Note 9**).
7. Purify the double-stranded cDNA on a Sephacryl S400 spin column. This step eliminates the excess of UPdN$_6$ primers, the salts, and the proteins. (The resulting volume is about 50 μL.)

### 3.1.2. cDNA Amplification

The following protocol is for a 50 μL reaction. For larger amounts, it can be scaled up to 200 μL per reaction or multiple reactions can be run in parallel. With one RNA sample, it is possible to obtain 500 μL of amplificate.

1. Prepare a mix combining in the followed order (*see* **Note 10**):
   <u>Final concentrations</u>
   34 μL millipore water

   | | |
   |---|---|
   | 5 μL 10X buffer (provided with the enzyme) | 1X |
   | 3 μL 25 mM MgCl$_2$, (*see* **Note 11**) | 1.5 m*M* |
   | 1 μL dNTP mix | 0.2 m*M* |
   | 2 μL UP primer | 2 μ*M* |
   | 5 μL purified double-stranded cDNA | |
   | 0.04 μL Goldstar *Taq* DNA polymerase (*see* **Note 1**) | 0.2 U |

2. Include two negative controls: one reaction without cDNA template and one without primer.
3. Cover each reaction mixture with a drop of mineral oil.
4. Program the thermocycler for the following steps: denaturation at 95°C for 2 min, then 30 cycles of denaturation (94°C, 1 min), hybridization (55°C, 1 min), elongation (72°C, 1.5 min).
5. Control the amplification by electrophoresis of a 5-µL aliquot of the PCR mixture on an 1% agarose gel, stained by ethidium bromide. The visualization is done at 362 or 254 nm (*see* **Note 12**).

## 3.2. cDNA Synthesis and Amplification from Total RNA

### 3.2.1. cDNA Synthesis

1. *See* introduction of **Subheading 3.1.1.**
2. Mix about 1–5 µg of total RNA, 100 ng of NIIIdTVV primer with RNAse free water to obtain a final volume of 7 µL.
3. *See* **steps 2–5** of **Subheading 3.1.1.**
4. Add 5 U of Large Fragment of Klenow enzyme, 5 µL of 10X DNA polymerase buffer (provided with the enzyme), 0.6 µL of dNTP mix, 25 ng of UPdNprimer, and adjust the final volume to 50 µL with millipore water. Incubate the mixture at 37°C, for 30 min, then heat at 70°C, for 10 min to inactivate the enzyme (*see* **Note 13**).
5. *See* **step 7** of **Subheading 3.1.1.**

### 3.2.2. cDNA Amplification

1. Prepare a mix combining in the followed order (*see* **Note 10**):
   <u>Final concentrations</u>

   | | |
   |---|---|
   | 34 µL millipore water | |
   | 5 µL 10X buffer (provided with the enzyme) | 1X |
   | 3 µL 25 m*M* MgCl, (*see* **Note 10**) | 1.5 m*M* |
   | 1 µL dNTP mix | 0.2 m*M* |
   | 1 µL UP primer | 1 µ*M* |
   | 1 µL NIII primer | 1 µ*M* |
   | 5 µL purified double stranded cDNA | |
   | 0.04 µL Goldstar *Taq* DNA polymerase (*see* **Note 1**) | 0.2 U |

2. *See* **steps 2–5** of **Subheading 3.1.2.**

## 3.3. Dot Blots

From a 50 µL PCR reaction, several dot blots can be made. Usually 2–5 µL of the amplificate per dot is enough to achieve a good sensitivity of the detection for a medium-abundant messenger (*see* **Note 14**). For higher sensitivity, the amount to be dotted can be increased.

1. For each dot, dilute 2 µL of the PCR reaction in 150 µL of TE buffer. Denature the DNA by heating the solution to 100°C, for 5 min, and then cool it rapidly on ice.

2. Wet the membrane with TE buffer, and place it on the hybridot (according to the manufacturer's instructions).
3. Fill the wells of the hybridot apparatus with the denatured samples, and apply the vacuum until the wells are empty (*see* **Note 15**), usually a 10 mbar vacuum is sufficient.
4. Transilluminate the membrane at 254 nm for 3 min to covalently fix the DNA to the membrane.

### 3.4. Hybridization and Detection of Digoxigenin-Labeled Probe

In order to evaluate the expression level of a particular messenger in the whole cDNA amplicons, the cDNA dot blots are hybridized with the corresponding probe and the hybridization signal is measured by densitometry. The total amount of dotted cDNA is taken as a standard. This amount is evaluated by hybridization with a total cDNA probe.

All the volumes are for a 100 cm$^2$ membrane.

1. Place the membrane in an hybridization bottle or a polypropylene tube, add 30 mL of hybridization buffer, and incubate at 68°C, for at least 2 h in the hybridization oven.
2. Dilute the digoxigenin-labeled probe in 10 mL of hybridization buffer to a final concentration of 2.5 ng/mL. For a PCR-labeled probe, 2 μL of the PCR product in 10 mL of hybridization buffer is usually adequate. Heat the diluted probe to 100°C for 10 min in a boiling water bath, and cool rapidly on ice (*see* **Notes 16–20**).
3. Pour off the prehybridization buffer, and replace it by the diluted probe. Perform the overnight hybridization at 68°C.
4. Wash the membrane with 30 mL of wash buffer at 68°C for 20 min. Repeat this step twice.
5. Wash the membrane with 10 mL of B1 buffer for 5 min at room temperature with constant shaking.
6. Incubate the membrane in 15 mL of B2 buffer for at least 1 h at room temperature.
7. Dilute 75 U/mL of AP-antidigoxigenin Fab fragment in 10 mL of B2 buffer. Incubate the membrane for 30 min at room temperature.
8. Wash the membrane in 30 mL of B1 buffer for 15 min. Repeat this step twice.
9. Equilibrate the membrane in 10 mL of B3 buffer for 5 min. Eliminate the excess of buffer from the membrane by briefly placing it on a Whatman paper.
10. Dilute to 1:100 the CSPD substrate in 1 mL of B3 buffer. On a sheet of plastic, scatter drops of the substrate solution, and place the membrane on the drops. Cover the membrane by another sheet of plastic, and incubate 5 min in the dark.
11. Eliminate the excess of substrate on a Whatman paper and place the membrane in a sealed plastic bag (*see* **Notes 17** and **21**).
12. Incubate the membrane at 37°C for 10 min to activate the alkaline phosphatase enzyme.
13. Place an X-ray film on the membrane for 20 min to 12 h (*see* **Note 22**).
14. After revelation, the signals are recorded by a camera or a scanner, and the image can be processed for the signal quantitation.

### *3.5. Dehybridization and Rehybridization*

1. Rinse the membrane in 30 mL of millipore water for 5 min.
2. Incubate the membrane in 50 mL of 0.1 % SDS, at 100°C for 7 min in a polypropylene tube.
3. Incubate the membrane in a flat dish containing 100 mL of 0.1% SDS, 50 m$M$ NaOH for 10 min twice, at room temperature with constant shaking.
4. Neutralize by incubation in B1 buffer for 10 min.
5. The membrane can be rehybridized with the total cDNA probe for standardizing the signals and processed as above. (*see* **Note 23**).

### *3.6. Densitometry Using NIH-Image and Excel Software*

Both images of dot blot hybridizations (obtained with the specific probe and with the total cDNA probe) must be processed in the same way (*see* **Note 5**).

1. Load the macro "gel plotting macro" of the NIH-Image software.
2. The dot to be quantified must be on horizontal lines in the image.
3. Select the first horizontal area containing the dots.
4. Press "Z" on the keyboard and enter the number of lines to be processed.
5. Press "P" on the keyboard to plot the density profile of the selected line.
6. In the plot window, draw the base line of the plot profile, and measure the area under the curve with the "auto measure tool." Results are outlined in the results window. To visualize it, select "show results" in the "analyze" menu. Be careful to the measure units, which must be in pixels.
7. Save the results as a text file and process the values in Microsoft Excel.
8. Microsoft Excel or any other software allow calculation of the ratio between the mRNA species signals and the total cDNA signal and creation of graphics of the results.
9. **Figure 2** shows an example of results obtained in this technique. The HL60 cell line was induced to differentiate along the monocytic pathway and randomly primed cDNA were prepared from the cells at different times after the induction (t = 0, 1, 24, and 48 h). The amplified cDNA were dotted and probed with a myeloperoxydase cDNA probe, dehybridized, and reprobed with the total cDNA. In agreement with previous results *(8)*, the amount of myeloperoxydase mRNA gradually decreases between the promyelocytic state and the macrophage state of the HL60 cells.

## 4. Notes

1. The very good efficiency of the Goldstar *Taq* DNA polymerase allows to decrease the amount of enzyme to 0.2 U/50 µL reaction. If another brand of *Taq* DNA polymerase is used, the usual amount is 1 U/reaction. Too high an amount of enzyme may lead to nonspecific priming.
2. DNA adsorption during dot blotting is optimal on positively charged membrane because of electrostatic interactions. The quality of the membrane greatly influences the background of a subsequent chemiluminescent detection. Care must be taken not to use "old membranes," which lead to poor adsorption of the

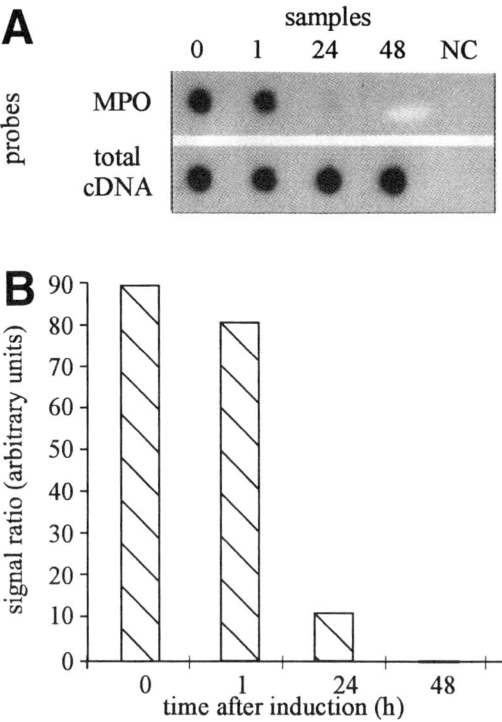


Fig. 2. Application to the measure of the decrease in myeloperoxidase mRNA level during the differentiation of HL 60 cells: **(A)** Poly (A)+ RNA was extracted from HL 60 cells at different times after treatment of the cells with 12-O-tetradecanoylphorbol-13-acetate. Double-stranded cDNA was prepared and amplified. Amplicons from each time point (0, 1, 24, 48 h) and a negative control (NC, pBS plasmid) were dotted on a nylon membrane and hybridized to a myeloperoxidase probe (MPO). The membrane was dehybridized and rehybridized to total cDNA (total cDNA) probe for normalization. **(B)** Normalized hybridization signals.

    sample and high background. We use Appligene or Boehringer Mannheim membranes, but other brands are adequate.

3. The digoxigenin used for labeling must be alkali labile to allow an efficient dehybridization of the first probe before rehybridization with the whole cDNA probe.

4. Be careful of the buffer sterility, in particular those containing Blocking Reagent, because bacterial contamination can lead to a high background.

5. Any image processing software can be used, but NIH-Image is very simple of use. It is available on the web (address: http://rsb.info.nih.gov/NIH-IMAGE/).

6. When magnetic beads are used, it is often necessary after the purification to heat the mRNA to 70°C for 5 min and rapidly cool the solution on ice, then centrifuge

the sample at 12,000$g$ for 5 min to pellet the remaining beads and carefully pipet the supernatant. We prefer to use Qiagen oligotex mRNA isolation kits.

7. Heating destroys RNA secondary structures. During the cooling, the UPdN$_6$ oligonucleotides randomly anneal along the RNA molecule, so that the whole mRNA sequence can be reverse transcribed.

8. Heating denatures RNA/cDNA hybrids.

9. The UPdN$_6$ molecules still present in the mixture anneal with the first-strand cDNA and the DNA polymerase uses them as primers for the second-strand cDNA extension. As a consequence, both extremities of the second-strand cDNA molecules are fitted with defined oligodeoxynucleotides sequences, making them amplifiable with UP primer.

10. The mix and the sample must be kept on ice until the denaturation step. This prohibits the formation of primer dimers during the PCR.

11. Add MgCl$_2$ to a final concentration of 1.5 m$M$ if it is not included in the *Taq* DNA polymerase buffer.

12. The length of amplified cDNA molecules usually ranges from 1000–200 bp.

13. The double–stranded cDNA molecules carries the UP primer at the 5' end and the complementary sequence of NIII primer at the 3' end. In contrast, to the use of purified mRNA, only the 3' sequences of the mRNA are present in this population. It is important to remember this fact for the subsequent choice of the probes.

14. With one RNA sample, it should be possible to get at least 100 dots. If more sequences are to be probed, it is possible to split the first amplicons and run a second run of amplification. This should yield enough material for several hundred measures.

15. The hybridot apparatus allows the dotting of the amplified cDNA on the nylon membrane. The DNA containing solution is filtrated through the membrane, and the DNA molecules are adsorbed on the membrane by electrostatic and hydrophobic binding. The filtration must be slow. The filtration should take at least 5 to 10 min for 150 µL. Faster filtration may result in poor DNA binding to the membrane.

16. Probe labeling with digoxigenin can be achieved by incorporation of dig-dUMP during the elongation step of PCR amplification. For labeling specific clone insert, one can use the vector sequences flanking the MSC (T3, T7, or SP6 primers). For labeling the whole cDNA population, the PCR conditions are the same as as for **Subheading 3.2.2.** In both cases, the PCR mix includes 4 n$M$ dig-dUTP and dTTP concentration is decreased to 0.1 m$M$. Purified inserts can also be labeled by random priming, in particular, if PCR primers are not available.

17. Usually a probe concentration around 2.5 ng/mL is suitable. Rather wide variations are tolerable. However too high a concentration generates high backgrounds and too low a concentration dramatically decreases the sensitivity. It may be necessary to test several probe concentrations to optimize the signal/ background ratio. The effective probe concentration is roughly determined by comparison with a series of standards provided by the manufacturer.

18. After cooling on ice, the SDS of the hybridization buffer can precipitate. If this is the case, briefly heat the solution to dissolve it.
19. Do not allow the membrane to dry after the prehybridization step and during the detection procedure.
20. The diluted probe can be reused several times without loss of efficiency. After hybridization, collect it in a polypropylene tube and store it at –20°C. The probes are stable for at least 6 mo when stored at –20°C. Reheat at 100°C for 10 min before used.
21. An excess of buffer will also give a high background.
22. Any X-Ray autoradiographic films are suitable for chemiluminescence detection. For the quantitative treatment of the results, it is necessary to remain below the saturation level of the film. At low exposure, the signal can be considered as proportional to the amount of probe on the filter. To set the signals in this range, we suggest to dot a series of dilutions of the purified target sequence probed at the same time as the cDNA dots. This allows to constitute a reference scale and to evaluate the range of hybridization signal linearity. Only signals within this range will give a valid quantitation. In case of saturating signal, one can decrease the exposure time or dilute the cDNA samples before dotting.
23. Contrary to the use of radioactive probes, it is not possible to repeatedly dehybridize and reprobe the blot here.

## Acknowledgments

We thank B. Mila for his skillfull technical assistance. This work has been supported by the University of Bordeaux II, the French MSER, INSERM (CRE N°920613) and La Ligue contre le Cancer.

## References

1. Ko, M. S. H., Ko, S. B. H., Takahasi, N., Nishigushi, K., and Abe, K. (1990). Unbiased amplification of a highly complex mixture of DNA fragments by "lone linker" tagged PCR. *Nucleic Acids Res.* **18,** 4293–4294.
2. Reyes, G. R. and Kim, J. P. (1991) Sequence independent single primer amplification (SISPA) of complex DNA populations. *Mol. Cell. Probes* **5,** 473–481.
3. Domec, C., Garbay, B., Fournier, M., and Bonnet, J. (1990) cDNA construction from small amounts of unfractionated RNA: association of cDNA synthesis with polymerase chain reaction amplification. *Anal. Biochem.* **188,** 422–426.
4. Brunet, J.-F., Shapiro, E., Foster, S. A., Kandel, E. R., and Iino, Y. (1991) Identification of a peptide specific for aplysia sensory neurons by PCR differential screening. *Science* **252,** 856–859.
5. Liang, P. and Pardee, A. B. (1992) Differential display of eukaryotic messenger RNA by means of the polymerase chain reaction. *Science* **257,** 967–971.
6. Lisitsyn, N., Lisitsyn, N., and Wigler, M. (1993) Cloning the difference between two complex genomes. *Science* **259,** 946–951.

7. Froussard, P. (1992) A random-PCR method (rPCR) to construct whole cDNA library from low amounts of RNA. *Nucleic Acids Res.* **20,** 2900.
8. Meyer, R. W., Chen, T., Mathews, S., Niklaus, G., and Tobler, A. (1992) The differentiation pathway of HL60 cells is a model system for studying the specific regulation of some myeloid genes. *Cell Growth Differ.* **3(10),** 663–669.

# Index